International Series in Operations Research & Management Science

Volume 158

For further volumes:
http://www.springer.com/series/6161

Jeff Kennington • Eli Olinick • Dinesh Rajan
Editors

Wireless Network Design

Optimization Models and Solution Procedures

Springer

Editors
Jeff Kennington
Department of Engineering Management,
Information, and Systems
Bobby B. Lyle School of Engineering
Southern Methodist University
PO Box 750123
75275-0123 Dallas Texas
USA
jlk@lyle.smu.edu

Eli Olinick
Department of Engineering Management,
Information, and Systems
Bobby B. Lyle School of Engineering
Southern Methodist University
PO Box 750123
75275-0123 Dallas Texas
USA
olinick@lyle.smu.edu

Dinesh Rajan
Department of Electrical Engineering
Bobby B. Lyle School of Engineering
Southern Methodist University
PO Box 750338
75275-0338 Dallas Texas
USA
rajand@lyle.smu.edu

ISSN 0884-8289
ISBN 978-1-4614-2763-6 ISBN 978-1-4419-6111-2 (eBook)
DOI 10.1007/978-1-4419-6111-2
Springer New York Dordrecht Heidelberg London

Printed on acid-free paper

Springer is part of Springer Science+Business Media (www.springer.com)

Preface

This book surveys state-of-the-art optimization modeling for design, analysis, and management of wireless networks, such as cellular and wireless local area networks (LANs), and the services they deliver. The past two decades have seen a tremendous growth in the deployment and use of wireless networks. The current-generation wireless systems can provide mobile users with high-speed data services at rates substantially higher than those of the previous generation. As a result, the demand for mobile information services with high reliability, fast response times, and ubiquitous connectivity continues to increase rapidly. The optimization of system performance has become critically important both in terms of practical utility and commercial viability, and presents a rich area for research.

In our previous work on traditional wired networks, we have observed that designing low cost, survivable telecommunication networks involves extremely complicated processes. Commercial products available to help with this task typically have been based on simulation and/or proprietary heuristics. As demonstrated in this book, however, mathematical programming deserves a prominent place in the designer's toolkit. Convenient modeling languages and powerful optimization solvers have greatly facilitated the implementation of mathematical programming theory into the practice of commercial network design.

These points are equally relevant and applicable in todays world of wireless network technology and design. But there are new issues as well: many wireless network design decisions, such as routing and facility/element location, must be dealt with in innovative ways that are unique and distinct from wired (fiber optic) networks. The book specifically treats the recent research and the use of modeling languages and network optimization techniques that are playing particularly important and distinctive roles in the wireless domain.

To cover the broad range of important topics in the design and optimization of wireless networks, we invited experts from academia and industry to contribute chapters in their particular areas of specialization. We thank them for their fine contributions to this work and for their patience during the editing and revising process of putting this book together. We also thank Dr. Ronald Dearing for his typesetting assistance. The work of the three editors was partly supported by ONR contract N00014-10-1-0330.

Dallas
August 2010

Jeff Kennington
Eli Olinick
Dinesh Rajan

Contents

List of Contributors

Khaldoun Al Agha
LRI, Bâtiment 490, Université Paris-Sud, 91405 Orsay cedex, France, e-mail: alagha@lri.fr

Marina Aguado
ETSI Bilbao, Depto. Electrónica y Comunicaciones, Al. Urquijo S/N, CP. 48013 Bilbao, Spain, e-mail: marina.aguado@ehu.es

Jasone Astorga
ETSI Bilbao, Depto. Electrónica y Comunicaciones, Al. Urquijo S/N, CP. 48013 Bilbao, Spain, e-mail: jasone.astorga@ehu.es

Orhan Dengiz
DnD Technical Solutions, Tunus Cad. 87-5, Kavaklidere Ankara 06680, Turkey, e-mail: orhan.dengiz@gmail.com

Li Guo
Cisco Systems, 2200 East President George Bush Highway, Richardson, Texas 75082-3550, USA, e-mail: liguo2@cisco.com

K. V. S. Hari
Department of ECE, Indian Institute of Science, Bangalore 560012, India, e-mail: hari@ece.iisc.ernet.in

Dag Haugland
Department of Informatics, University of Bergen, PB. 7800, N-5020 Bergen, Norway, e-mail: Dag.Haugland@ii.uib.no

Karla Hoffman
George Mason University, Systems Engineering and Operations Research Department, Fairfax, Virginia 22030, USA, e-mail: khoffman@gmu.edu

Hang Jin
Cisco Systems, 2200 East President George Bush Highway, Richardson, Texas 75082-3550, USA, e-mail: hangjin@cisco.com

Abhay Karandikar
Department of Electrical Engineering, Indian Institute of Technology Bombay, Mumbai, 400076, India, e-mail: karandi@ee.iitb.ac.in

Jeff Kennington
Department of Engineering Management, Information, and Systems, Bobby B. Lyle School of Engineering, Southern Methodist University, Dallas, TX 75275-0123, USA, e-mail: jlk@smu.edu

Abdullah Konak
Department of Information Sciences and Technology, Penn State Berks, Tulpehocken Road P.O. Box 7009, Reading, PA 19610-6009, USA, e-mail: auk3@psu.edu

Jason Kratz
Department of Engineering Management, Information, and Systems, Bobby B. Lyle School of Engineering, Southern Methodist University, Dallas, TX 75275-0123, USA, e-mail: jkratz@smu.edu

Khiem Le
Research In Motion, 5000 Riverside Drive, Irving, TX 75039, USA, e-mail: khiem.le@ieee.org

Steven Martin
LRI, Bâtiment 490, Université Paris-Sud, 91405 Orsay cedex, France, e-mail: smartin@lri.fr

Jon Matias
ETSI Bilbao, Depto. Electrónica y Comunicaciones, Al. Urquijo S/N, CP. 48013 Bilbao, Spain, e-mail: jon.matias@ehu.es

Syam Menon
School of Management, University of Texas at Dallas, Richardson, TX 75083, USA, e-mail: syam@utdallas.edu

Eli Olinick
Department of Engineering Management, Information, and Systems, Bobby B. Lyle School of Engineering, Southern Methodist University, Dallas, TX 75275-0123, USA, e-mail: olinick@smu.edu

Dinesh Rajan
Department of Electrical Engineering, Bobby B. Lyle School of Engineering, Southern Methodist University, Dallas, TX 75275-0340, USA, e-mail: rajand@smu.edu

Nitin Salodkar
Department of Computer Science and Engineering, Indian Institute of Technology Bombay, Mumbai, 400076, India, e-mail: nitinsalodkar@gmail.com

Alice E. Smith
Department of Industrial and Systems Engineering, 3301 Shelby Center, Auburn University, Auburn, AL 36849, USA, e-mail: smithae@auburn.edu

J. Cole Smith
University of Florida, Department of Industrial and Systems Engineering, Gainesville, FL 32611, USA, e-mail: cole@ise.ufl.edu

Sibel B. Sonuc
University of Florida, Department of Industrial and Systems Engineering, Gainesville, FL 32611, USA, e-mail: sibel.bilge@ufl.edu

Gheorghe Spiride
Department of Engineering Management, Information, and Systems, Bobby B. Lyle School of Engineering, Southern Methodist University, Dallas, TX 75275-0123, USA, e-mail: gspiride@smu.edu

Nerea Toledo
ETSI Bilbao, Depto. Electrónica y Comunicaciones, Al. Urquijo S/N, CP. 48013 Bilbao, Spain, e-mail: nerea.toledo@ehu.es

Di Yuan
Department of Science and Technology, Linköping University, SE-601 74 Norrköping, Sweden, e-mail: diyua@itn.liu.se

Acronyms

AMPS	Advanced Mobile Phone Service
AP	Access Point
BPSK	Binary Phase Shift Keying
CDMA	Code Division Multiple Access
DS-CDMA	Direct Sequence CDMA
FCC	Federal Communications Commission
FDMA	Frequency Division Multiple Access
FH-CDMA	Frequency Hopping CDMA
GSM	Global System for Mobile Communications
IEEE	Institute of Electrical and Electronics Engineers
INFORMS	Institute for Operations Research and the Management Sciences
IS-95	Interim Standard 95
LTE	Long Term Evolution
MAC	Medium Access Control
MANET	Mobile Ad-Hoc Network
MNO	Mobile Network Operator
OFDM	Orthogonal Frequency Division Multiplexing
PSTN	Public Switched Telephone Network
QAM	Quadrature Amplitude Modulation
QoS	Quality of Service
RF	Radio Frequency
SIM	Subscriber Identity Module
SMS	Short Message Service
TACS	Total Access Communication System
TDMA	Time Division Multiple Access
UMTS	Universal Mobile Telecommunication System
VoIP	Voice over Internet Protocol
WiMAX	Worldwide Interoperability for Microwave Access
WLAN	Wireless Local Area Network
WPAN	Wireless Personal Area Network
WWAN	Wireless Wide Area Network

Chapter 1
Introduction to Optimization in Wireless Networks

Jeff Kennington, Eli Olinick, and Dinesh Rajan

Wireless communications have progressed tremendously from the days of Marconi and Tesla. It is hard to imagine life without the comforts enabled by wireless technologies, which has impacted several facets of society.

The growth of cellular technologies worldwide over the last few years has been phenomenal. Nearly 5 billion people currently receive wireless services. The development of cellular technologies is commonly categorized into various generations. The first generation (1G) of cellular networks were those installed in the period from the late 70's and 80's and were primarily designed to provide only voice services. These networks are analog in nature and the quality of service was mediocre. Examples of first generation systems include Advanced Mobile Phone Service (AMPS) and Total Access Communication System (TACS). The second generation networks were installed in the 90's and were digital systems. Low rate data services were first introduced at the final stages of 1G networks. However, it was second generation (2G) systems that made data services more prevalent. Examples of 2G technologies include the popular Global System for Mobile Communications (GSM) and IS-95 systems. This period also witnessed the rapid growth of the world wide web, which spawned several new applications and widespread adoption in society. These changes created the need for mobile access to data at high speeds. The third generation (3G) cellular technologies adopted around the start of this century were designed to address these needs. Universal Mobile Telecommunication System (UMTS) and CDMA-2000 are examples of 3G technologies. The planned fourth

Jeff Kennington
Department of Engineering Management, Information, and Systems, Bobby B. Lyle School of Engineering, Southern Methodist University, Dallas, TX 75275-0123, USA, e-mail: jlk@smu.edu

Eli Olinick
Department of Engineering Management, Information, and Systems, Bobby B. Lyle School of Engineering, Southern Methodist University, Dallas, TX 75275-0123, USA, e-mail: olinick@smu.edu

Dinesh Rajan
Department of Electrical Engineering, Bobby B. Lyle School of Engineering, Southern Methodist University, Dallas, TX 75275-0340, USA, e-mail: rajand@smu.edu

J. Kennington et al. (eds.), *Wireless Network Design: Optimization Models and Solution Procedures*, International Series in Operations Research and Management Science 158, DOI 10.1007/978-1-4419-6111-2_1,

generation (4G) technologies including Long Term Evolution (LTE) and Worldwide Interoperability for Microwave Access (WiMAX) are currently being rolled out in many parts of the world and promise unprecedented data rates and quality of service in fixed wireless and mobile settings.

The rise of WLANs has mirrored the growth in wide area cellular networks. A major reason for this growth is the inexpensive computational power that leads to affordable WLAN adapters and access points. The technologies and algorithms used in the various generations of WLANs have also contributed to a progressive improvement in their performance. The popular 802.11 based WLANs were first standardized in 1997. Since then several versions of the standards (including 802.11b, 802.11g and 802.11n) have been released and successfully commercialized.

In both cellular and WLAN networks, there is typically a backbone network that is composed of fixed nodes that act as a gateway to the legacy public switched telephone network (PSTN) or the Internet. With the emergence of voice over IP (VoIP) technology and provisioning of Internet access over PSTN networks, the boundary between the PSTN and Internet is disappearing. Equally impressive has been the explosion in the number of wireless personal area network (WPAN) devices. For instance, in the commercial setting, Bluetooth based devices have become popular in a variety of short range wireless products including, hands free headsets, video game consoles, and wireless keyboards. In many of these WPAN networks, a small group of nodes operate in an ad-hoc manner without any backbone network support. While ad-hoc networks have promised spectacular results in several settings, their widespread use has not materialized. The major problems have been with scaling the number of nodes and also a lack of standardization efforts which have restricted interoperability. Nevertheless, recent advances in a wide range of disciplines including semiconductors, distributed algorithms, and optimization raises the hope that ad-hoc networks will become ubiquitous.

One of the common key factors in the design of these various networks is the use of rigorous mathematical models to formulate the system design as an optimization problem and utilize tools from optimization theory to solve them efficiently. Very sophisticated tools for solving many types of optimization problems are currently available for both research investigations and industrial applications. Many of the software firms offering optimization products advertise in ORMS Today, a monthly publication of the Institute for Operations Research and the Management Sciences (INFORMS). In addition, ORMS Today routinely publishes software surveys for various problem classes. The June 2009 issue provides a comprehensive survey of linear programming solvers. In addition, the Computational Infrastructure for Operations Research (COIN-OR) organization is a group of volunteers devoted to creating and disseminating open-source software for the operations research (OR) community. Hosted by INFORMS, COIN-ORs Web site, http://www.coin-or.org/ is currently hosting over 30 projects. The NEOS site, http://www-neos.mcs.anl.gov/ can be used to test various software systems prior to purchase. The user simply provides the proper input to the site and the solution is returned by e-mail. Using these two sites, users can find solvers for network programming, linear programming, integer programming, nonlinear programming, global optimization, mixed integer nonlinear programming, stochastic programming, and metaheuristics.

Several software tools have been developed to assist in network optimization. Examples of such tools include Aruba, Opnet, S^4W and Wireless Valley's Site Planner. While some of these tools are used in the network design phase prior to deployment, some tools can also be used to fine tune the performance post-deployment based on measured performance. Indeed, optimization techniques have emerged as a key tool in the design and operation of wireless networks.

This book is broadly organized into four parts. The first part includes Chapters 2, 3, and 4 which provide background material, respectively, on wireless communications, channel modeling, and optimization techniques. The second part of the book focuses on optimization in planning of networks with a backbone structure including cellular and WLAN networks. Chapters 5, 6, 7, and 8 fall into this category. The third part of the book comprises Chapters 9, 10, and 11 which deal with optimization problems in ad-hoc networks that operate without any backbone support. The final part of the book includes Chapters 12, 13, 14, and 15 and addresses optimization problems that arise in operation of wireless networks in general. Next, we provide a brief preview of these chapters.

Chapter 2 provides a brief introduction to the area of wireless communication. This chapter discusses the basics of modulation and coding for transmission of digital data. Advanced multiuser communication strategies including code division multiple access (CDMA) and orthogonal frequency division multiplexing (OFDM) are also covered in this chapter. This chapter is followed by Chapter 3 which discusses the various models for wireless channels. Wireless channel modeling is an area with a rich history and several types of models have been proposed over the years. These models include the simple flat fading models for narrow band channels and the frequency selective wide band channels. The effect of various system settings such as mobile speed, carrier frequency, bandwidth, and context information (building/wall details in indoor settings, downtown/suburban settings in outdoor scenarios) are addressed in this chapter. These channel models describe the effect of these parameters on the received signal. For instance, wireless signals propagate over longer distances at lower bands such as 900 MHz than at higher frequencies such as 60 GHz. Together, Chapters 2 and 3 provide the reader with sufficient detail to understand the basic wireless system including, transmitter, channel and receiver design, analysis, and operation. Chapter 4 provides an introduction to the fundamental concepts, theory and methods of mathematical optimization (also known as mathematical programming) with an emphasis on integer and large-scale linear models that have proven to be particulary useful for solving telecommunications network design problems. The chapter is written primarily to prepare readers who have little or no background in OR for the optimization models and solution techniques used in most of the chapters in the remainder of the book.

Chapters 5 and 6 discuss, respectively, the optimization of 3G cellular and WLAN networks. In a cellular network, coverage and connectivity depend critically on the location and placement of the base stations (BTS). Similarly, indoor coverage in WLANs depends on the location of the access points (AP). Optimizing the location of these BTS and APs is the primary focus of Chapters 5 and 6. While these two problems appear to be similar, there are several crucial differences: i) the quality of service requirements depends on the specific applications supported by the networks and may be different in the cellular and WLAN cases, ii) the signal propagation in indoor and outdoor channels is very different which requires the use of different models, and iii) constraints used in formulating the optimization problem are different in the two cases. For instance, the cost of a BTS is significantly higher than the cost of APs, and is a primary concern for cellular operators[1]. The primary focus of Chapter 5 is the design of 3G cellular networks based on CDMA technology. The primary focus of Chapter 6 is the design of 802.11 based WLANs.

One common and critical ingredient in all wireless networks is the requirement for radio frequency (RF) spectrum. Networks can be classified into two types based on the spectrum being used: licensed and unlicensed networks. The use of a licensed spectrum is regulated in each country by an agency such as the federal communications commission (FCC) in the United States. Examples of networks using licensed spectrum include the cellular networks at 1800 and 1900 MHz. Similarly, networks that use unlicensed spectrum include the 802.11 based WLANs and Bluetooth networks that operate in the 2.4GHz industrial, scientific, medical (ISM) band. The advantage of using the unlicensed band is that the spectrum is free to use. However, interference from other devices which use the same frequency band can cause disruptions in wireless service. While the use of the licensed spectrum is more controlled, obtaining a license to operate a device in a licensed spectrum can be expensive. In most countries, spectrum is allocated to various operators via an auction process. Chapter 7 gives a history of spectrum allocation with a focus on the Simultaneous Ascending Auction (SAA). In the past, agencies such as the FCC have raised billions of dollars in revenue by selling spectrum in SAAs. A potential risk to wireless service providers bidding for spectrum in an SAA is that they might win only part of what they require for a particular network design and overpay for the resulting "sub-package" of their desired frequencies. Chapter 7 discusses newer designs that eliminate this risk by allowing bids on "all or nothing" packages of licenses. An understanding of the formal optimization framework described in Chapter 7 for the design of spectrum auctions can help operators plan their strategy for obtaining this crucial resource.

Chapter 8 discusses the importance of designing cellular networks with increased reliability. Specifically, redundancy is built into the cellular backbone network to overcome connectivity problems arising out of node and link failures. Several flavors of this problem and solution methodologies are addressed in this chapter.

Chapters 9, 10, and 11 are focused on the area of optimizing the performance of ad-hoc wireless networks. A mobile ad-hoc network (sometimes referred to as MANET) is a collection of nodes that communicate with each other using a multihop forwarding mechanism without any backbone network support. One of the primary challenges in a MANET is the node routing problem, where each node tries to establish a route to send packets to its intended destination. Numerous routing

[1] The emergence of inexpensive pico base stations is beginning to create a new dimension to consider for cellular operators.

protocols have been proposed that are solutions to optimization problems with different objective functions and constraints. For instance, a routing protocol might be optimized to minimize the overall energy consumption in the network. Chapter 9 discusses in detail the routing and the associated quality of service provisioning problems in MANETs.

Chapter 10 considers power optimization for a specific problem in ad-hoc networks. This chapter considers a broadcast/multicast scenario in which one transmit node desires to send its data to a collection of nodes. Several compact integer programs to solve problems of this flavor are addressed in Chapter 10. As noted before, in a MANET, node mobility can pose a significant challenge in ensuring a fully connected network at all times. Chapter 11 addresses the issue of maintaining network connectivity under mobility conditions. The main idea in Chapter 11 is the use of designated mobile nodes (called agents) which predict the movement of the other nodes (based on prior observations) and select their own path to ensure that the network remains fully connected. A particle swarm optimization is proposed to select the positions for the mobile agents at each time interval.

With the growth in the size and complexity of the networks, analyzing their performance is a challenging task. Fortunately, the advances in semiconductor technology have made computing power and memory inexpensive and widely available. Simultaneous to this growth, novel tools and techniques for simulating networks have also been developed. Chapter 12 provides an overview of the major techniques used to design networks and analyze their performance using simulation methods.

WiMAXWorldwide Interoperability for Microwave Access has emerged as a serious contender for providing ubiquitous broadband services in both fixed wireless and mobile settings over a broad service area. Chapter 13 is focused on operational time optimization procedures used to improve the performance of WiMAX networks. These WiMAX networks are based on the OFDM concept and this chapter explores the optimization issues that arise specifically in that setting. The focus of the chapter is on optimizing the radio resource allocation and multiuser packet scheduling within the OFDM framework.

One of the challenges of communicating over the wireless channel is the time varying nature of the channel. These time variations pose a serious challenge in ensuring that the desired quality of service is maintained without interruption. Simultaneously, the system must leverage these variations to improve the long term performance of the system. Channel aware scheduling has emerged as a key concept in achieving superior performance over a wireless channel. Chapter 14 provides a comprehensive overview of the main scheduling methods and their performance. The scheduling problem becomes even more relevant in a multiuser setting where the channels of the different users vary independently of each other. The scheduling framework should strike a balance between optimizing network performance (such as throughput) with the requirement for ensuring fair allocation of resources to the various users. Chapter 14 discusses several fairness metrics and how they can be incorporated as constraints in the scheduler optimization framework. Finally, we conclude this book with Chapter 15, which looks at the major trends in the field of wireless communications. In addition to a historical perspective of the growth of the field, some pointers to future development are also provided.

The book is at a level that is appropriate for a senior undergraduate or first year graduate course for students with an engineering background. A typical course would cover the first introductory chapters in the first half of the course. Selected topics from the remainder of the chapters can be covered in the second half of the course. The later chapters could also form the basis for semester long projects. We also hope that this book will be helpful to researchers and practitioners working at the intersection of OR and wireless networking.

Part I
Background

Chapter 2
Introduction to Wireless Communications

Dinesh Rajan

Abstract This chapter provides an introduction to the basic concepts of a wireless communication system. First, a simple point-to-point wireless link is considered and the key techniques at the transmitter and receiver that are required for a successful communication are discussed. The wireless medium being inherently a shared medium, multiple access methods to support numerous users in a given area is critical. This chapter also discusses several multiuser communication schemes. Specifically, two popular multiuser transmission strategies - CDMA and OFDM, are described and their merits and challenges discussed. We also provide a brief introduction to Shannon's theoretic capacity concept and discuss simple coding methods used in wireless channels.

2.1 Introduction

Digital communication systems have advanced significantly in the past few decades. Coupled with the progress in the semiconductor industry which obeys Moore's law, inexpensive and power efficient mobile communication devices have become widely prevalent. Cellular networks have become ubiquitous in most of the developed world and continues to see exponential growth in developing countries. Wireless local area networks (WLANs) based on the 802.11 standard have opened new venues for high speed data access in homes, offices and hot-spots. Bluetooth based commercial short range personal area networks have enabled several new commercial technologies such as hands-free cellular headsets and remote controllers for video game consoles. Further, the data rates for these technologies are constantly increasing and enabling several new applications as evident from the hugely popular *app stores* from Apple and Android.

Dinesh Rajan
Department of Electrical Engineering, Bobby B. Lyle School of Engineering, Southern Methodist University, Dallas, TX 75275-0340, USA, e-mail: rajand@smu.edu

J. Kennington et al. (eds.), *Wireless Network Design: Optimization Models and Solution Procedures*, International Series in Operations Research and Management Science 158,
DOI 10.1007/978-1-4419-6111-2_2,

The goal of this chapter is to introduce the basic components of a wireless communications system. To understand the functioning of digital communication devices, we first introduce the layering principle that is commonly used in the design of these systems.

2.1.1 Layered Architecture

The task of transmitting information from a transmitter device to a receiver device is a complex process that requires careful design, analysis, and implementation. This challenging task is usually accomplished with a layered architecture such as the open systems interconnection (OSI) model proposed by the international standards organization (ISO), which is shown in Figure 2.1. An OSI model, typically consists of 7 layers which function together to accomplish the required data transmission task. The 7 layers and a brief description of their functionality are:

- *Physical Layer*: This layer performs the basic task of converting digital bits into signals that may be sent and recovered over the transmission medium.
- *Data Link Layer*: This layer provides a certain amount of data reliability over a particular link and also assists in separating the signals from the various users that are transmitted over the same channel.
- *Network Layer*: The network layer provides functionality such as routing to enable data packets to be efficiently sent over multiple hops in a network.
- *Transport Layer*: This layer provides a mechanism to ensure end-to-end reliability of data. The transport layer is also tasked with creating congestion control mechanisms that reduces packet losses in a network occurring due to buffer overflows.
- *Session Layer*: The session layer provides services such as directory assistance to enable efficient and reliable data delivery.
- *Presentation Layer*: This layer typically provides data encryption and compression functionality.
- *Application Layer*: This layer includes all other tasks involved in the data transmission process as well as preparing the data for user or machine consumption.

The advantages of a layered architecture is that the design and implementation of each layer is simplified and can be done independent of the implementation of the other layers, as long as the interfaces with the layer above and below it are standardized. This freedom in implementation allows many of these layers to be reused in many different systems. For instance, the transmission control protocol (TCP) functionality (which forms the backbone for the Internet) could be the same irrespective of what physical layer transmission scheme is used. Also, changes in the implementation of one layer are transparent to the other layers. One of the disadvantages of a layered architecture is that the resulting end-to-end system performance may not be optimized. For instance, the use of TCP functionality which was originally developed for wired networks, in the wireless domain results in severe throughput degradation due to the time varying nature of the wireless channel [4].

Most systems however, use the simplified Internet layering scheme which was proposed in a Request for Comments (RC) document of the Internet Engineering Task Force (IETF) [11]. This scheme broadly consists of 4 layers: link layer, Internet layer, transport layer, and application layer [11]. Note that several variations of these layering methods exist and different authors interpret the RFCs in a slightly different manner [42].

Next, we look at the main elements of a digital communication system.

2.1.2 Digital Communication System

The basic functional blocks in a digital communication system are shown in Figure 2.2. The digital signal to be transmitted is first processed through a source encoder [37]. Most common digital data files have a significant amount of redundancy.

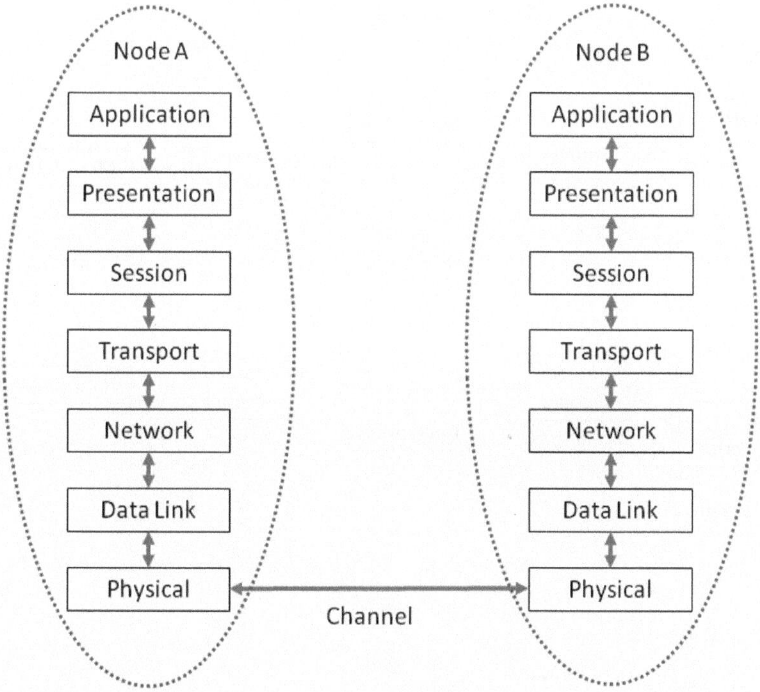

Fig. 2.1 Layered architecture

The source encoder leverages this redundancy to compress the data to the fewest number of bits required to meet the desired fidelity. For instance, text and document files could be subject to a lossless compression using a utility such as WinZip; image files could be subject to a lossy compression such as JPEG. If the signal to be transmitted is an analog signal such as voice, this signal is first processed through an analog-to-digital conversion using a sampling and quantization device before feeding it to the source encoder. In certain systems, the output of the source encoder is passed through an encryption mechanism to ensure data integrity and security [44].

The compressed (and possibly encrypted) data is then passed through a channel encoder. The primary goal of a channel encoder is to add redundancy to the data to protect it against errors experienced when transmitted over the channel [26]. Channel coding may also be used to detect errors in data packets using simple parity check mechanisms. Error detection allows the higher layers of the system to request a retransmission of the erroneous packets. Channel coding schemes are further discussed in Section 2.4.

The output of the channel encoder is passed through a modulator which effectively converts the digital bit stream to an analog signal that is appropriate for transmission over the channel. Examples of popular modulation schemes and their performance are discussed in Section 2.2. The modulated signal is then transmitted

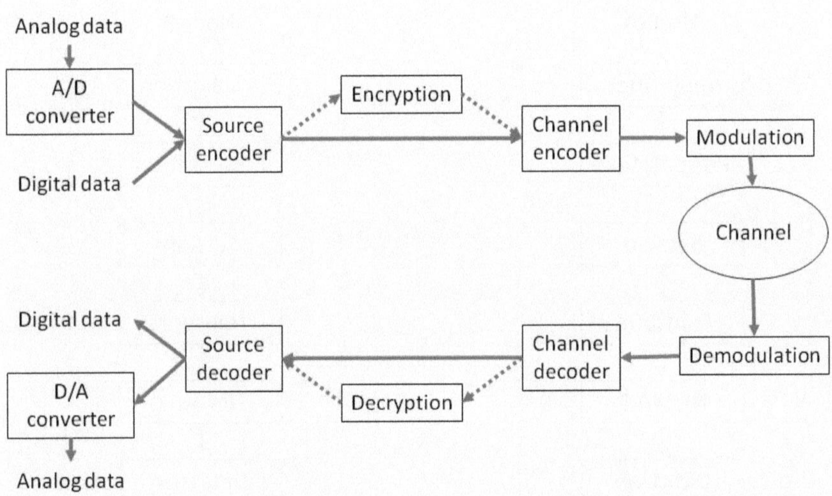

Fig. 2.2 Block diagram of digital communication system

over the desired medium. For instance, in a wireless channel this signal is radiated out from an antenna.

At the receiver, these blocks are processed in the exact reverse order. First, the received signal is demodulated, *i.e.* the noisy analogy received signal is converted to a sequence of bits. These bits are then passed through a channel decoder, which leverages the redundant bits to reduce the effective errors in the data bits. If channel encryption is used at the transmitter, then the signal is passed through a equivalent decryptor. Finally, the output bits from the channel decoder (or decryptor) is processed in a source decoder which converts it back into the original signal that is desired to be transmitted. While in legacy systems, *hard* decisions about the detected bits are sent to the channel decoder, the use of *soft* information about the bits (e.g. bit probabilities) was later shown to improve performance significantly [33]. With the increased capabilities of processing elements, sophisticated receiver techniques have also been proposed. For instance, the use of iterative detection and decoding using soft information is proposed in [50]. In such iterative processing, the output of the decoder is fed back to the bit detection block to reduce the overall bit error rate.

The separation of the source and channel encoding as indicated in Figure 2.2 has certain advantages. The design of the source encoder can be optimized to the characteristics of the source, independent of the channel over which it is transmitted. Similarly, the channel encoder can be optimized for the channel of interest, independent of the source that is transmitted. One potential disadvantage of this separation principle could be the loss in optimality of performance as compared to a system with a single joint source-channel encoder. Fortunately, one of the results in Shannon's pioneering work [39] is that the separation of the source and channel encoding does not entail any loss in asymptotic performance. As a result, the source coding and channel coding research developed in parallel for many years without many significant interactions. Note, however, that Shannon's separation principle only proves optimality under certain assumptions including processing over very large data lengths, which could significantly increase the delay, memory, and computational complexity requirements [46]. For a system with finite resources (memory and processing power) it turns out that joint source-channel coders do outperform separable source and channel coders [20, 35]. Moreover, Shannon's separation principle does not hold in multiuser settings [22]. Further details on source coding and joint source-channel coding is available in [37, 40].

Since the focus of this chapter is on wireless systems, we discuss the basics of digital modulation in Section 2.2. We introduce the Shannon theoretic concept of channel capacity in Section 2.3. The channel capacity is defined as the maximum rate at which data can be *reliably* transmitted over the channel. The channel capacity concept however does not provide practical coding schemes that achieve this capacity. Practical code design and analysis is covered in Section 2.4. We then focus attention on multiuser communication systems in Section 2.5. Specifically, in Section 2.5.1 we consider random access methods enabled by algorithms at the data link layer. Common wideband multiuser methods at the physical layer, including direct sequence (DS) code division multiple access (CDMA), frequency hopping (FH) CDMA, and orthogonal frequency division multiplexing (OFDM), are covered in Sections 2.5.2.1, 2.5.2.2, and 2.5.2.3, respectively. Finally, we discuss several advanced wireless transceiver techniques in Section 2.6.

Due to the wide range of topics involved in the design of wireless systems, in this chapter, we discuss only the key results and the underlying intuition. Detailed mathematical proofs and derivations of the main results are omitted due to space constraints; relevant references are provided as appropriate. To stay in focus we only consider the transmission of digital data. As and when possible, we provide performance metrics that can be easily used by network designers to ensure that desired quality of service guarantees are provided by their designs.

2.2 Digital Modulation in Single User Point-to-point Communication

Consider the transmission of digital data at rate R bits per second. Let T_s denote the symbol period in which M bits are transmitted. Then, $R = M/T_s$. The information sequence to be transmitted is denoted by a sequence of real symbols $\{I_n\}$. This sequence could be the output of the channel coding block shown in Figure 2.2. A traditional modulation scheme consists of transmitting one of 2^M different signals every T_s seconds. In orthogonal modulation the signal $s(t)$ is given as

$$s(t) = \sum_n I_n q_n(t), \tag{2.1}$$

where $q_n(t)$ form a set of orthonormal pulses, *i.e.* $\int q_i(t)q_j(t)dt = \delta_{i-j}$, where δ_m is the Kronecker delta function which is defined as follows: $\delta_0 = 1$, and $\delta_m = 0, m \neq 0$.

In many traditional modulation schemes these orthonormal waveforms are generated as time-shifts of a basic pulse $p(t)$. Examples of $p(t)$ include the raised cosine and Gaussian pulse shapes. For example, in pulse-amplitude modulation (PAM), the transmit signal $s(t)$ equals

$$s(t) = \sum_n I_n p(t - nT_s). \tag{2.2}$$

In (2.2) for a 2-PAM system, $I_n \in \{-A, +A\}$, where the amplitude A is selected to meet average and peak power constraints at the transmitter. Similarly, for a 4-PAM system $I_n \in \{-3A, -A, +A, +3A\}$.

Consider the transmission of data through an ideal additive white Gaussian noise (AWGN) channel, in which the received signal, $r(t)$ is given by,

$$r(t) = s(t) + n(t), \tag{2.3}$$

where input signal $s(t)$ has an average power constraint P and the noise $n(t)$ has a constant power spectral density of $N_0/2$. The additive noise channel is one of the simplest and most widely analyzed channel. In practical systems, the noise at the receiver is a combination of several independent effects. Hence, using the central limit theorem (CLT) [31], the noise distribution is well approximated by a Gaussian density. Realistic wireless channel models are more sophisticated and are discussed in Chapter 3 and also in several references [29, 36].

At the receiver, the signal $r(t)$ given in (2.3) is first passed through a correlator to obtain

$$y_n = I_n \int_{T_s} p(t-nT_s)p(t-mT_s)dt + \int_{T_s} n(t).p(t-mT_s)dt. \tag{2.4}$$

It is known that the output of the correlator is a sufficient statistic for bit detection, *i.e.* all information about I_n that is contained in $y(t)$ is preserved in y_n. The orthonormality condition requires that the choice of the pulse shape should satisfy:

$$\int p(t-nT_s)p(t-mT_s)dt = \delta_{n-m}. \tag{2.5}$$

Under this condition, the output $y_n = I_n + z_n$, where it can be easily shown that z_n is a zero mean, Gaussian random variable with variance $N_0/2$. Under this condition, the output also does not have any inter-symbol interference.

The detection process involves making a decision on which symbol was transmitted given y_n is received. The detector design depends on the performance metric that one desires to optimize. For instance, the detector that minimizes the probability of error in the detection process, is the maximum a posteriori probability (MAP) detector. The MAP detector selects the $\hat{I}_n$ that maximizes $p(\hat{I}_n/y_n) = \frac{p(y_n/\hat{I}_n)p(\hat{I}_n)}{p(y_n)}$. If all possible symbols I_n are *a priori* equally likely, then the MAP detector is the same as the maximum likelihood (ML) receiver. The ML receiver maximizes $p(y_n/\hat{I}_n)$. Since the effective noise z_n is Gaussian, the MAP receiver also reduces to the minimum Euclidean receiver, *i.e.*, the constellation point that is the closest to the received signal is selected as an estimate for the transmitted signal. The decision region corresponding to this Euclidean distance based receiver is simple to derive and is shown in Figure 2.3 for two different modulation schemes.

The probability of error in this case can be computed as the tail probability of the Gaussian noise variable. For instance, the conditional probability of symbol error given symbol *"10"* is transmitted can be evaluated as the area under the tails of the Gaussian probability density function (pdf) as shown in Figure 2.3. The width of the Gaussian pdf equals the variance of the noise in the system. The overall probability of symbol error can be calculated by averaging over all possible transmit symbols. This probability of symbol error for M-ary PAM is given by

$$P_e \approx 2Q\left(\sqrt{\frac{6\log_2(M)SNR_{bit}}{M^2-1}}\right), \tag{2.6}$$

where, $SNR_{bit} = \frac{E_b}{N_0}$ is the SNR per bit and Q is the tail probability of the standard normal distribution [53]. The average bit energy $E_b = P_{avg}T_s$ where P_{avg} is the average power of the transmit symbols. A plot of this error probability as a function of SNR is given in Figure 2.4 for different values of M. Clearly, we can see that for a given SNR the error rate is higher for larger M. Also, for a given M, the error probability decreases with increasing SNR. In the presence of multiple users and other interference sources, the probability of error would depend on the signal to interference ratio (SIR) instead of just the SNR. The error performance can be further decreased at a given SNR using error control codes as discussed in Section 2.4. The probability of bit error can be evaluated from the probability of symbol error as follows. Typically, bits are assigned to symbols using a process known as *gray* coding in which neighboring symbols differ in only 1 bit as shown in Figure 2.3. With such a bit assignment process, a symbol error will cause only a single bit error with high probability. Hence the probability of bit error $P_{bit-error} \approx \frac{1}{\log_2(M)} P_e$.

The transmitted signal in the above case is actually a signal in baseband. Typically, this signal is modulated up to a pass-band frequency, f_c before transmission. This pass band signal could be obtained as $s(t)e^{-j2\pi f_c t}$ which is however, a complex signal. To make the transmitted signal real, we can add its complex conjugate to obtain $s_p(t) = s(t)e^{-j2\pi f_c t} + s(t)e^{j2\pi f_c t}$.

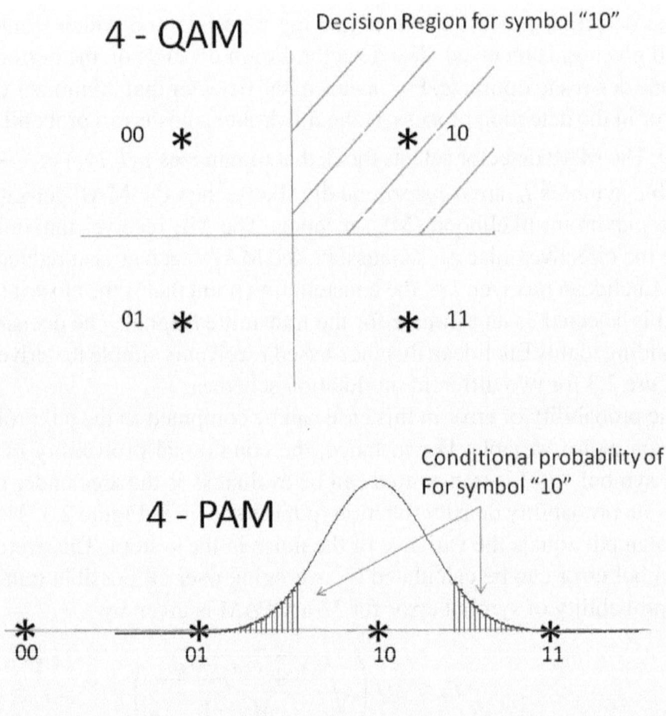

Fig. 2.3 Euclidean distance based detection and error calculation for various modulation schemes. The figure also demonstrates the concept of gray coding, in which neighboring symbols only differ in 1 bit.

The bandwidth of the signal is defined as the range of positive frequencies that are transmitted. It can easily be seen that the bandwidth of $s_p(t)$ equals twice the bandwidth of $s(t)$; this process is referred to as double sideband modulation. Clearly, this process is wasteful in bandwidth.

One way to use the bandwidth more efficiently is to use a complex signal constellation $u(t)$ instead of a real $s(t)$. Thus the transmitted passband signal is now $u(t)e^{-j2\pi f_c t} + u^*(t)e^{j2\pi f_c t}$ where $u^*(t)$ is the complex conjugate of $u(t)$.

Examples of such complex constellation are the quadrature amplitude modulation (QAM) signals given in Figure 2.5. The error performance of QAM signals can be derived in a manner similar to that for PAM signals and is given by [34]

$$P_e \leq 4Q\left(\sqrt{\frac{3\log_2(M)SNR_{bit}}{M-1}}\right). \tag{2.7}$$

It turns out that for high SNR and large values of M this upper bound is quite tight. The probability of error versus SNR for QAM is given in Figure 2.4.

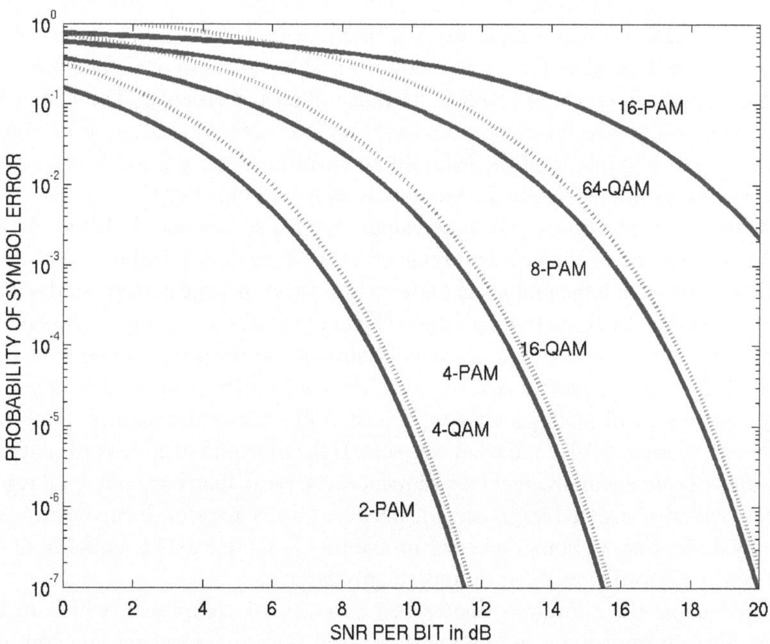

Fig. 2.4 Probability of symbol error with specific PAM and QAM schemes.

2.3 Information Theoretic Capacity and Coding

Consider the discrete time additive Gaussian noise channel given by

$$y_n = x_n + z_n \tag{2.8}$$

where, transmit signal x_n has power constraint P and Gaussian noise z_n has 0 mean and variance of σ^2. A fundamental question that arises is to find the maximum rate at which data can be reliably transmitted over this channel. This question is answered in an emphatic manner by Shannon's landmark paper [39]. For this ideal discrete time real AWGN channel the Shannon capacity [39] is given by

$$C_{ideal} = 0.5\log(1+SNR) \text{ bits per channel use} \tag{2.9}$$

The interpretation of this capacity is that for transmission at any rate $R > C_{ideal}$, the probability of error in decoding that message at the receiver is necessarily bounded away from 0. Shannon also showed that information can be transmitted at any rate $R < C_{ideal}$ with arbitrarily small probability of error.

This capacity can be achieved by using a *random* Gaussian codebook as follows. To transmit R bits per use of a channel, a codebook is constructed containing $M = 2^{nR}$ codewords that each spans n symbols. Each symbol in the codebook is selected as an independent Gaussian random variable with zero mean and variance N. This codebook is revealed to both the transmitter and receiver. The codeword to be transmitted is selected at random (with uniform density) based on nR information bits. The n symbols of the selected codeword are transmitted in n uses of the channel. An example of such a codebook is shown in Figure 2.6.

At the receiver, decoding is done using a typical set decoder [14, 54]. A typical set for a sequence of independent realizations of a random variable is the set of all sequences for which the empirical entropy is close to the true entropy of the random variable. In this decoding method, the receiver compares the received sequence with all possible 2^{nR} codewords to see which pairs of the received sequence and codewords belong to the *jointly typical set*. If the transmitted codeword is the one and only one codeword that is jointly typical with the received sequence, then the decoder can "correctly" decode what was sent. If the transmitted codeword sequence is not jointly typical with the received sequence or if more than one codeword is jointly typical with the received sequence, the decoder makes an error. It can be proved that for any desired upper bound on error probability ε, there exists a sequence of codes that achieve error below ε for asymptotically large n.

The typical set decoding process has exponential complexity which makes it impractical to implement in real systems. Real systems use some low complexity decoders based typically on the maximum likelihood (ML) criterion or maximum a posteriori probability (MAP) criterion as discussed in Section 2.4. Further, rather than using *random* codes, practical systems use codes with structure. The fundamental challenge in code design is to find codes with *just enough* structure to enable design of a low complexity decoder and *just enough* randomness to ensure good performance.

A plot of the capacity versus SNR is given in Figure 2.7. As expected, the capacity shows a linear increase with SNR for very small SNR values and a logarithmic increase for higher SNR's. Also shown in the figure is the achievable rate for various PAM schemes. This achievable rate is calculated as the mutual information between an equiprobable PAM constellation input to the channel and the output of the channel [16]. As expected the rate for the M-PAM is upper bounded by $\log_2(M)$.

In addition to the complexity issue mentioned before, another challenge in using the random coding formulation in practical systems is that it does not give any indication of the size of codebook required to achieve a desired error probability. To provide some theoretical guidance on the performance of the best coding scheme for a given length, the theory of error exponents has been developed [18]. For a given code of length n, the error exponent is the logarithm of the error probability. Specifically, for an ensemble of random codes, it can be shown that the average error probability is upper bounded as,

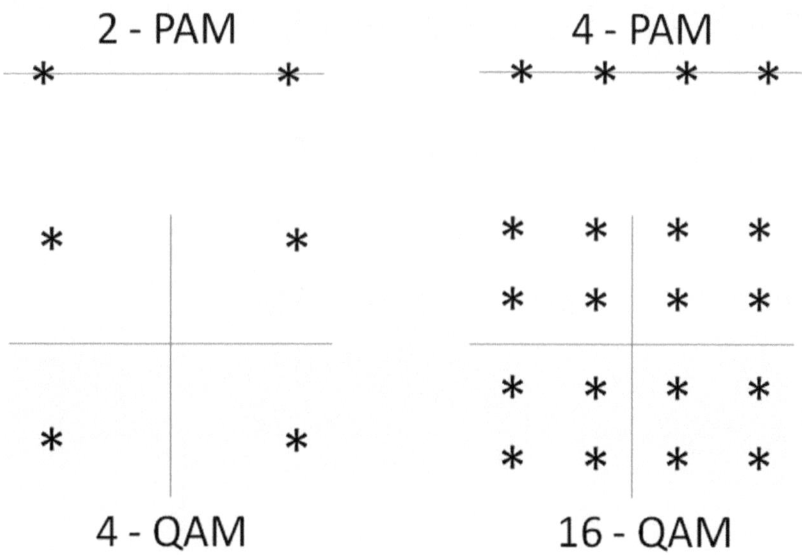

Fig. 2.5 Different digital modulation schemes

$$P_{error} \leq e^{-nE_{random}(R)} \tag{2.10}$$

where, $E_{random}(R)$ is the random coding exponent and R is the rate of transmission. This random coding exponent can be calculated based on the channel characteristics. For the Gaussian noise channel,

$$E_{random}(R) = 1 - \beta + \frac{SNR}{2} + 0.5\log(\beta - \frac{SNR}{2}) + 0.5\log(\beta) - R, \tag{2.11}$$

$$\text{where } \beta = 0.5[1 + \frac{SNR}{2} + \sqrt{1 + \frac{SNR^2}{4}}].$$

This expression for the error exponent is valid for transmission rates $R < 0.5\log[0.5 + \frac{SNR}{4} + 0.5\sqrt{1 + \frac{SNR^2}{4}}]$. For rates above this threshold and below the capacity, the expression for the error exponent is slightly different and is given in [18]. Further, a lower bound on the decoding error probability for any code over a particular channel is given by the sphere packing error exponent, which can be computed in a manner similar to the random coding error exponent [18].

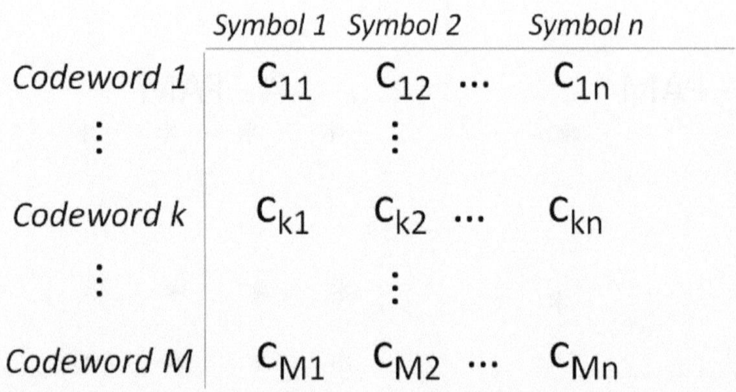

Fig. 2.6 Illustration of a random codebook used to prove achievability of Shannon's channel capacity.

Similar to the discrete time channel, for a continuous time additive white Gaussian noise channel with bandwidth W the capacity can be calculated as

$$C_{AWGN} = W\log(1+P/N_0W)\text{bits/sec} \tag{2.12}$$

Further, the capacity of wireless fading channels has been well investigated under several conditions [10]. The concept of channel capacity has also been extended to multiuser channels. We illustrate the idea in 2 different settings.

1. Uplink Channel: Consider a simple uplink channel in which multiple users are communicating with a single base station. Again, consider the simple Gaussian noise channel in which the received signal $y_{bs}(n)$ at the base station is given by

$$y_{bs}(n) = \sum_{i=1}^{K} x_i(n) + z_{bs}(n) \tag{2.13}$$

where $x_i(n)$ is the signal transmitted by user i and z_{bs} is the effective noise at the base station receiver, which is modeled as having a Gaussian pdf. Similar to the capacity of a single user channel, in this case, the capacity region is defined as the set of all $K-$tuples of achievable rates for the various users.

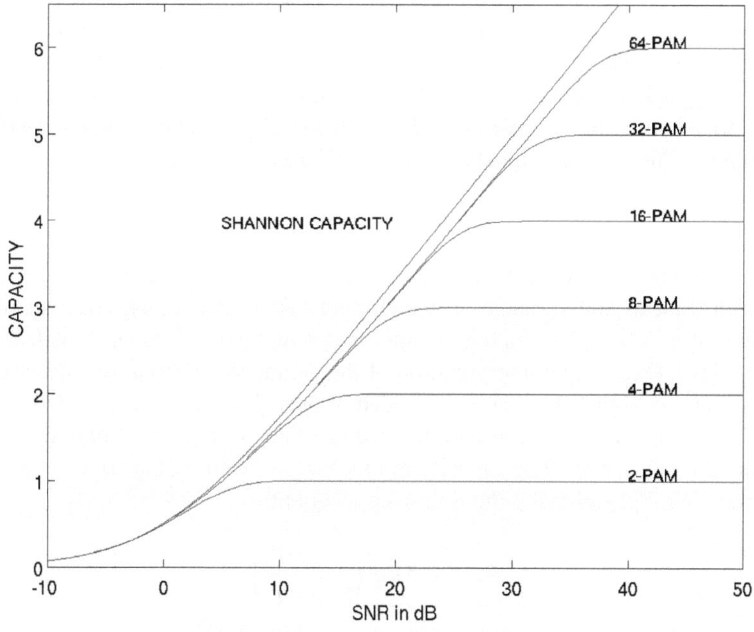

Fig. 2.7 Capacity of AWGN channel and maximum possible transmission rates with various modulation schemes.

For instance, for the $K=2$ user channel, this capacity is given by the convex hull of rate pairs (R_1,R_2) satisfying the following inequalities:

$$\begin{aligned} R_1 &\le 0.5\log\left(1+\frac{P_1}{N}\right) \\ R_2 &\le 0.5\log\left(1+\frac{P_2}{N}\right) \\ R_1+R_2 &\le 0.5\log\left(1+\frac{P_1+P_2}{N}\right) \end{aligned} \tag{2.14}$$

where P_i is the average power constraint at transmitter i and N is the variance of the noise at the receiver. This capacity region is illustrated in Figure 2.8. The interpretation of this capacity is that any rate pair within the boundary of the pentagon capacity region can be simultaneously achieved for both users with any desired error probability. Further, any rate pair outside the boundary cannot be simultaneously supported for both users and still ensure low probability of error. Also shown in the figure is the achievable rate with *naive* TDMA, in which user i exclusively access the channel for a fraction α_i of the time, where $\sum_i \alpha_i = 1$. The achievable rates with FDMA is also shown in the figure. It can be seen that FDMA achieves the boundary of the capacity region at one point; this point is the case when the whole bandwidth is allocated to the different users in proportion to their powers. Recognize that in a TDMA scheme, user i is only active for a portion α_i of the time, where $\sum_{k=1}^{K} \alpha_k = 1$. Hence, if we adjust the transmission power of user i to $\frac{P_i}{\alpha_i}$ its performance will improve. The performance of this variation of TDMA is identical to the FDMA scheme shown in Figure 2.8.

2. Downlink Channel: Now, consider a downlink channel in which a base station is transmitting to multiple users. Let $x_{bs}(n)$ be the signal transmitted by the base station. The received signal $y_i(n)$ at the ith user is given by

$$y_i(n) = x_{bs}(n) + z_i(n) \tag{2.15}$$

where $z_i(n)$ is the noise at the receiver and is modeled as being Gaussian with 0 mean and variance of N_i. Without loss in generality, consider the case that $N_i < N_j \, \forall \, i < j$. Such a channel is denoted as a degraded broadcast channel [14]. One simple interpretation of the degraded Gaussian broadcast channel is that any signal that can be decoded at user j can also be decoded at user i, since user i has lower noise on the average than user j. For simplicity, consider the 2-user channel. The capacity in this case is given by the convex hull of rate pairs (R_1,R_2) satisfying the following inequalities:

$$\begin{aligned} R_1 &\le 0.5\log\left(1+\frac{\alpha P}{N_1}\right) \\ R_2 &\le 0.5\log\left(1+\frac{(1-\alpha)P}{N_2+\alpha P}\right) \end{aligned} \tag{2.16}$$

where P is the power constraint at the base station. This capacity region is illustrated in Figure 2.9. Again, the achievable rate with FDMA is also shown in the figure.

These theoretical capacity bounds are very useful for network designers in planning and resource allocation issues.

2.4 Channel Coding

Channel codes, also referred to as forward error correction (FEC) codes, are used to reduce the effective bit error rate when data is sent over a lossy medium such as a wireless channel. There are two main categories of FEC codes - block codes and convolutional codes. We now briefly discuss the properties of these codes. For simplicity, we only discuss binary codes in this chapter. Codes over other alphabets are discussed in several references such as [26, 51].

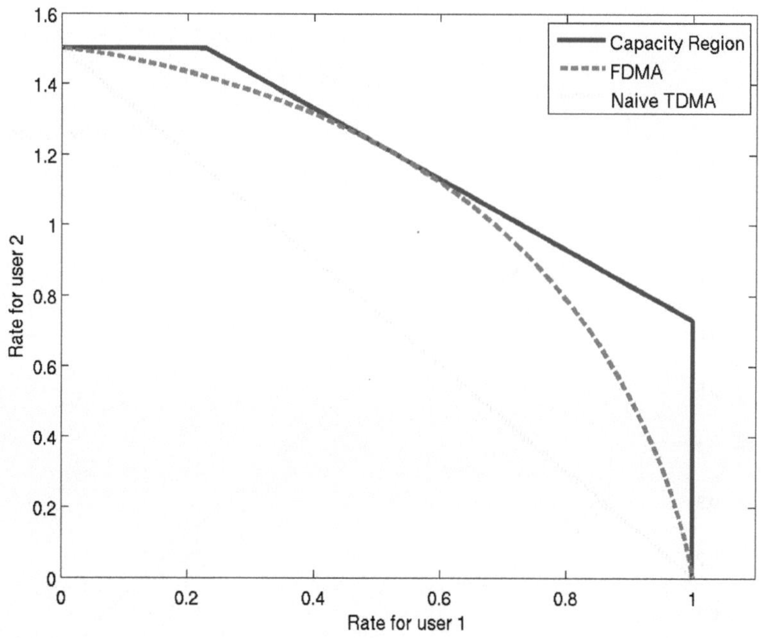

Fig. 2.8 Illustration of capacity of an uplink channel

2.4.1 Block Codes

In a simple block code, the information data stream is parsed into non-overlapping groups of a fixed size, say k bits. Each group of k bits is then mapped to a group of $n > k$ bits independent of the bits in the other groups. This mapping into a higher dimensional space essentially defines the code and is responsible for the error correction and detection capabilities of the code. Since each group of n bits are essentially independent of other groups, this type of coding is sometimes referred to as memoryless. The rate, r, of the channel code is defined as $r = \frac{k}{n}$. The random coding described in Section 2.3 and also depicted in Figure 2.6 is a block code (over a non-binary alphabet) with $k = \log_2(M)$ and n coded symbols.

The channel encoding can be achieved using a generator matrix $\mathbf{G}$ of size $k \times n$ which uniquely defines the code. Let $\mathbf{x}$ and $\mathbf{y}$ denote, respectively, the $1 \times k$ and $1 \times n$ vectors of the information bits and coded bits. Then $\mathbf{y} = \mathbf{x}\mathbf{G}$. For each code, there is also a unique parity check matrix $\mathbf{H}$ of size $n - k \times n$ such that $\mathbf{y}\mathbf{H}^T = 0$, for all valid codewords $\mathbf{y}$, which also implies that $\mathbf{G}\mathbf{H}^T = 0$.

There are several decoding algorithms for block codes depending on the type of code and the channel over which the code is used. These algorithms offer the system designer several options to trade-off the complexity of the decoding process

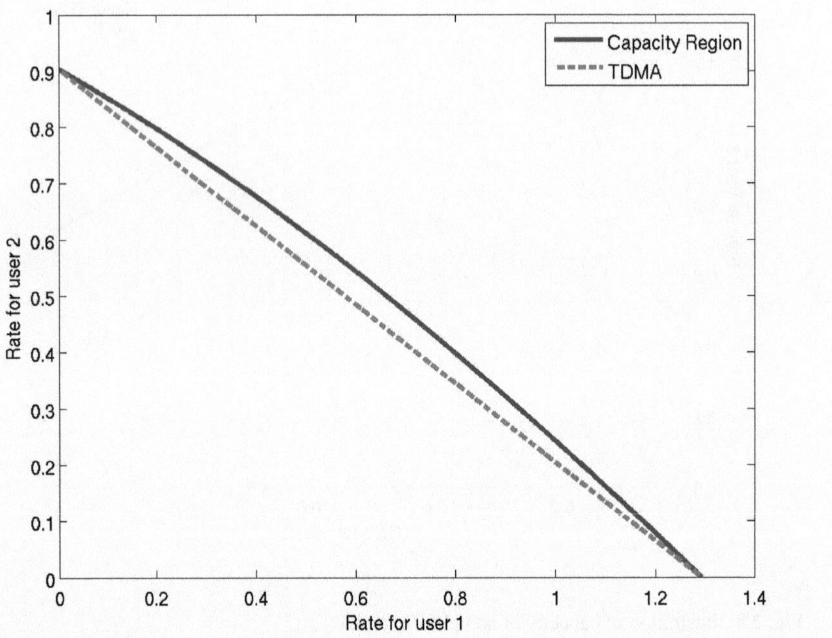

Fig. 2.9 Illustration of capacity of a downlink channel

with the performance achieved. For example, consider a simple binary symmetric channel (BSC) which randomly flips each of the coded bits with a certain probability. Hence, the received vector $\mathbf{b} = \mathbf{y} + \mathbf{e}$, where $\mathbf{e}$ is a binary error vector in which a value of 1 indicates that the corresponding bit is flipped.

The maximum likelihood decoding in this case selects the codeword that is closest to $\mathbf{b}$ in terms of the Hamming distance,

$$\hat{\mathbf{y}} = \arg\min d_H(\mathbf{c}, \mathbf{b}) \tag{2.17}$$

This ML decoding can be implemented using a standard array decoder [27]. The standard array decoder is similar to a look up table based decoder, with a separate entry for each possible received vector $\mathbf{b}$. The standard array gives the ML estimate of the error vector for each of the 2^n possible received vectors. Thus, the decoder simply looks for the error corresponding to the vector $\mathbf{b}$ and then subtracts the error vector from $\mathbf{b}$ to obtain the ML estimate of the transmitted codeword.

A major problem with this type of decoding is that the memory requirements for storing the standard array becomes prohibitively large for even reasonable size codebooks. Further, the search procedure to find the correct element in the array is also computationally intensive. To reduce these complexities, another decoding process called syndrome decoding is sometimes used. The syndrome, $\mathbf{s}$ of vector $\mathbf{b}$ is defined as $\mathbf{s} = \mathbf{b}\mathbf{H}^T = (\mathbf{y} + \mathbf{e})\mathbf{H}^T$. Since $\mathbf{y}\mathbf{H}^T = 0$ it turns out that the value of the syndrome only depends on the error vector. Thus, in syndrome decoding a table of all 2^{n-k} possible syndromes and the error vectors associated with them are stored. Decoding is achieved by simply computing the syndrome for the vector $\mathbf{b}$ and then looking up the associated error vector from the table. It turns out that in many cases the complexity and memory requirements of syndrome decoding are still prohibitively high. In a typical minimum distance decoder, which is often used over Gaussian wireless channels, the decoder calculates the distance of the received signal to each of the 2^k possible transmit sequences and selects the one with the minimum distance. Decoders with lower memory and complexity requirements have been derived which effectively exploit the structure in the codes [23].

The error correction and detection performance of a channel code depends on the *distance spectrum* of the code. Essentially, the distance spectrum is an indication of the separation between the 2^k codewords of length n. The calculation of this distance spectrum is, however, not an easy task in many scenarios. Fortunately, the performance of the code can be approximated using a single quantity referred to as minimum distance, typically denoted by d. This minimum distance d equals the minimum of the Hamming distance between all pairs of codewords. The Hamming distance between two binary strings is the number of positions in which the two strings differ. For the special case of linear block codes, the minimum distance of the code also equals the minimum (non-zero) weight of the code.

The weight distribution of codes is conveniently represented in polynomial form as $B(z) = \sum_{i=0}^{n} B_i z^i$, where B_i is the number of codewords with Hamming weight i. For example, the weight distribution of Hamming codes is given by

$$B(z) = \frac{1}{n+1}\left[(1+z)^n + n(1-z)(1-z^2)^{(n-1)/2}\right] \tag{2.18}$$

The calculation of this minimum distance is a non-trivial task in many cases. There are several options available to a system designer in selecting the appropriate code.

1. Use a code from an existing family of codes for which the minimum distance has already been calculated. Examples of such codes are the Hamming codes which have $k = 2^m - m - 1, n = 2^m - 1$ for any $m \geq 3$. It is well known that the minimum distance for all Hamming codes is $d = 3$. The generator matrix $\mathbf{G}$ for a Hamming code, has a specific structure. The columns of $\mathbf{G}$ are the $2^m - 1$ binary codewords of length k excluding the all zero codeword. The order of these columns does not affect the properties of the code.
2. Use bounds to approximate the Hamming distance. For instance, for a (n,k,d) code, the Hamming bound gives the minimum amount of redundancy $(n-k)$ required as
$$n - k \geq \log_2 V(n, \lfloor\frac{d-1}{2}\rfloor) \tag{2.19}$$
where $V(n,t) = \sum_{j=0}^{t} C_j^n$ is the number of binary vectors of length n that are at most Hamming distance t from a given vector.
3. Evaluate the minimum distance using numerical simulations.

There is a basic trade-off between code rate r and minimum distance d. The smaller the code rate, the larger is the minimum distance and vice-versa. Given a code with a certain minimum distance d, a simple bound on the probability of decoding error for a bounded-distance based decoder is given by

$$P_{decoding-error} \leq \sum_{j=t}^{n} \binom{n}{j} \alpha^j (1-\alpha)^{n-j} \tag{2.20}$$

where α is the crossover probability of the BSC and $t = \lfloor\frac{d-1}{2}\rfloor$. The probability of bit error can be easily lower and upper bounded as $\frac{1}{k}P_{decoding-error} \leq P_{bit-error} \leq P_{decoding-error}$.

2.4.2 Convolutional Codes

In a convolutional code, similar to a block code, the information bits are parsed into non-overlapping groups of k bits and each group is mapped to $n > k$ bits. However, unlike block codes, the mapping is dependent on the bits in the prior groups which introduces memory in the system. The n coded bits thus depends both on the input k bits as well as the state of the system. The constraint length, K, of the code is an indicator of the effective memory in the system.

One advantage of a convolutional code is that the encoder can be implemented as a digital filter, using shift registers and adders. An example of such an encoder is shown in Figure 2.10. While there are several analytical representations for convolutional codes, for simplicity we only present a couple of graphical representations.

Consider for instance the rate 1/2 code shown in Figure 2.10. In this case, for each input bit x_i, the encoder produces two output bits, $y_i(1) = x_i + x_{i-2}$ and $y_i(2) = x_i + x_{i-1}$. The state diagram corresponding to this code is also shown in Figure 2.10. The advantage of this approach is that it clearly represents the relationship between the current state and next state as a function of the input and the output bits. One disadvantage of the state diagram approach is that the temporal sequence of coded bits is difficult to represent with this approach. For this reason, the equivalent trellis is used and is shown in Figure 2.11.

There are two main types of decoders for convolutional codes. The first type of decoder is the maximum likelihood sequence decoder which is typically implemented using a Viterbi algorithm [48]. The second decoder is the maximum a posteriori probability decoder which is implemented using the BCJR algorithm [3]. The Viterbi algorithm has much lower complexity than the BCJR algorithm whereas the performance of both are nearly similar. The description of these algorithms are fairly complex and hence relegated to the appropriate references [33]. It turns out that the trellis description shown in Figure 2.11 is useful in describing the Viterbi decoding algorithm. The enumeration of the performance of convolutional codes has received significant attention in the literature [9, 27] and is not duplicated here for conciseness.

The performance of convolutional codes over an AWGN channel is shown in Figure 2.12 for three different codes with rates $\frac{1}{2}$, $\frac{1}{3}$ and $\frac{1}{4}$. The constraint length of these codes were set to $K = 3$. Clearly, we can see that the effective bit error rate reduces as the rate of the code decreases. Also, in this case, the performance of the soft decoder is about 2dB better than the hard decoder.

Convolutional codes are also the basic elements of the hugely popular Turbo codes [5]. The original Turbo codes are essentially a concatenation of two convolutional codes acting on different pseudo-random permutations of the input bits. The performance of Turbo codes is close to the best possible theoretical limits.

The choice of whether to use block codes or convolutional codes in a specific scenario has been a subject of extensive debate between code designers. Some coding theorists believe that convolutional codes offer superior performance over block codes for the same complexity. Many real systems use more sophisticated coding methods such as trellis coded modulation [45] and concatenated codes [15].

2.5 Multiuser Communication

One of the features of the wireless medium is its broadcast nature which makes simultaneous use of the spectrum by multiple users in a certain region challenging. This multiple access is accomplished using techniques at physical and/or data link layers. Two of the simplest and intuitive methods of multiple access is to assign the different users non-overlapping time-slots and/or frequency bands. Such schemes are referred to as time-division multiple access (TDMA) and frequency division multiple access (FDMA). The popular GSM standard uses a combination of time and frequency division multiplexing to support multiple users. For instance, the basic GSM standard specifies different groups of carrier frequencies for each base station. Within each base station, each carrier frequency is shared among 7 different users in a time-division manner.

2.5.1 Data Link Layer Random Access Methods

Traditional data networks such as Ethernet which primarily support packet data traffic (which is bursty in nature) tend to enable random access using a data link layer protocol. The basic scenario being considered in the design of these protocols is as

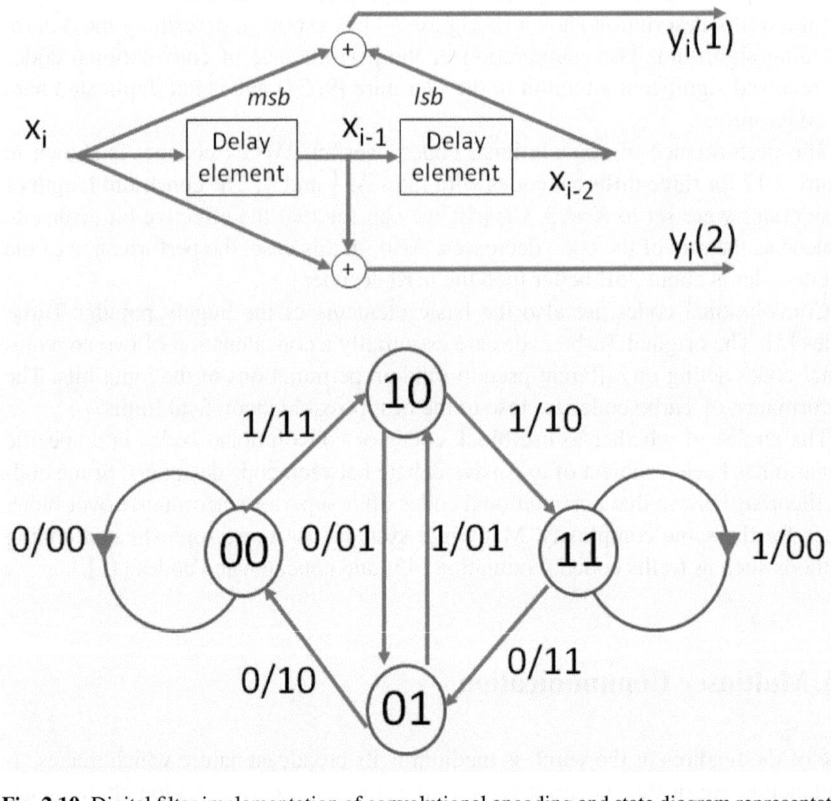

Fig. 2.10 Digital filter implementation of convolutional encoding and state diagram representation of convolutional codes

follows. Let there be N distributed nodes that are sharing the wireless medium. Traffic arrivals at each node is modeled as a random process, such as a Poisson point process. Some basic assumptions are typically made to enable simple analysis of these protocols. For instance, if two or more nodes transmit a packet at the same time, a *collision* is said to occur and the receiver will be unable to accurately decode either of these packets.[1] It is also assumed that some form of instantaneous feedback mechanism exists from the receiver to the transmitter indicating whether a packet is received successfully or not. This feedback in turns enables the transmit node to retransmit packets that are not received correctly.

The earliest and simplest form of multiple access is the well known Aloha method [1]. In an Aloha system, users simply transmit their packets whenever they have data to be sent. Clearly, such a scheme might cause significant amount of collisions if multiple packets are sent simultaneously. Further, if a collision is detected, the nodes which are involved in the collision, delay their subsequent retransmissions by random amounts. This randomization will allow the multiple competing nodes to reduce their probability of collision in their retransmissions. It turns out that the maximum throughput of Aloha is severely limited. Specifically, if the wireless chan-

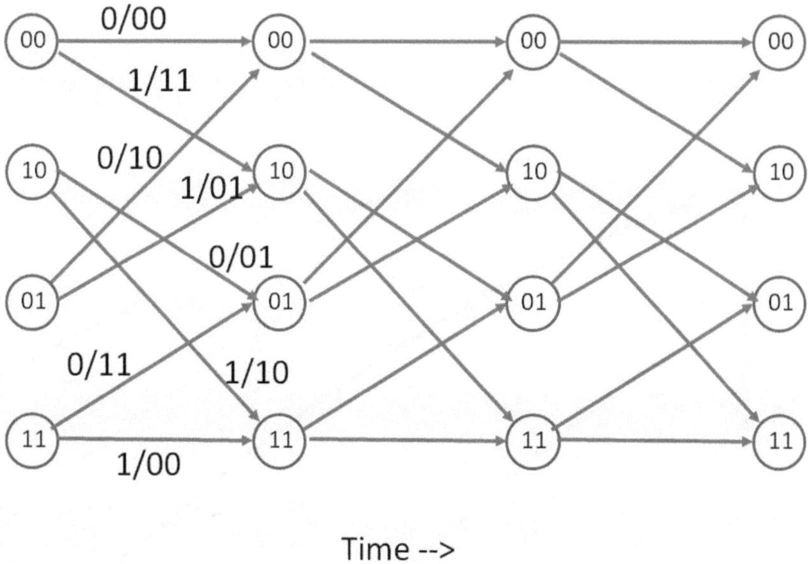

Fig. 2.11 Depiction of convolutional code using a trellis.

[1] This assumption has been modified in recent years to address the issue of *capture*, which is discussed later.

nel can support a peak data rate of R, then with random multiple access using the Aloha protocol, the maximum effective data rate is only $\frac{1}{2e}R$, or approximately 18% of data rate. However, for lightly loaded systems Aloha is a simple strategy that offers reasonable performance. An excellent historical perspective of Aloha written by the original inventor including recent application areas is given in [2].

Several variations of Aloha including slotted-Aloha have also been proposed which improve the performance of the basic scheme [7]. In a slotted Aloha system, the entire time is discretized into time-slots of fixed duration and nodes are allowed to transmit packets only at the beginning of a time-slot. Further, each node that has a packet to transmit will only transmit it with a finite probability q at the beginning of the next time-slot. These changes to Aloha protocol, reduces the number of collisions and increases the overall throughput. The maximum data rate of a slotted Aloha system can be shown to equal $\frac{1}{e}R$.

Many recent multiple access (MA) strategies are based on the idea of carrier sense multiple access (CSMA) [7, 25]. Several modifications of CSMA have been proposed in the literature. In a CSMA scheme, each node listens to the medium to see if the medium is free or if there exists any other signal in the medium. If the wireless medium is believed to be free then the node transmits its data after waiting for a random amount of time. If the medium is sensed as being busy, then

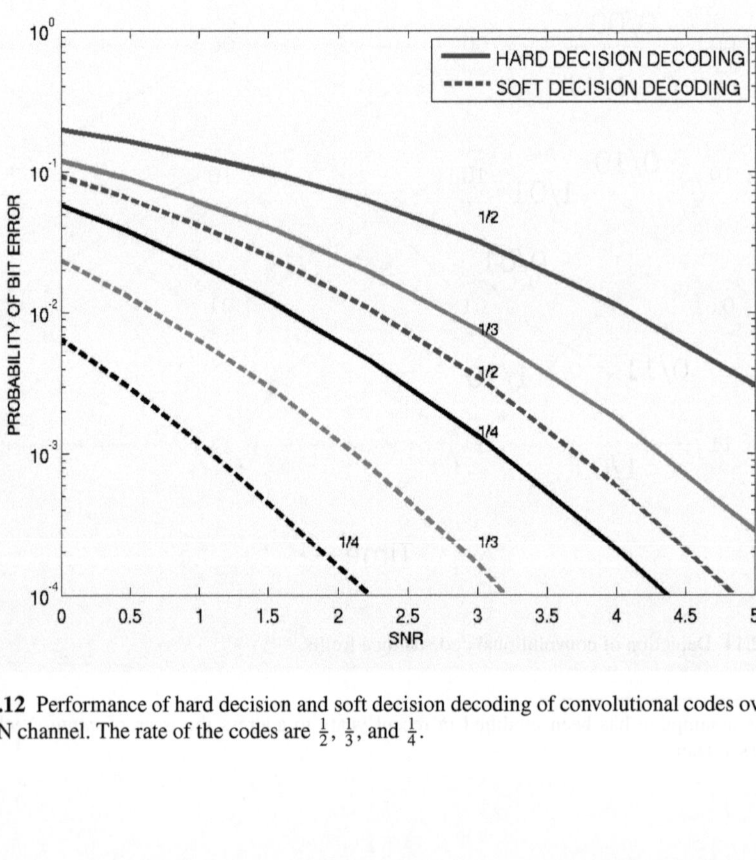

Fig. 2.12 Performance of hard decision and soft decision decoding of convolutional codes over an AWGN channel. The rate of the codes are $\frac{1}{2}$, $\frac{1}{3}$, and $\frac{1}{4}$.

the node backs off from transmission for a random amount of time and then senses the channel again. However, CSMA still suffers from collisions if multiple nodes are constantly monitoring a busy medium waiting for it to become free. As soon as the medium becomes free all these nodes might begin transmission simultaneously resulting in a packet collision.

While CSMA offers significant improvement in performance over Aloha type protocols, it suffers from the hidden node and exposed node problems. Consider the scenario shown in Figure 2.13, where nodes B and C are trying to communicate with node A. If nodes B and C are not within range of each other, they cannot detect the presence of the other node's signal and might try to simultaneously communicate with A. These transmissions would likely result in a collision and node A may not be able to decode either of these signals. This problem is commonly referred to as the hidden node problem.

Now consider, the same scenario as shown in Figure 2.13. If node A is trying to send data to node B, then node C is prevented from transmitting due to carrier sense aspect. However, it is possible for node C to simultaneously transmit data to a different node since its signal is not received at node B and hence will not cause any interference. This problem is called the exposed node problem.

A related problem that sometimes occurs in wireless networks is the issue of *capture*. For example, if signals from two different transmitters are received at a particular node at significantly different power levels, then the transmitter with the stronger signal will *capture* the medium. This capture will prevent the weaker transmitter from sending data and lead to severe unfairness in the resource allocation.

Virtual carrier sense is proposed as a solution to overcome the hidden node problem [19]. In virtual carrier sense, each node that wants to transmit a packet first sends a short request-to-send (RTS) packet to the intended destination. If the destination node receives this packet successfully and determines that there are no other active signals in its vicinity, it replies with a short clear-to-send (CTS) packet. These RTS and CTS packets also include a field indicating the amount of time that the medium is expected to be busy. All nodes that receive either the RTS or CTS packet will not attempt transmission during the period of time specified in these packets. The use of RTS/CTS mechanism can also solve the exposed node problem under certain scenarios; for instance, using the MACAW protocol [8]. The fairness issues associated with capture can also be solved using more sophisticated back-off mechanisms.

While the use of RTS/CTS coupled with a CSMA variant scheme reduces collisions significantly, it can add significant overhead especially if the packet size is small. The other problem is that in some scenarios nodes might back-off unnecessarily thereby reducing the number of simultaneous transmissions in the network and effectively reducing the network throughput. The popular 802.11 based standard uses a combination of CSMA with the RTS/CTS mechanism for multiple access.

2.5.2 *Physical layer Multiuser Signaling Schemes*

Traditional cellular networks which primarily support voice tend to also build multiple access capabilities in the physical layer signaling. Next generation wideband networks tend to uses a variation of CDMA or OFDM, which are described next.

2.5.2.1 Direct Sequence Code Division Multiple Access DS-CDMA

In a direct sequence CDMA system, all users use the same frequency and time to communicate. User separation is achieved by modulating the bits of each user with a distinct spreading sequence. By a careful selection of these spreading sequences, it is possible to recover the bits of each user from a linear combination of their transmit sequences. In many systems, these spreading codes are generated to form an orthonormal set. Although we only discus uplink communications in this section, CDMA can also be used in downlink channels. The CDMA2000 standard specified by the International Telecommunication Union (ITU) is based on DS-CDMA technology.

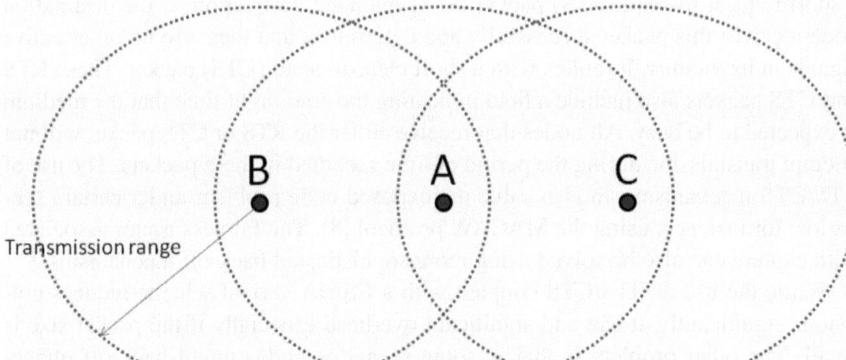

Fig. 2.13 Topology showing the hidden node problem. If both nodes *B* and *C* transmit at the same time, their signals collide at node *A*.

Consider an uplink CDMA system with K users. A simplified block diagram of the entire uplink CDMA system is shown in Figure 2.14. The transmit signal $x_k(t)$ for user k at time t is given by,

$$x_k(t) = \sqrt{E_k} \sum_{m=0}^{\infty} b_k^{(m)} \sum_{l=0}^{N_c-1} c_{k,l}\, p(t - mT_s - lT_c), \tag{2.21}$$

where, $b_k^{(m)}$ is the mth bit of user k, E_k is the energy, $c_{k,l}$ is the spreading sequence and $p(t)$ is the normalized pulse of duration T_c. The chip period and symbol period are represented by T_c and T_s, respectively. The ratio $\frac{T_s}{T_c}$ is referred to as the spreading gain of the DS-CDMA system. The received signal, $y(t)$, at the base station equals,

$$y(t) = \sum_k h_k x_k(t) + n(t), \tag{2.22}$$

where, h_k is the channel between user k and the base-station and $n(t)$ is the additive noise at the base-station receiver. For simplicity, we have not shown the temporal or spectral variations of the channel h_k. Detailed models for h_k are discussed in Chapter 3.

This received signal, $y(t)$, is passed through a bank of parallel filters that are matched to the spreading codes of the various users. The output, w_k of the kth

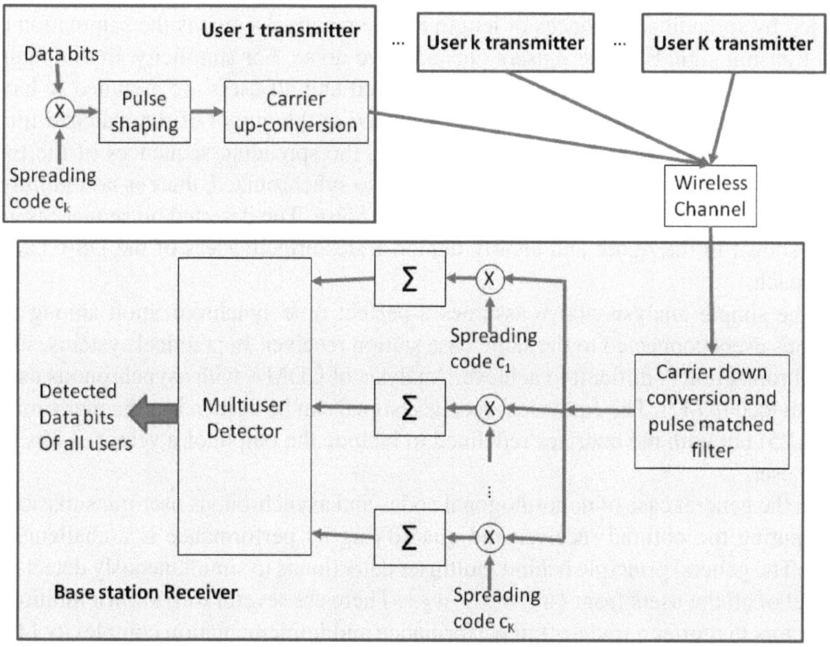

Fig. 2.14 Basic block diagram of an uplink DS-CDMA system.

matched filter for the mth time period equals

$$w_k^{(m)} = \sqrt{E_k} b_k^{(m)} h_k < c_k, c_k > + \sqrt{E_k} b_k^{(m)} \sum_{j \neq k} h_j < c_k, c_j > + n_{eff} \tag{2.23}$$

where $< ., . >$ represents the inner product of the two vectors. In the special case that the spreading codes for the various users are selected to belong to an orthonormal set (*i.e.* $< c_k, c_j > = 0$ and $< c_k, c_k > = 1$, for $k \neq j$), then (2.23) reduces to the form

$$w_k^{(m)} = \sqrt{E_k} h_k b_k^{(m)} + n_{eff} \tag{2.24}$$

Effectively, the output of the matched filter in (2.24) equals the scaled signal of user k corrupted by additive white Gaussian noise. The effect of the multiuser interference is completely eliminated due to the orthogonality of the codes selected.

In the general case of non-orthogonal codes, an equivalent discrete time model for the output of the matched filters for bit m of all users can be written in matrix form as

$$\mathbf{w}^{(k)} = \mathbf{RAb}^{(k)} + \mathbf{n} \tag{2.25}$$

where $\mathbf{w}^{(k)} = [w_1^{(k)}, w_2^{(k)}, \ldots, w_K^{(k)}]$, $\mathbf{b} = [b_1^{(k)}, b_2^{(k)}, \ldots, b_K^{(k)}]$ and $\mathbf{n} = [n_1, n_2, \ldots, n_K]$. Matrix $\mathbf{R}$ is the cross-correlation between the various spreading sequences and matrix $\mathbf{A}$ is a diagonal matrix of the channel gains $\{h_k\}$.

Figure 2.15 illustrate the various signals in a DS-CDMA system. The figure shows the first 5 bits in a simple 2-user CDMA system. The bit sequences are modulated by spreading sequences of length 8. The received signal is the summation of the transmit signals of the 2 users plus additive noise. For simplicity, in this figure the attenuation in the channel is not considered and all users are assumed to have the same transmit power. The signals recovered at the output of the matched filter are also shown in the figure. In this example, the spreading sequences of the two users are orthogonal and since the users are also synchronized, there is no multiuser interference and the bits are only corrupted by noise. The detected bit sequences are also shown in the figure and clearly demonstrates effectiveness of the DS-CDMA approach.

The simple analysis above assumes a perfect time synchronization among the various users connected to the same base station receiver. In practical systems, such synchronization is difficult to achieve. Analysis of CDMA with asynchronous users is provided in [47]. The equivalent received signal can be written in a manner similar to (2.25) but with the matrices redefined to include the output of a vector of bits for each user.

In the general case of nonorthogonal codes and asynchronous user transmissions, computing the optimal receiver and quantifying its performance is a challenging task. The general principle behind multiuser detection is to simultaneously detect the signal of all the users from $\{w_1, w_2, \ldots w_K\}$. There are several well known multiuser detectors that offer a trade-off in performance and implementation complexity [47]. Typically, the performance of a MUD depends on the signal-to-interference (SIR) ratio before bit detection.

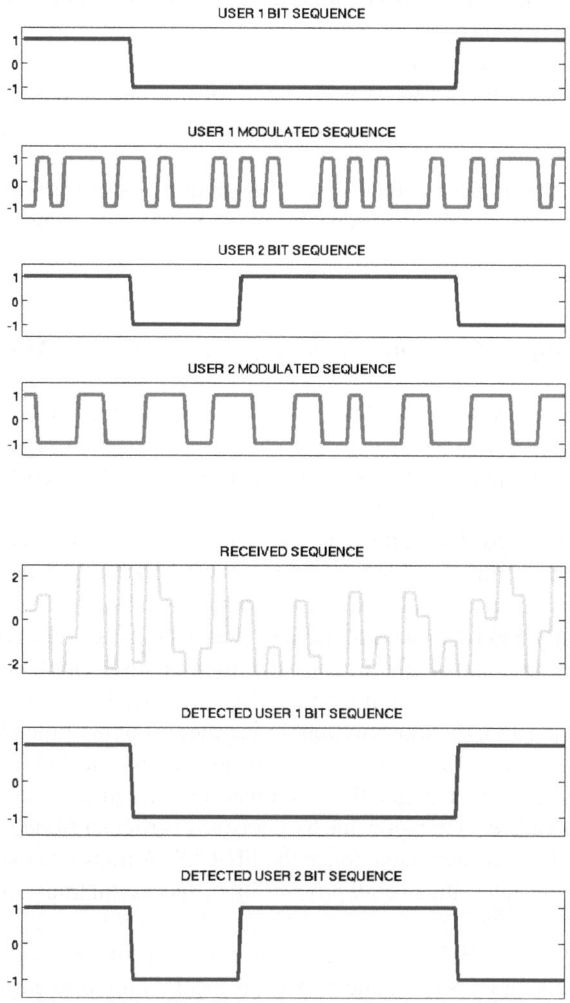

Fig. 2.15 Simple illustration of DS-CDMA signals. The bit sequences for users 1 and 2 are [1 -1 -1 -1 1] and [1 -1 1 1 -1], respectively. The corresponding spreading codes are [-1 1 -1 1 1 1 -1 1] and [1 -1 -1 -1 1 1 -1 -1]. In this case, the two spreading codes are orthogonal and hence the multiuser interference is completely eliminated at the receiver and all bits for both users are detected correctly.

In summary, the main advantages of CDMA are: i) efficient utilization of frequency spectrum with a reuse factor of 1 in cellular systems, which leads to increased system throughput [49], ii) efficient allocation of resources (spreading codes) leads to better utilization in bursty traffic scenarios, and iii) better immunity against narrow band interference.

One of the main challenges with DS-CDMA is the near-far problem, *i.e.*, the received power for various users can vary by orders of magnitude depending on the relative distances of the users from the base station and the fading conditions. This large difference in received power will cause the signal of certain users to completely swamp out the signals of other users. The solution proposed to overcome this problem is to carefully control the transmit power of the users using a power control algorithm [52].

2.5.2.2 Frequency Hopping Code Division Multiple Access (FH-CDMA)

The basic idea in a frequency hopping CDMA system is to temporally vary the frequency used by each user in a pseudo random manner as illustrated in Figure 2.16. Since each user selects the pattern in a pseudo random manner, the probability of more than one user selecting the same frequency in a given time-slot is low. Thus packet collisions are reduced. Depending on the rate at which users change the frequency, the hopping is termed as slow-hopping or fast hopping. In addition to providing multiple access capability, FH-CDMA also has the advantage that it provides frequency diversity and robustness to narrow band interference. For instance, if another system is using a narrow frequency band, then the user employing FH-CDMA will only be affected for a small fraction of the time when its hop-frequency matches the interference band; the transmission during the rest of the time-slots are not affected by the interference. Similarly, in a wideband wireless channel (formally defined in Chapter 3) the channel attenuation is a function of frequency. So, even if the channel encounters severe attenuation due to fading at a particular frequency, the other frequencies may be less attenuated. Since the FH-CDMA transmits over multiple time-slots and frequencies, the overall probability of successful transmission is increased.

One of the challenges in a FH-CDMA system is to ensure that the transmitter and intended receiver hop using the same pattern. The protocol required to ensure that these sequences match is fairly sophisticated. The popular Bluetooth standard [12] is based on FH-CDMA technology. The standard specifies an elaborate sequence of steps that nodes need to follow in order to discover and share a frequency hopping pattern based on the unique Bluetooth identity of each node and their system clocks. The standard also employs an adaptive frequency hopping protocol. In adaptive frequency hopping, the hopping pattern that each node uses can be varied with time depending on the interference encountered [21, 32].

2.5.2.3 Orthogonal Frequency Division Multiplexing (OFDM)

OFDM is essentially a combined method for modulating and multiplexing multiple signals onto a wideband channel. The primary objective of OFDM is an efficient system to transmit information over parallel channels. Consider for instance a simple frequency division multiplexing (FDM) system in which the available bandwidth B Hz is divided into $m = B/b$ non-overlapping segments of width b Hz. A traditional modulation pulse and carrier can be used in each of these segments. However, in an OFDM system, the carrier frequency on each of these smaller segments (typically referred to as subcarriers) are selected in such a manner that they are orthogonal with each other as shown in Figure 2.17. As can be seen from the figure, at the center frequency of each subcarrier, the contributions from the other subcarriers is 0. This orthogonality is the key to the superior spectral efficiency of OFDM since it allows numerous subcarriers to be used within the given band without causing any interference between subcarriers.

There are several variations of OFDM [24] and it has become widely used in next generation wireless standards such as WiMAX. OFDM based signaling is also used in wire-line systems where it is frequently referred to as discrete multitone modulation (DMT). In this chapter, we only focus on a very basic OFDM system and provide the key insights into its performance.

Consider the basic blocks in an OFDM system as shown in Figure 2.18. The input data stream is first parsed into groups of N bits. Without loss in generality consider the transmission of one input vector $\mathbf{X}$ of length N. This input is first passed through

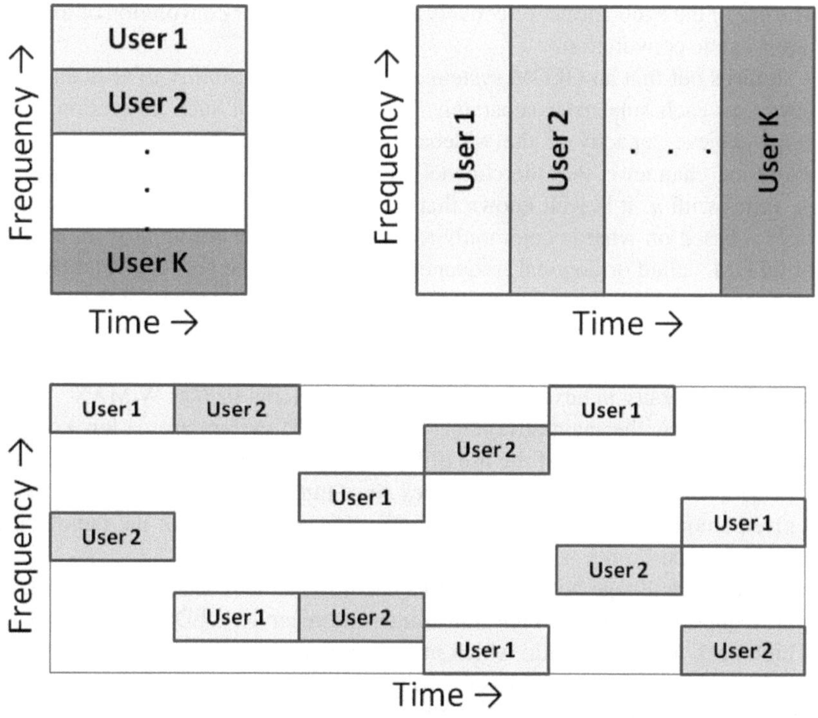

Fig. 2.16 Simple illustration of FDMA, TDMA and frequency hopping CDMA

a length N Inverse Discrete Fourier Transform (IDFT) resulting in output $\mathbf{x}$. A cyclic prefix which is a replica of the first c symbols is appended to the end of the symbol stream. This sequence is then transmitted over the channel.

Let the channel be represented by a linear, time-invariant (LTI) system with impulse response g_n. Consequently, the received signal y_n is given by the convolution of the input signal and the channel response, as

$$y_n = \sum_k g_k x_{n-k} + z_n \tag{2.26}$$

where z_n is the additive Gaussian noise. At the receiver the portion corresponding to the cyclic prefix is first removed. Then the resulting signal is passed through a Discrete Fourier Transform (DFT). The output, Y of the DFT block is given by

$$\begin{aligned} Y = DFT(y_n) &= DFT(g_n).DFT(y_n) + DFT(z_n) \\ &= h_n.X_n + Z_n \end{aligned} \tag{2.27}$$

Effectively, the signal Y can be treated as the output of a set of parallel channels with gains h_n and noise Z_n. The properties of the effective discrete wideband channel (represented by g_n and h_n) are discussed in Chapter 3 of this book. As long as the length of the cyclic prefix c is greater than the spread of the channel, represented by the size of g_n, there is no intersymbol interference or interchannel interference. The use of the cyclic prefix effectively converts the linear convolution of the channel into a cyclic convolution.

It turns out that an OFDM system also offers the flexibility to alter the rate and power on each subcarrier separately. The advantage of such adaptation is that it helps achieve capacity of the wideband channel. One of the characteristics of a wideband channel is that the channel gain is different on each subcarrier, *i.e.* the h_n varies with n. It is well known that the optimal power allocation across subcarriers is based on what is commonly referred to as water-filling [14]. In a variation of OFDM, called orthogonal frequency division multiple access (OFDMA), multiple users are assigned non-overlapping groups of subcarriers within the entire band. The allocation of the subcarriers can be fixed or adaptively vary based on channel measurements. Such adaptive subcarrier allocation plays a critical role in achieving multiuser diversity in next generation wireless systems such as WiMAX.

In summary, the main advantages of an OFDM system are: i) low complexity implementation because of the use of Fast Fourier Transform (FFT) to compute the DFT [30, 41], ii) high spectral efficiency since multiple signals are superimposed in a given amount of bandwidth, and iii) effective equalization of the fading channel characteristics.

A major challenge with an OFDM system is that it is sensitive to timing errors and frequency offsets between transmitter and receiver. OFDM systems also have a high peak-to-average ratio which makes implementation a challenge. The power amplifier used before transmission typically has a narrow operating range where its performance is linear. Thus, the high peak-to-average ratio poses a challenge in ensuring efficient operation of the power amplifier.

2.5.3 Power Control

Determining the transmission power to use in a wireless system is a crucial task which affects the overall performance. The transmission power is controlled to achieve a variety of desired objectives, such as: i) provide performance guarantees even in the presence of a time varying channel attenuation, ii) conserve battery power in mobile devices by regulating the transmission power to the bare minimum required to provide desired level of service, or iii) reduce the amount of interference caused to other users in a multiuser system.

The variety and complexity of power control problems that arise in wireless systems are too numerous to cover in this chapter. For conciseness, we provide two examples of power control problems.

Cellular Uplink: We will consider the uplink of a cellular network in which the mobile users attempt to minimize their transmission powers to meet their desired performance level. To keep the discussion general we will consider both orthogonal and non-orthogonal signaling schemes used by multiple users.

Let set M and set B represent, respectively, the set of mobile users and base stations in the network. Mobile user $m \in M$ is connected to base station $b_m \in B$. Let set $K_m \subset M$ denote the set of mobile users whose transmissions interfere with

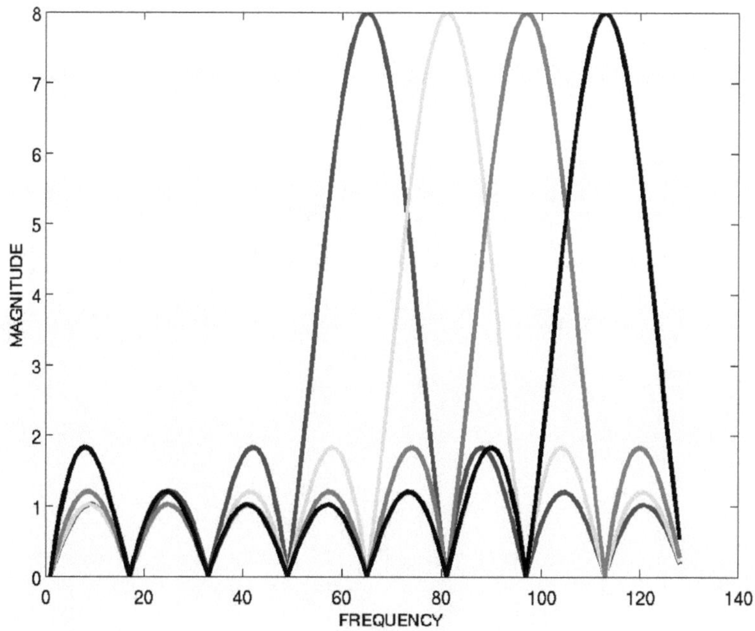

Fig. 2.17 Simple illustration orthogonality in an OFDM system

the transmission of user m. For instance, in a CDMA network using non-orthogonal codes and a frequency reuse of 1, $K_m = M - m \forall m \in M$. In a GSM system, K_m will only include mobiles from cells other than b_m and which use the same frequency as m.

The average channel attenuation between mobile $m \in M$ and base station $b \in B$ is given by h_{mb}. Let the transmit power of mobile m equal P_m and the receiver noise at base station b be denoted by N_b. Thus, the received signal-to-interference ratio SIR_{mb} for mobile m at base station b is given by,

$$SIR_{mb} = \frac{P_m |h_{mb}|^2}{\sum_{i \in K_m} P_i |h_{ib}|^2 + N_b} \tag{2.28}$$

As noted before, the obtained average SIR is proportional to a variety of performance metrics including throughput, and bit error rate. Thus, a typical performance requirement might be to ensure a desired minimum level of signal-to-interference ratio, $\bar{SIR}$ is achieved for each user. There is clearly a conflicting requirement for each user. For instance, increasing P_m results in an increase in the ratio SIR_{mb} but also decreases the ratio SIR_{jk} for all users j such that $m \in K_j$. Thus, if each user employed a greedy approach of increasing their power when their received SIR is

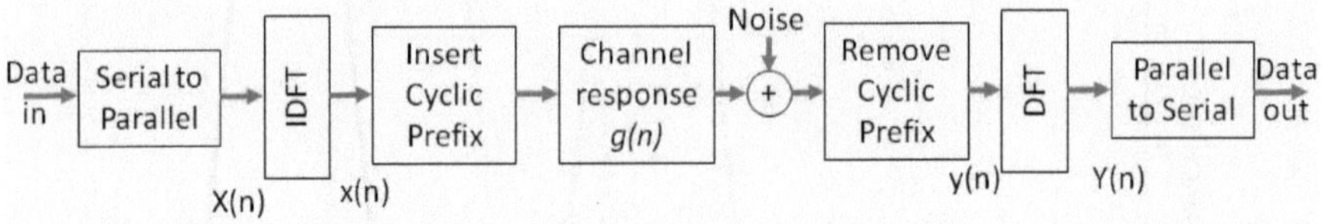

Fig. 2.18 Basic block diagram of an OFDM system.

lower than $\bar{SIR}$ and not changing their power when their SIR is greater than $\bar{SIR}$ all user powers would increase to infinity and the system would become unstable.

However, it can be proved that in this case, there exists an optimum transmit power level for each user that results in a stable operating point. This optimum operating point can be obtained as the solution to the following optimization problem:

$$\min \quad \sum_{m \in M} P_m \tag{2.29}$$

$$s.t. SIR_{mb} \geq \bar{SIR} \tag{2.30}$$

As the keen reader might recognize, (2.29) is a centralized problem and requires global knowledge of all parameters to solve. However, it turns out that there exists a simple distributed and iterative solution to this problem. This solution only relies on each link observing its own SIR during iteration n and modifying its transmission power in iteration $n+1$, as follows:

$$P_m(n+1) = \frac{\bar{SIR}(m)}{SIR_{mb}(n)} P_m(n) \tag{2.31}$$

One of the features of this solution is that the updates for each user can occur independent of the other users and in a completely asynchronous manner. In most cases, the number of iterations required to achieve convergence is small. It is typically assumed that all system parameters including channel gains, network topology variations due to mobility, variation in the number of users occur on a time-scale that is larger than what is required to obtain convergence.

Several variations of this problem and resulting solutions have been studied in the literature. For instance, problems that make the solution robust, opportunistic, and couple power control with beam-forming and scheduling have been investigated [13].

Ad-hoc Networks: The second example of power control is from ad-hoc networks. An ad-hoc network is a collection of nodes that communicate with each other in a multihop manner without the support of a backbone network. In an ad-hoc network, power control is also used for network topology control (for instance to achieve a fully connected network), or to improve the efficiency of the routing protocol or the MAC contention process. Further, the diverse nature of the nodes might cause a link to not have bidirectional connectivity.

Let c_{ij}^1 be the indicator function for first order or direct connectivity between nodes i and j. That is, $c_{ij}^1 = 1$ if node i can directly communicate with node j. Set M denotes the set of all nodes. In the case of asymmetric links, $c_{ij}^1 \neq c_{ji}^1$. Let c_{ij}^k indicate whether a $k-$hop connection exists between nodes i and j. Then,

$$c_{ij}^k = I[\{\max_{0 \leq n \leq k, l \in M, l \neq i,j} (c_{il}^n + c_{lj}^{k-n})\} > 2] \tag{2.32}$$

where $I[x]$ is the indicator function that equals 1 if the condition x is true, and equals 0, otherwise. Let the power of node i be denoted by P_i. In a more general setting, a node may choose to use a different power level to communicate with each of its neighboring nodes. At the other extreme, the protocol design may restrict all nodes in the network to use the same transmission power. Recognize that c_{ij}^1 depends on the transmission power used by node i. Computing the minimum network power that ensures a fully connected network can be formally posed as the following optimization problem:

$$\min \sum_{i \in M} P_i \tag{2.33}$$

$$s.t. \quad P_i < P_i^{max} \tag{2.34}$$

$$c_{ij}^1(P_i) = c_{ji}^1(P_j) \tag{2.35}$$

$$\exists \quad k : c_{ij}^k = 1 \forall i, j \in M, i \neq j \tag{2.36}$$

Recognize that (2.35) imposes the condition that each link is bidirectional and (2.36) ensures that the network is fully connected.

Several protocols have been developed to solve problems of this nature in a distributed manner. We now describe a popular protocol, COMPOW [28], which selects a common transmit power for all nodes, thereby ensuring each link is bidirectional ($P_i = P_j, \forall i, j$).

COMPOW computes and maintains multiple routing tables at each node, one for each possible transmit power level. A routing table is essentially a look up table that a node uses to determine the next hop to forward the packets for each possible destination. COMPOW then determines the smallest power level at each node such that the list of nodes that can be reached (using a finite number of hops) from that particular node using the selected power, equals the list of nodes that can be reached from that node using the highest transmission power. In other words, network connectivity is not sacrificed in reducing the transmit power levels.

The key features of COMPOW are: i) the authors show that there is no significant loss in asymptotic (large number of nodes) network capacity by using a common transmit power for all nodes, ii) existence of low power routes using a low common transmit power for all nodes, and iii) the contention at the data link layer is also minimized.

2.6 Advanced Transceiver Methods

In this section we discuss selected advanced techniques used in state of the art wireless systems. These techniques are also an area of active research and form a critical component of future wireless networks.

2.6.1 Multiple Antenna Systems

One of the most promising methods to improve the performance of wireless systems is the use of multiple antennas at the transmitter and receiver. Such systems are broadly classified as multi-input-multi-output (MIMO) systems. The capacity of a MIMO system with n_t transmit and n_r receive antennas is given by [17, 43]

$$C_{MIMO} = \log\det(\mathbf{I}_{n_r} + \mathbf{HPH}^*), \tag{2.37}$$

where, $\mathbf{H}$ is the $n_r \times n_t$ matrix of the channel gains between the various users, $\mathbf{I}_{n_r}$ is the identity matrix of size n_r, $\mathbf{P}$ is the diagonal matrix of size n_t with the diagonal elements being the power transmitted on the various transmit antennas and $\mathbf{H}^*$ denotes the complex conjugate of the matrix $\mathbf{H}$. For simplicity, the noise at the receiver is normalized to 1 in (2.37). For channels with certain types of fading characteristics, this capacity grows asymptotically with SNR as

$$C_{MIMO} \approx \min(n_t, n_r)\log(SNR). \tag{2.38}$$

Thus, the capacity grows linearly with the number of antennas, if both the number of transmit and receive antennas increases.

The multiple antennas can also be used for improving the diversity of the system. The diversity of the system is defined as the exponent of the SNR in the expression for probability of error. For instance, the maximum diversity of a Gaussian noise channel is infinite. This diversity can be derived from the expression for probability of error, similar to (2.6) or (2.7). By approximating the Q-function with an exponential function, the asymptotic fall off of error rate with SNR can be computed. In contrast, the diversity order in fading channels is finite. For the MIMO system with an independent Rayleigh fading distribution between each pair of transmit and receiver antennas, the maximum diversity equals $d_{max} = n_t n_r$.

While it is desirable to use multiple antennas to increase the throughput as well as reduce the error probability, there is a fundamental trade-off between these two objectives. This trade-off is quantified by the diversity-multiplexing trade-off [55]. Specifically, the maximum diversity $d(r)$ that can be obtained when the throughput equals $r\log(SNR)$ is given by the piecewise linear boundary specified by

$$d(r) = (n_r - r)(n_t - r), \forall 0 \le r \le \min\{n_t, n_r\}, r \in \mathbb{Z}^+ \tag{2.39}$$

The multiple antennas can also be used for beam-forming. In digital beam-forming, the transmit node (for instance a cellular base station) uses a group of antennas to transmit and receive signals. The signal for each user is premultiplied by a different complex number at each antenna in order to shape the beam towards the direction of that user. This beam-forming is useful in reducing the interference experienced by other users in the system. Recent advances in antenna technology and signal processing methods have enabled adaptive beam-forming methods to be implemented in which the beams are adaptively designed in real-time depending on user locations and the desired performance metric to optimize. A similar adaptive weighting of the signals received from multiple antennas can be used to enhance the signal of one user and minimize the interference from the other users.

2.6.2 Scheduling, Rate, and Power Control

One of the promising areas of recent research is in the area of cross layer optimization [6, 38]. The essential idea in cross layer optimization is to modify the design of the protocol/algorithm in one layer based on input from a different layer. Specifically, the area of cross layer scheduling, rate and power control has emerged as a useful tool to improve performance significantly.

Consider a multiuser wireless network in which each of the users encounters a different time varying wireless channel gain. Further, each user supports a different type of traffic and consequently has different rate and delay requirements. The task of the scheduler is to optimally allocate the resources (power, modulation, coding rate, time-slots, frequency bands) to the various users to optimize a desired global network level metric while also satisfying some fairness constraints. These fairness constraints are required to ensure that no single user is starved of network resources. These scheduling issues are covered in detail in Chapter 14 for general wireless systems and in Chapter 13 for WiMAX systems.

2.7 Conclusions

In this chapter, we considered the basics of a digital wireless communication system. The focus is on introducing the fundamental concepts of modulation and coding in a manner that is broadly accessible. Basic multiple access methods including popular CDMA and OFDM are also introduced.

We did not consider the various models for wireless channels. As noted before, several detailed and extensive models for the wireless channel are available and are briefly discussed in Chapter 3.

Acknowledgements The work of the author was supported in part by the Office of Naval Research through a research grant.

References

1. Abramson, N.: The ALOHA system: another alternative for computer communications. In: Proceedings of Fall Joint Computer Conference (ACM), pp. 281–285. Houston, TX (1970)
2. Abramson, N.: The Alohanet - surfing for wireless data. IEEE Comm. Mag. **47**(12), 21–25 (2009)

3. Bahl, L., Cocke, J., Jelinek, F., Raviv, J.: Optimal decoding of linear codes for minimizing symbol error rate. IEEE Transactions on Information Theory **20**(2), 284–287 (1974)
4. Balakrishnan, H., Padmanabhan, V., Seshan, S., Katz, R.H.: A comparison of mechanisms for improving tcp performance over wireless links. IEEE/ACM Transactions on Networking **5**(6), 756–769 (1997)
5. Berrou, C., Glavieux, A., Thitimajshima, P.: Near Shannon limit error–correcting coding and decoding: Turbo-codes. In: Proceedings of ICC, pp. 1064–1070. Geneva, Switzerland (1993)
6. Berry, R., Yeh, E.: Cross layer wireless resource allocation - fundamental performance limits. IEEE Signal Processing Magazine **21**(5), 59–68 (2004)
7. Bertsekas, D.P., Gallager, R.: Data Networks. Prentice Hall (1992)
8. Bhargavan, V., Demers, A., Shenker, S., Zhang, L.: MACAW: A media access protocols for wireless LANs. In: Proceedings of ACM Sigcomm. London, England (1994)
9. Biglieri, E.: Coding for Wireless Channels. Springer, New York, NY (2005)
10. Biglieri, E., Proakis, J.G., Shamai, S.: Fading channels: Information-theoretic and communications aspects. IEEE Transactions on Information Theory **44**(6), 2619–2692 (1998)
11. Braden, R.: RFC 1122 Requirements for Internet Hosts – Communication Layers. In: Internet Engineering Task Force, Request for Comments (RFC) document. Network Working Group (1989)
12. Bray, J., Sturman, C.F.: Bluetooth 1.1: Connect Without Cables. 2nd edn. Prentice Hall, Upper Saddle River, NJ (2001)
13. Chiang, M., Hande, P., Lan, T., Tan, C.W.: Power control in wireless cellular networks. Foundations and Trends in Networking **2**, 381–533 (2008)
14. Cover, T.M., Thomas, J.A.: Elements of Information Theory. 2nd edn. Wiley, New York (2006)
15. Forney, D.G.: Concatenated Codes. The MIT Press (1966)
16. Forney, D.G., Ungerboeck, G.: Modulation and coding for linear Gaussian channels. IEEE Transactions on Information Theory **44**(6), 2384–2415 (1998)
17. Foschini, G.J., Gans, M.J.: On limits of wireless communications in a fading environment when using multiple antennas. Wireless Personal Communication **6**, 311–355 (1998)
18. Gallager, R.G.: Information Theory and Reliable Communication. John Wiley and Sons, Inc. (1968)
19. Gast, M.: 802.11 Wireless Networks: The Definitive Guide. 2nd edn. O'Reilly Media, Sebastopol, CA (2005)
20. Gastpar, M., Rimoldi, B., Vetterli, M.: To code, or not to code: Lossy source-channel communication revisited. IEEE Transactions on Information Theory **49**(5), 1147–1158 (2003)
21. Golmie, N., Rebala, O., Chevrollier, N.: Bluetooth adaptive frequency hopping and scheduling. In: Proceedings of MILCOM, pp. 1138–1142. Boston, MA (2003)
22. Gunduz, D., Erkip, E., Goldsmith, A., Poor, H.V.: Source and channel coding for correlated sources over multiuser channels. IEEE Transactions on Information Theory **55**(5), 3927–3944 (2009)
23. Guruswami, V., Sudan, M.: Improved decoding of Reed-Solomon codes and algebraic geometry codes. IEEE Transactions on Information Theory **45**(6), 1757–1767 (1999)
24. Hanzo, L., Munster, M., Choi, B.J., Keller, T.: OFDM and MC-CDMA for Broadband Multi-User Communications, WLANs and Broadcasting. Wiley-IEEE Press, West Sussex, England (2003)
25. Keshav, S.: An Engineering Approach to Computer Networks. Addison-Wesley Longman, Inc. (1997)
26. Lin, S., Costello, D.J.: Error Control Coding: Fundamentals and Applications. Prentice Hall, second edition (2004)
27. Moon, T.: Error Correction Coding: Mathematical Methods and Algorithms. Wiley-Interscience, Hoboken, NJ (2005)
28. Narayanaswamy, S., Kawadia, V., Sreenivas, R.S., Kumar, P.R.: Power control in ad-hoc networks: Theory, architecture, algorithm and implementation of the COMPOW protocol. In: Proceedings of European Wireless 2002. Next Generation Wireless Networks: Technologies, Protocols, Services and Applications, pp. 156–162. Florence, Italy (2002)

29. Oestges, C., Clerckx, B.: MIMO wireless communications: from real-word propagation to spacetime code design. Academic Press: Oxford (2007)
30. Oppenheim, A.V., Schafer, R.W.: Discrete Time Signal Processing. Prentice Hall, Englewood Cliffs, NJ (1989)
31. Papoulis, A., Pillai, U.S.: Probability, Random Variables and Stochastic Processes. Mc-Graw Hill Inc., New York, NY (2002)
32. Popovski, P., Yomo, H., Prasad, R.: Dynamic adaptive frequency hopping for mutually interfering wireless personal area networks. In: Proceedings of ACM Mobihoc, vol. 4, pp. 199–209. Tokyo, Japan (2004)
33. Proakis, J.G.: Digital Communications, Fourth Edition. Mc-Graw Hill Inc., New York, NY (2001)
34. Proakis, J.G., Salehi, M.: Fundamentals of Communication Systems. Prentice Hall, Upper Saddle River, NJ (2004)
35. Ramstad, T.: Insights into mobile multimedia communications, chap. Combined Source Coding and Modulation for Mobile Multimedia Communication, pp. 415–430. Academic Press Signal Processing And Its Applications Series. Academic Press Professional, Inc., San Diego, CA (1998)
36. Rappaport, T.: Wireless Communications: Principles and Practice. 2nd edn. Prentice Hall, Upper Saddle River, NJ (2001)
37. Sayood, K.: Introduction to Data Compression. Morgan Kauffman, San Francisco, CA (2005)
38. Shakkottai, S., Rappaport, T.S., Karlsson, P.C.: Cross-layer design for wireless networks. IEEE Communications Magazine **41**(10), 74–80 (2003)
39. Shannon, C.E.: A mathematical theory of communication. Bell Systems Technical Journal **27**, 379–423 (1948)
40. Sherwood, P.G., Zeger, K.: Progressive image coding on noisy channels. In: Proceedings of Data Compression Conference, pp. 72–81. Snowbird, UT (1997)
41. Soliman, S., Srinath, M.D.: Continuous and Discrete Signals and Systems. Prentice Hall, Upper Saddle River, NJ (1998)
42. Tanenbaum, A.S.: Computer Networks. Prentice Hall, Upper Saddle River, NJ (2002)
43. Telatar, I.E.: Capacity of multi-antenna Gaussian channels. European Transactions on Telecommunications **10**(6), 585–596 (1999)
44. Trappe, W., Washington, L.C.: Introduction to Cryptography with Coding Theory. Prentice Hall (2005)
45. Ungerboeck, G.: Channel coding with multilevel/phase signals. IEEE Transactions on Information Theory **25**(1), 55–66 (1982)
46. Vembu, S., Verdu, S., Steinberg, Y.: The source-channel separation theorem revisited. IEEE Transactions on Information Theory **41**(1), 44–54 (1995)
47. Verdu, S.: Multiuser Detection. Cambridge University Press, Cambridge, UK (1998)
48. Viterbi, A.J.: Error bounds for convolutional codes and an asymptotically optimum decoding algorithm. IEEE Transactions on Information Theory **13**(2), 260–269 (1967)
49. Viterbi, A.J.: CDMA : Principles of Spread Spectrum Communication. Addison-Wesley Publishing Company, Reading, MA (1995)
50. Wang, X., Poor, V.: Iterative (Turbo) soft interference cancellation and decoding for coded CDMA. IEEE Transactions on Communications **47**(7), 1046–1061 (1999)
51. Wicker, S.B.: Error Control Systems for Digital Communication and Storage. Prentice Hall, Upper Saddle River, NJ (1995)
52. Yates, R.: A framework for uplink power control in cellular radio systems. IEEE Journal on Selected Areas in Communications **13**(7), 1341–1348 (1995)
53. Yates, R., Goodman, D.J.: Probability and Stochastic Processes: A Friendly Introduction for Electrical and Computer Engineers. John Wiley and Sons Inc., Hoboken, NJ (2004)
54. Yeung, R.W.: A First Course in Information Theory. Kluwer Academic/Plenum publishers, New York, NY (2002)
55. Zheng, L., Tse, D.: Diversity and multiplexing: A fundamental tradeoff in multiple antenna channels. IEEE Transactions on Information Theory **49**(5), 1073–1096 (2003)

Chapter 3
Channel Models for Wireless Communication Systems

K. V. S. Hari

3.1 Introduction

Wireless communication has evolved significantly, over the past several decades, to meet the ever-growing demand for high data rates over the wireless medium. Systems have been designed for indoor applications and outdoor applications where mobility is a very important aspect of system specification. One of the key elements in the design of such systems is understanding the characteristics of the wireless channel [7, 10, 13, 16, 22, 23, 25, 28, 31, 30, 34].

3.2 Wireless Channel

The wireless signal propagates in space, based on the laws of physics. An electromagnetic Radio Frequency (RF) signal which travels in a medium suffers an attenuation (path loss) based on the nature of the medium. In addition, the signal encounters objects and gets reflected, refracted, diffracted, and scattered. The cumulative effect results in the signal getting absorbed, signal traversing multiple paths, signal's frequency being shifted due to relative motion between the source and objects (Doppler effect), thus getting modified significantly. It is clear that the RF signal is a space-time-frequency signal and can be represented as $s(x,y,z,t,f)$ where x,y,z are the space variables and t,f are time and frequency variables, respectively. If we represent the transmitted signal as $s_T(x_T,y_T,z_T,t,f_T)$ and the received signal as $s_R(x_R,y_R,z_R,t,f_R)$, one can relate both space-time-frequency signals as $s_R(x_R,y_R,z_R,t,f_xxR) = H(s_T(x_T,y_T,z_T,t,f_T))$. $H(.)$ is a function which can be called the *Wireless Channel*. The wireless channel is also dependent on factors like

K. V. S. Hari
Department of ECE, Indian Institute of Science, Bangalore 560012, India, e-mail: hari@ece.iisc.ernet.in

J. Kennington et al. (eds.), *Wireless Network Design: Optimization Models and Solution Procedures*, International Series in Operations Research and Management Science 158,
DOI 10.1007/978-1-4419-6111-2_3,

terrain of operation, atmospheric conditions, relative mobility of transmitters and receivers, types of antennas used, heights of antennas, and other practical parameters.

3.3 Propagation Model for Wireless Channel

The propagation model for the wireless channel is characterized by path loss, multipath fading and Doppler spread (or Doppler spectrum). All these characteristics are affected by the physical environment between the transmitter and receiver as well as system dependent parameters such as antenna heights, antenna beam-widths, antenna polarization, and mutual coupling between multiple antennas.

Path Loss characterizes the attenuation between the transmitted and received signals. This is typically based on physical phenomena like *absorption, diffraction, scattering, reflection and refraction.* Absorption of RF (microwave frequency band) signals in the atmosphere is due to the phenomenon of molecular resonance due to the presence of water and oxygen molecules. Absorption, therefore, is higher during rain and snowfall. Absorption also occurs when the propagation is through solids. Buildings with metal attenuate between 10-25 dB depending on the thickness of the walls, trees attenuate about 10-12 dB and metals attenuate the most. Diffraction around objects occurs when the wavelength of the signal is comparable to the size of the object. In the microwave region, for example, a 2.5 GHz signal has a wavelength of 12 cm. Diffraction occurs when the signal encounters objects of these sizes and if there are sharp edges, the signal 'diffracts' across the edges. Diffraction is also observed in a city with buildings. Lower frequency signals can diffract over hilltops. Scattering is a phenomenon which occurs when the signal encounters a cluster of objects and multiple reflections from these objects causes the random scattering effect. An example of a scattering environment is foliage. Trees with lots of branches and moving leaves (due to wind) scatter the signal significantly.

Path loss is a function of the frequency of operation, the distance between the transmitter and receiver, terrain of operation, and system parameters like antenna heights and antenna characteristics.

Fading is a phenomenon which occurs when the scattering environment between the transmitter and the receiver varies with time. This change in the scatterers alters the signals being added constructively or destructively at the receiver location, as a function of time. This change in the signal level is called fading. It is obvious that the terrain of operation (open spaces, dense foliage, tall buildings, hilly areas) will determine the level of fading. Microscopic fading occurs when the receiver receives multiple copies of signals due to scatterers close to the receiver. This type of fading is also called flat fading. Macroscopic Fading occurs when the receiver receives multiple delayed copies of the transmitted signal from a collection of scatterers, which are distributed in space over a large region. This type of fading is also called frequency selective fading. Frequency selective fading is characterized by a parameter called delay spread of a channel [40]. It indicates the statistical nature of the delays introduced by the scattering environment which can cause significant distortion in the signal. The relative velocity of the scattering environment between the transmitter and receiver determines the extent of shift in the frequency of the received signal, due to Doppler effect. This causes a change in the frequency spread of the signal [50].

A channel has coherence if it does not show changes in its characteristics along the space, time and frequency dimensions. The different types of coherence are explained below. Consider a narrow band signal and fixed in space. Then the temporal coherence of a channel can be expressed as $|h(t)| \approx constant$, for $|t - t_0| \leq T_c/2$ where T_c is the temporal coherence time. It means that the temporal characteristics of the channel do not change within the temporal coherence time. The frequency coherence of a channel is defined when the magnitude of the channel does not change over a certain frequency band of interest. The form of the expression is similar to the previous case and is given by $|h(f)| \approx constant$ for $|f - f_c| \leq B_c/2$. As in the previous cases, spatial coherence can be defined as $h(r) \approx constant$ for $|r - r_0| \leq DD_c/2$ with r_0 is an arbitrary position in space, D_c is the spatial coherence. All the above coherence parameters are very important for system design. For example, the knowledge of the value of spatial coherence will help mitigate fading. To explain further, if two antennas of the receiver are separated by a distance more than the spatial coherence distance, it is likely that one of the antennas could receive a stronger signal compared to the other antenna. Thus, one can combine the signals from both antennas in an effective manner to increase the signal strength. In other words, the principle of 'diversity' is employed. Similar techniques can be used in the context of temporal and frequency coherence to improve signal quality.

3.4 Mathematical Modeling of Wireless Channels

For the design of wireless systems, where the signal is distorted due to physical phenomena, it is necessary to characterize the channel, using mathematical models. It is clear that the scattering environment would be different for different locations of the transmitter and receiver. The best framework is to model the channel as a random quantity and characterize it appropriately. In order to characterize the channel, covering most of the possible cases, two approaches are used.

- Statistical model using experimental data : This model is based on measurements which are carried out using transmitters and receivers in different terrain locations, using different antennas, and other relevant experimental parameters. It is clear that an exhaustive measurement campaign to cover all possibilities, is not possible. Representative sets of experiments are carried out and the variations are modeled in terms of a stochastic framework. Mathematical models are developed based on the analysis of data. This is also called the classical modeling approach.
- Models using principle of geometry: In this technique, the scattering environment is characterized by geometrical models where the signal paths are assumed to be rays traveling between the transmitter and receiver [32, 36, 46]. The level of

fading is calculated based on the number of rays and the distance the rays travel. In addition, the angles of arrival (AOA), and Doppler spread are characterized using the information about array geometry. The channel is then numerically simulated and compared with measurements obtained through experiments to validate the models.

3.4.1 Experiments to Characterize Wireless Channels

Wireless channels are characterized by conducting experiments [5, 6, 8, 9, 11, 14, 15, 17, 26, 27, 29, 35, 38, 39, 40, 47, 49] where real transmitters and receivers are used to measure the received signal characteristics. Outdoor and Indoor measurements are carried out for different terrain conditions, different weather conditions etc, to collect data for modeling.

3.4.1.1 Path Loss Models

Path loss is defined as the difference in power levels (in dB) of the received signal and the transmitted signal due to the overall effect of 'path' or distance between the transmitter and receiver. The path loss for a line-of-sight (LOS) scenario where free-space path loss is assumed, is given by

$$PL(dB) = 32.44 + 20\log_{10}(f_c) + 20\log_{10}(d) \tag{3.1}$$

where d is the distance between the transmitter and receiver in km, f_c is the frequency of operation in MHz. It may be observed that the path loss increases with frequency and distance. Some terrain locations provide additional path loss due to large structures (hills, buildings etc) which is called the shadow fading loss.

In the case of a multi-path fading environment and shadowing, several experiments were carried out by Okumura [51] and the data was later analyzed by Hata [19] to develop path loss models characterizing the statistical nature of the channel. The mean path loss is usually obtained using curve fitting techniques. In the logarithmic domain, these models tend to be linear as a function of distance. As an example, some of the models for different terrains proposed by Hata are given below (Note: All path loss models need to be used with caution. If the equation yields a loss less than the free-space path loss, one must use the free-space path loss value.)
Typical Urban

$$\begin{aligned} PL(dB) &= 69.55 + 26.16\log_{10} f_c \\ &= (44.9 - 6.55\log_{10} h_b) log_{10} d - 13.82\log_{10} h_b - a(h_m) \end{aligned} \tag{3.2}$$

where f_c is the carrier frequency in MHz, d is the distance in km, h_b is the height of the base-station in meters, and $a(h_m)$ is a correction term for the receiver antenna height (h_m) given by for large cities:

$$a(h_m) = 8.29(\log_{10}(1.54h_m))^2 - 1.1 \text{ for } f_c \leq 200\text{MHz} \quad (3.3)$$

$$a(h_m) = 3.2(\log_{10}(11.75h_m))^2 - 4.97 \text{ for } f_c \geq 400\text{MHz} \quad (3.4)$$

for small cities:

$$a(h_m) = (1.1\log_{10}(f_c) - 0.7)h_m - (1.56\log_{10}(f_c) - 0.8) \quad (3.5)$$

Typical sub-urban

$$\begin{aligned} PL(dB) &= 69.55 + 26.16\log_{10} f_c \\ &= (44.9 - 6.55\log_{10}\Delta h_b)log_{10}d - 13.82\log_{10} h_b \\ &- a(h_m)\,2(\log_{10}(f_c/28))^2 - 5.4. \end{aligned} \quad (3.6)$$

Rural area

$$\begin{aligned} PL(dB) &= 69.55 + 26.16\log_{10} f_c \\ &= (44.9 - 6.55\log_{10} h_b)log_{10}d - 13.82\log_{10} h_b \\ &-a(h_m) - 4.78(\log_{10} f_c)^2 + 18.33\log_{10}(f_c) - 40.94. \end{aligned} \quad (3.7)$$

AT&T Bell Labs had made extensive measurements in the 1.9 GHz frequency band for an outdoor fixed wireless system. For a given close-in distance d_0, the median path loss (PL in dB) is given by

$$PL_a = A + 10\gamma\log_{10}(d/d0) + s for d > d0, \quad (3.8)$$

where $A = 20\log_{10}(4\pi d0/\lambda), \lambda$ being the wavelength in meters, s is the shadow fade margin, γ is the path-loss exponent with $\gamma = (abh_b + c/h_b)$ for h_b between 10 m and 80 m d_0 = 100m and a, b, c are constants which depend on the terrain as given in Table 3.1. The shadowing margin, s takes a value between 8.2 dB and 10.6 dB

Table 3.1 Values of a,b,c for different terrains

Model parameter	Terrain Type A	Terrain Type B	Terrain Type C
a	4.6	4	3.6
b	0.0075	0.0065	0.005
c	12.6	17.1	20

depending on the terrain.

The above path loss model is for receive antenna heights close to 2 m. In order to use the model for other frequencies and for receive antenna heights between 2 m and 10 m, several experiments were conducted at Stanford University and correction terms have been proposed [11]. The path loss model (in dB) with the correction terms would be

$$PLmodified = PL_a + \Delta PL_f + \Delta PL_h \quad (3.9)$$

where ΔPL_f (in dB) is the frequency correction term given by

$$\Delta PLf = 6log(f/2000) \quad (3.10)$$

where f is the frequency in MHz, and ΔPL_h (in dB) is the receive antenna height correction term given by $\Delta PL_h = -10.8log(h/2)$ for Categories A and B and $\Delta PL_h = -20log(h/2)$ for Category C h is the receive antenna height between 2 m and 10 m.

Path Loss Models for indoor stationary/low-mobility applications: As an example, if the transmitter and receiver are both indoors, the indoor path loss model (dB) for the ITU M.1225 model is based on the COST 231 model[1, 2] and is given below.

$$PL(dB) = 37 + 30\log_{10} d + 18.3n^{\left(\frac{n+2}{n+1} - 0.46\right)} \quad (3.11)$$

where $d(m)$ is the transmitter-receiver separation and n is number of floors in the path. A log-normal shadow fading standard deviation of 12 dB can be expected in such an environment.

Similarly, for the case where the receiver is indoor and the transmitter is indoor, the path loss model is given by

$$PL(dB) = 40\log_{10} d + 30\log_{10} f + 49 \quad (3.12)$$

where $d(km)$ is the transmitter-receiver separation and f is in the 2GHz band. The model can be modified to include the antenna system parameters and expressed as

$$PL(dB) = 40(1 - 0.004\Delta h_b)\log_{10} d - 18\log_{10}\Delta h_b + 21\log_{10} f + 80 \quad (3.13)$$

where $d(km)$ is the transmitter-receiver separation, Δh_b is the base station antenna height (m) measured from the average rooftop level ($0 \leq \Delta h_b \leq 50$), f is in the 2GHz band.

3.4.2 Rayleigh and Ricean Flat Fading Channels

The analysis of experimental data led to the modeling of a wireless channel as a random process. The fact that the signal is a superposition of several faded signals and the use of the central limit theorem leads to the modeling of the channel as a Gaussian random process.

Rayleigh channel: The amplitude of a channel which is modeled as a *zero* mean complex Gaussian random process has a Rayleigh distribution. Such channels are called Rayleigh channels and occur when there is a non-line-of-sight (NLOS) communication between the transmitter and receiver. This type of fading is also called *Flat fading*.

Ricean channel: The amplitude of a channel which is modeled as a *non-zero* mean complex Gaussian random process has a Ricean distribution. Such channels are called Ricean channels and occur when there is Line-Of-Sight (LOS) communication between the transmitter and receiver. The Ricean pdf is given in [31].

The above channels are characterized by the mean and variance of the channel, which is modeled as a Gaussian random process. For the Ricean channel, an additional factor called the $k - factor$ is defined as the ratio of total power in the LOS path to the total scattered power received. If this ratio is zero, it implies Rayleigh fading.

3.4.3 Doppler Spread Due to Relative Motion Between Transmitter and Receiver and Coherence Time

Consider a mobile receiver moving at a velocity v and receiving multiple signals from various directions (due to obstacles). Consider the case where there is no direct LOS path and hence Rayleigh fading is considered. The signal is assumed to be a narrowband signal centered around the carrier frequency f_c. Considering the case that the signal is received from all directions and assuming that they are all of equal amplitude, the power spectrum of the Rayleigh fading channel is given by

$$S(f) = \frac{b_0}{\pi f_d \sqrt{1 - \{f/f_d\}^2}} \tag{3.14}$$

where f_d is the doppler frequency and the equation is valid for $f < f_d$. The shape of the spectrum is like a 'bathtub with infinitely high steep walls'. Jakes [50] proposed a technique to simulate a fading channel with the above doppler spectrum. The technique involves considering several oscillators and appropriately combined after suitable amplitude scaling. This is popularly known as the Jakes spectrum as shown in Figure 3.1.

Consider the case where the receiver and transmitter are stationary. In this case too, one gets a Doppler spread due to the movement of the scatterers (for example, branches of trees, moving traffic) between the transmitter and receiver. The spectrum of such a channel was obtained through measurements carried out at Stanford University in the 2.5GHZ band [11]. A mathematical model, obtained from the measurement data, is given by

$$S_{fixed}(f) = \sqrt{1 - \{f/f_m\}^2} \tag{3.15}$$

f_m is the maximum doppler frequency (typically 2Hz to 4Hz). This spectrum shape is an 'inverted cup' as shown in Figure 3.2.

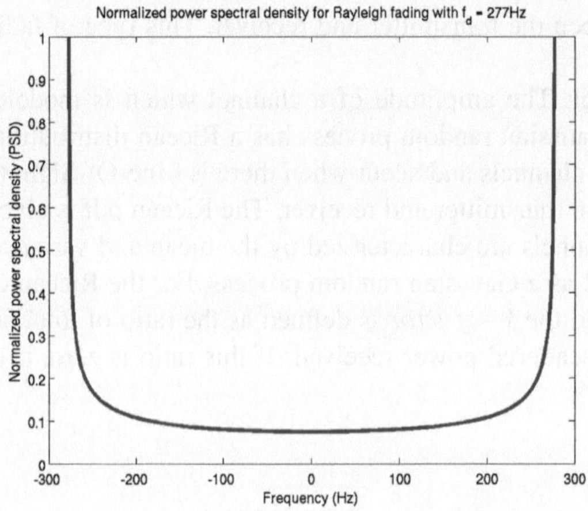

Fig. 3.1 Normalized power spectral density for Rayleigh fading with Doppler spread of 277 Hz.

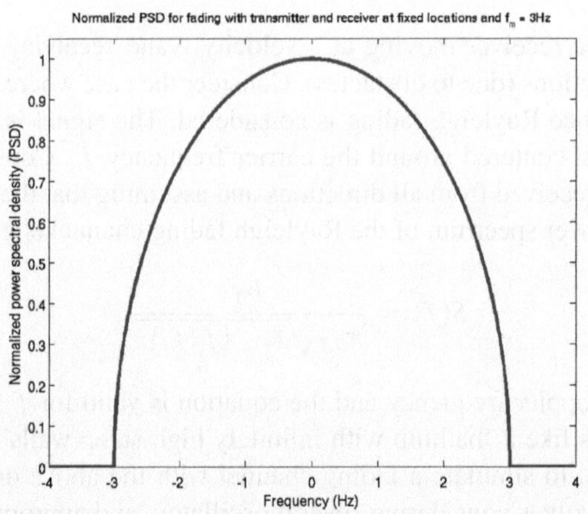

Fig. 3.2 Normalized power spectral density for fading channel with transmitter and receiver at fixed locations with Doppler spread close to 3Hz.

Coherence time, T_c, is the time during which the channel is assumed to be invariant with time. It can be approximately obtained from the Doppler spread characterization. One such approximation is given [34] as $T_c = 0.423/f_d$. where f_d is the maximum Doppler frequency.

3.4.4 *Finite Impulse Response Channel Model for Frequency Selective Fading Channels*

In most cases, the receiver receives multiple copies of attenuated and delayed copies of the transmitted signal. At the baseband level, the channel response can be modeled as a linear system where the transfer function is characterized by a Finite Impulse Response (FIR) with $L-1$ multipaths, given by

$$h(n) = \sum_{l=0}^{L-1} \alpha_l \, \delta(n - \tau_l) \tag{3.16}$$

where α_l is modeled as a random co-efficient with a certain probability density function and τ_l is the delay of the lth multipath, which need not be uniform.

Assuming that this FIR channel is ‘constant’ for a certain period of time, the channel behaves like an FIR filter. The transfer function of this channel, $H(z)$, is a finite length polynomial. In the frequency domain, $H(f)$ will have a response which will indicate the attenuation, the filter offers, to various frequencies. It implies that some frequencies of the signal could suffer heavy attenuation, indicating the frequency selective nature of the channel. If the zeros of $H(z)$ are on the unit circle, then the channel would null the frequency component in the signal, causing loss of valuable information. In the case of flat fading channels, it is equivalent to a single-tap FIR channel and all frequencies of the signal suffer the same attenuation. The following tables give the details about the impulse response coefficients for different indoor scenarios. The Doppler spectrum is assumed to be the classic spectrum for the outdoor scenario.

Table 3.2 Impulse response coefficients for two types of wireless indoor channels

Tap	Channel A Relative Delay(ns)	Channel A Relative Avg Pwr(dB)	Channel B Relative Delay (ns)	Channel B Relative Avg Pwr (dB)
1	0	0	0	0
2	50	-3.0	100	-3.6
3	110	-10.0	200	-7.2
4	170	-18.0	300	-10.8
5	290	-26.0	500	-18.0
6	310	-32.0	700	-25.2

3.4.5 *Delay Spread Parameters*

Average Delay: The average path delay is defined as $\tau_{avg} = \sum_{l=0}^{L-1} |\alpha_l|^2 \tau_l$. The Root Mean Square Delay (RMS Delay), τ_{rms} is given by

Table 3.3 Impulse response coefficients for the outdoor to indoor and pedestrian scenario

Tap	Channel A Relative Delay(ns)	Channel A Relative Avg Pwr(dB)	Channel B Relative Delay (ns)	Channel B Relative Avg Pwr (dB)
1	0	0	0	0
2	110	-9.7	200	-0.9
3	190	-19.2	800	-4.9
4	410	-22.8	1200	-8.0
5	–	–	2300	-7.8
6	–	–	3700	-23.9

Table 3.4 Impulse response coefficients for the vehicular scenario

Tap	Channel A Relative Delay(ns)	Channel A Relative Avg Pwr(dB)	Channel B Relative Delay (ns)	Channel B Relative Avg Pwr (dB)
1	0	0	0	-2.5
2	310	-1.0	300	0
3	710	-9.0	8900	-12.8
4	1090	-10.0	12900	-10.0
5	1730	-15.0	17100	-25.2
6	2510	-20.0	20000	-16.0

$$\tau_{rms}^2 = \sum_{l=0}^{L-1} P_l \tau_l^2 - \tau_{avg}^2 \tag{3.17}$$

where P_l is the power (variance) of the amplitude of the lth path. Some experiments were conducted to measure the power delay profiles [21, 48]

We now present some of the typical frequency selective channel model parameter values used in practice for the indoor channel case in the 2-2.5GHZ band.

Table 3.5 Delay spread values for indoor channels

	Channel A RMS(ns)	Channel B RMS(ns)
Indoor Office	35	100
Outdoor to Indoor and Pedestrian	45	750
Vehicular high antenna	370	4000

3.4.6 Decorrelation Distance of the Channel

The space-time wireless channel has some interesting properties with respect to the spatial dimension [3]. The spatial correlation function of the channel can be obtained as a function of the space dimension. The distance at which the magnitude of the correlation takes the value zero (in practical systems, close to zero) is the decorrelation distance. It means that if the receiver has multiple antennas spaced equal to the decorrelation distance, the signals received by the two antennas will be statistically uncorrelated. Measurements have found that the decorrelation distance is approximately equal to half the wavelength of the signal. This property of the channel is used to exploit spatial diversity of the channel and signal processing techniques are used to improve the performance of the receiver. It has also led to the deployment of multiple antennas at transmitters and receivers of wireless communication systems.

3.5 Channel Models for Systems using Multiple Antennas

Consider a system where the transmitter is transmitting using multiple antennas and the receiver is receiving signals using multiple antennas. The scattering environment is assumed to scatter the signal and the receiving antenna is assumed to receive several modified (in amplitude, phase, frequency and time) copies of the transmitted signals. Using this scenario, several spatial channel models have been developed using principles of geometry and a statistical framework [4, 12, 42, 43, 52].

3.5.1 Spatial Wireless Channel Models using Geometry

Consider a general representation of the wireless channel (extension of the FIR model given in a previous section) when one mobile receiver (having one antenna) is considered, as

$$h_1(t,\tau) = \sum_{l=0}^{L(t)-1} A_{l,1}(t)\exp(j\phi_{l,1}(t))\,\delta(t-\tau_{l,1}(t)) \tag{3.18}$$

where $L(t)$ is the number of multipath components, $\phi_{l,k}, A_{l,k}(t), \tau_{l,k}(t)$ are the phase shift, amplitude and delay due to the lth path induced in the kth receiver. Note that the above channel model assumes all parameters to be time varying. This model can be extended to the case of using multiple antennas by including the Angles of Arrival (AOA) into the model. It is given by

$$\underline{h}_1(t,\tau) = \sum_{l=0}^{L(t)-1} A_{l,1}(t)\exp(j\phi_{l,1}(t))\,\underline{a}(\theta_l(t))\,\delta(t-\tau_{l,1}(t)) \tag{3.19}$$

where $\underline{a}(\theta_l(t))$ is the array steering vector corresponding to the angle of arrival, $\theta_l(t))$ and is given by

$$\underline{a}(\theta_l(t)) = (\exp(-j\psi_{l,1}), \psi_{l,2}, \cdots, \psi_{l,m})^T \tag{3.20}$$

where m antennas are used and $\psi_{l,i} = (2\pi/\lambda)\,[x_i \cos(\theta_l(t)) + j\sin(\theta_l(t))]$, x_i, y_i are the in-phase and quadrature phase components. The angular spread is dependent on the scenario and the distribution is assumed accordingly.

Several models have been proposed. Lee [22] assumed that all the scatterers are uniformly placed in a circular fashion. For this configuration, the correlation function can be obtained [4]. But this was not found to be a satisfactory model. The model was extended [42, 43] to add additional rings of scatterers. This model was further modified [4] but all these models do not consider the delay spread or doppler spectrum.

3.5.2 Geometrically Based Single-Bounce Statistical Models

The use of spatial scatter density functions to characterize the channel model is developed using some geometry based principles and these models are called Geometrically based single-Bounce Statistical Models (GBSB). Using these models, the joint AOA and TOA probability density function is obtained (or numerically computed) and the communication system performance is studied using these channel models [20, 24].

3.5.2.1 Geometrically Based Single Bounce Models

The first model is called the Geometrically Based Single Bounce Circular Model (GBSBCM) where the scatterers are assumed to be in a circular arrangement around the transmitter and the other model is called the Geometrically Based Single Bounce Elliptical Model (GBSBEM) which assumes that scatterers are uniformly distributed within an ellipse, where the foci of the ellipse are the location of the transmitter and receiver. The joint AOA-TOA probability density function at the transmitter (base station) is given as

$$
\begin{aligned}
f_{\tau,\theta_T}(\tau,\theta_T) &= \frac{(d^2-\tau^2c^2)(d^2c+\tau^2c^3-2\tau c^2 d\cos(\theta_T))}{(4\pi r_R^2)(d\cos(\theta_T)-\tau c)^3} \\
&\quad \mathit{for}\ d^2-2\tau c d\cos(\theta_T)+\tau^2c^2/(\tau c-d\cos(\theta_T)) \le 2r_R \\
f_{\tau,\theta_b}(\tau,\theta_T) &= 0,\ \ \mathit{else}.
\end{aligned}
$$

Similarly, the joint AOA-TOA probability density function at the receiver (mobile) is given as

$$
\begin{aligned}
f_{\tau,\theta_R}(\tau,\theta_R) &= \frac{(d^2-\tau^2c^2)(d^2c+\tau^2c^3-2\tau c^2 d\cos(\theta_R))}{(4\pi r_R^2)(d\cos(\theta_R)-\tau c)^3} \\
&\quad \mathit{for}\ d^2-2\tau c d\cos(\theta_R)+\tau^2c^2/(\tau c-d\cos(\theta_R)) \le 2r_R \\
f_{\tau,\theta_R}(\tau,\theta_R) &= 0,\ \ \mathit{else}
\end{aligned}
$$

where θ_T, θ_R are the angles of arrival relative to the line of sight of the array at the transmitter and receiver respectively, c is the velocity of light and τ_R is the maximum delay at the receiver.

3.5.2.2 Gaussian Wide Sense Stationary Uncorrelated Scattering model

The next model is called the Gaussian Wide Sense Stationary Uncorrelated Scattering model (GWSSUS). This model assumes several clusters in space around the transmitter. Each cluster will have several scatterers. The scatters in each cluster are chosen/assigned such that the difference in delays with the scatterers are not resolvable with the signal transmission bandwidth. Each cluster has a mean AOA, θ_{0k} for the cluster. One can define

$$v_{k,b} = \sum_{i=1}^{N_k} \alpha_{k,i} \exp(j\phi_{k,i})\underline{a}(\theta_{0k} - \theta_{k,i}) \tag{3.21}$$

where N_k is the number of scatterers in the kth cluster, $\alpha_{k,j}, \phi_{k,j}, \theta_{k,j}$ are the amplitude, phaseshift and AOS of the jth scatterer in the kth cluster. Assuming that the scatterers are assigned to a cluster at least for b data bursts, the received signal over b data bursts is given by

$$\underline{x}_b = \sum_{k=1}^{N_c} \underline{v}_{k,b} s(t - \tau_k) \tag{3.22}$$

N_c being the number of clusters. If there are a large number of clusters, the Gaussian nature of $\underline{v}_{k,b}$ can be assumed and the mean and covariance matrix of this vector can be obtained easily.

3.5.2.3 Gaussian Angle Arrival Model

A special case of GWSSUS is when $N_c = 1$ and the density function of the mean AOA of the clusters is assumed to be Gaussian. The array covariance matrix is given by

$$R(\theta_0, \sigma_\theta) = P_r \underline{a}(\theta_0)\underline{a}^H(\theta_0) \odot B(\theta_0, \sigma_\theta) \tag{3.23}$$

where the klth element of the matrix is given by $B_{kl}(\theta_0, \sigma_\theta) = \exp(-2(\pi d_s(k - l))^2 \sigma_\theta^2 \cos^2\theta)$, with P_r as the received signal power, d_s is the element spacing, $\odot$ is the Schur (element-wise) product of matrices. This model is called the Gaussian Angle Arrival (GAA) model. The performance of the AOA estimation algorithms using this model have been studied in [18] [45].

3.5.2.4 Dominant Reflectors Model

In this model [33], it is assumed that there are L_r dominant reflectors and the received signal vector is defined as

$$\underline{x}(t) = \sum_{l=0}^{L_r(t)-1} \underline{a}(\theta)\alpha_l(t)s(t-\tau) + \underline{n}(t) \tag{3.24}$$

α_j is the complex amplitude of the lth dominant reflector given by

$$\alpha_l(t) = \beta_l(t)\sqrt{\Gamma_l \psi(\tau_l)} \tag{3.25}$$

where Γ_l denotes lognormal fading and $\psi(\tau_l)$ represents the power delay profile and

$$\beta_l(t) = K_r \sum_{r=1}^{N_r} C_r(\theta_l)\exp(j\omega_d \cos(\Omega_{r,l}t)) \tag{3.26}$$

where N_r is the number of signal components adding to the lth dominant reflector. $C_r(\theta_l)$ is the complex amplitude of the rth component of the lth dominant reflector. Simulations using this model have shown that the results are close to measured data.

3.5.2.5 Typically Urban (TU) and Bad Urban (BU) Models

TU and BU models are specifically specified for GSM and PCS systems where the scatterers are assumed to be away at a certain distance and the scatterer positions are kept fixed for a certain amount of time and then moved to a new position based on the velocity profile and random phases assigned to the scatterers. These models have been successfully used for practical system design [44]. Both models assume that the received array vector. In the TU model, 120 scatterers are assumed to be randomly placed 1km away. The positions of the scatterers are kept fixed for every interval of 5m movement of the mobile. During this time duration, random phases are assigned to the scatterers as well as randomized shadowing losses are assigned with a log-normal distribution. For the BU model, an additional cluster of 120 scatterers are assumed to be located at an angular offset of 45 degrees with respect to the first cluster of scatterers. The average power for this second cluster of scatterers is assumed to be 5dB lower than that of the first cluster. Thus, the BU channel introduces more angular spread and increases the delay spread of the channel. These models have been compared with actual measurements and found to be in close agreement with the models used by GSM system designers.

3.5.2.6 Indoor Models Based on Measurement Data

The clustering model is applied to the indoor environment and validated with measurements [37, 41]. The channel response is given by

$$h(t) = \sum_{k=0}^{\infty} \sum_{l=0}^{\infty} \alpha_{kl} \, \delta(t - \tau_{kl} - T_k) \tag{3.27}$$

where k denotes the cluster, l denotes the ray within a cluster, and α_{kl} is a Rayleigh distributed random variable. An extension to this model was proposed in [37]. Here, $h(t, \theta) = h(t)h(\theta)$ with the assumption that temporal and angular impulse responses are independent. The angular impulse response is given by

$$h(\theta) = \sum_{k=0}^{\infty} \sum_{l=0}^{\infty} \alpha_{kl} \, \delta(t - \omega_{kl} - \Theta_k) \tag{3.28}$$

with ω_{kl} is the ray angle and Θ_k is the mean angle of the kth cluster. The probability density function of ω_{kl} is given by $f(\omega) = (1/2\sqrt{\sigma} \exp(-|\sqrt{2}\omega/\sigma|))$ The Time-Angle FIR filter model [20] defines the channel impulse response as

$$h(t, \tau, \theta) = \sum_{q=1}^{Q} \alpha_q \delta(t - \tau_q) \, \delta(\theta - \theta_q) \tag{3.29}$$

where the FIR structure has Q taps. This is a simple extension of the temporal FIR model presented before.

3.6 Conclusion

The wireless channel models presented in previous sections are used for designing highly efficient communication systems which can provide huge data rates. The statistical nature of these models provides a means to cover most of the possible scenarios in the real-world through an exhaustive numerical simulation study. The choice of models is varied and an appropriate choice of the model can be made based on the specifications of the wireless system to be designed. This chapter provides a brief introduction to this topic of wireless channel models which have been used to design the present digital wireless communication systems.

References

1. Digital Land Mobile Radio Communications - COST 207", Commission of the European Communities, Final Report. Commission of the European Communities, Final Report, 14

March, 1984–13 September, 1988, Office for Official Publications of the European Communities, Luxembourg (1989)
2. EURO-COST-231 Revision 2, Urban transmission loss models for mobile radio in the 900 and 1800 MHz bands. Commission of the European Communities, Office for Official Publications of the European Communities, Luxembourg (1991)
3. Adachi, F., Feeney, M., Parsons, J., Williamson, A.: Crosscorrelation between the envelopes of 900 MHz signals received at a mobile radio base station site. In: IEE Proceedings F Communications, Radar and Signal Processing, vol. 133, pp. 506–512 (1986)
4. Asztély, D.: On antenna arrays in mobile communication systems: Fast fading and GSM base station receiver algorithms. Ph.D Thesis, Royal Institute of Technology, Stockholm, Sweden, IR-S3-SB-9611 (1996)
5. Bajwa, A., Parsons, J.: Large area characterization of urban UHF multipath propagation and its relevance to the performance bounds of mobile radio systems. IEE Proceedings on Communications, Radar and Signal Processing **132**(2 Part F), 99–106 (1985)
6. Blanz, J., Klein, A., Mohr, W.: Measurement-based parameter adaptation of wideband spatial mobileradio channel models. In: IEEE 4th International Symposium on Spread Spectrum Techniques and Applications Proceedings, 1996., vol. 1 (1996)
7. Clarke, R.: A statistical theory of mobile-radio reception. Bell Syst. Tech. J **47**(6), 957–1000 (1968)
8. Cox, D., Leck, R.: Distributions of multipath delay spread and average excess delay for 910-MHz urban mobile radio paths. IEEE Transactions on Antennas and Propagation **23**(2), 206–213 (1975)
9. Devasirvatham, D.: A comparison of time delay spread and signal level measurements within two dissimilar office buildings. IEEE Transactions on Antennas and Propagation **35**(3), 319–324 (1987)
10. Durgin, G.: Space-Time Wireless Channels. Prentice Hall PTR, NJ, USA (2003)
11. Erceg, V., Hari, K. V. S., Smith, M., Baum, D., Sheikh, K., Tappenden, C., Costa, J., Bushue, C., Sarajedini, A., Schwartz, R., Branlund, D., Kaitz, T., D., T.: Channel Models for Fixed Wireless Applications. IEEE 802.16 Standards (2001)
12. Ertel, R.: Vector channel model evaluation. SW Bell Tech. Res., Tech. Rep (1997)
13. Ertel, R., Cardieri, P., Sowerby, K., Rappoport, T., Reed, J.: Overview of Spatial Channel Models for Antenna Array Communication Systems. IEEE Personal Communications magazine pp. 10–22 (1998)
14. Feuerstein, M., Blackard, K., Rappaport, T., Seidel, S., Xia, H.: Path loss, delay spread, and outage models as functions of antennaheight for microcellular system design. IEEE Transactions on Vehicular Technology **43**(3 Part 1), 487–498 (1994)
15. Fuhl, J., Rossi, J., Bonek, E.: High-resolution 3-D direction-of-arrival determination for urbanmobile radio. IEEE Transactions on Antennas and Propagation **45**(4), 672–682 (1997)
16. Gans, M.: A power-spectral theory of propagation in the mobile-radio environment. IEEE Transactions on Vehicular Technology **21**(1), 27–38 (1972)
17. Greenstein, L., Erceg, V., Yeh, Y., Clark, M.: A new path-gain/delay-spread propagation model for digital cellularchannels. IEEE Transactions on Vehicular Technology **46**(2), 477–485 (1997)
18. Hari, K. V. S., Ottersten, B.: Parameter Estimation using a Sensor Array in a Ricean Fading channe. Sadhana, Journal of Indian Academy of Sciences **23**(1), 5–15 (1998)
19. Hata, M.: Empirical formula for propagation loss in land mobile radio services. IEEE Transactions on Vehicular Technology **29**(3), 317–325
20. Klein, A., Mohr, W.: A statistical wideband mobile radio channel model including the directions-of-arrival. In: IEEE 4th International Symposium on Spread Spectrum Techniques and Applications, pp. 102–06 (1996)
21. Kozono, S., Taguchi, A.: Mobile propagation loss and delay spread characteristics with a low-base station antenna on an urban road. IEEE Transactions on vehicular technology **42**(1), 103–109 (1993)
22. Lee, W.: Mobile communications engineering. McGraw-Hill, Inc. New York, NY, USA (1982)

23. Lee, W.: Mobile cellular telecommunications. McGraw-Hill Professional (1989)
24. Liberti, J., Rappaport, T.: A geometrically based model for line-of-sight multipath radiochannels. In: IEEE 46th Vehicular Technology Conference, vol. 2, pp. 844–848 (1996)
25. Lu, M., Lo, T., Litva, J.: A physical spatio-temporal model of multipath propagation channels. In: IEEE 47th Vehicular Technology Conference, vol. 2, pp. 810–814 (1997)
26. Maeyama, T., Ikegami, F., Kitano, Y.: Analysis of mountain-reflected signal strength in digital mobile radio communications. IEICE Transactions on Communications **76**(2), 98–102 (1993)
27. Michelson, D., Erceg, V., Greenstein, L.: Modeling diversity reception over narrowband fixed wireless channels. In: IEEE MTT-S Symposium on Technologies for Wireless Applications, pp. 95–100 (1999)
28. Monzingo, R., Miller, T.: Introduction to adaptive arrays. John Wiley and Sons, NY, USA (1980)
29. Moriyama, E., Iwama, T., Saruwatari, T.: Experimental investigation of 1.5 GHz, 2.3 GHz and 2.6 GHz band land mobile radio propagation in urban and rural areas . In: IEEE Vehicular Technology Conference, pp. 311–315 (1989)
30. Parsons, J.: The mobile radio propagation channel. Wiley New York (1992)
31. Paulraj, A.J., Nabar, R., Gore, D.: Introduction to space-time wireless communications. Cambridge University Press (2003)
32. Petrus, P., Reed, J., Rappaport, T.: Geometrically Based Statistical Channel Model for Macrocellular Mobile Environments. In: IEEE Proc. GLOBECOM, pp. 1197–1201 (1996)
33. Raleigh, G., Paulraj, A.: Time varying vector channel estimation for adaptive spatialequalization. In: IEEE Global Telecommunications Conference (GLOBECOM), vol. 1 (1995)
34. Rappaport, T.: Wireless communications: principles and practice. Prentice Hall PTR, New Jersey, USA (2002)
35. Rappaport, T., Seidel, S., Singh, R.: 900-MHz multipath propagation measurements for U.S. digital cellular radiotelephone. IEEE Transactions on Vehicular Technology **39**, 132–139 (1990)
36. Rossi, J., Levy, A.: A ray model for decimetric radiowave propagation in an urban area. Radio Science **27**(6), 971–979 (1992)
37. Saleh, A., Valenzuela, R.: A Statistical Model for Indoor Multipath Propagation. IEEE Journal on Selected Areas in Communications **5**(2), 128–137 (1987)
38. Seidel, S., Rappaport, T.: Site-specific propagation prediction for wireless in-buildingpersonal communication system design. IEEE Transactions on Vehicular Technology **43**(4), 879–891 (1994)
39. Seidel, S., Rappaport, T., Jain, S., Lord, M., Singh, R.: Path loss, scattering and multipath delay statistics in four european cities for digital cellular and microcellular radiotelephone. Vehicular Technology, IEEE Transactions on **40**(4), 721 –730 (1991)
40. Sousa, E., Jovanovic, V., Daigneault, C.: Delay spread measurements for the digital cellular channel inToronto. IEEE Transactions on Vehicular Technology **43**(4), 837–847 (1994)
41. Spencer, Q., Rice, M., Jeffs, B., Jensen, M.: A statistical model for angle of arrival in indoor multipath propagation. In: IEEE Vehicular Technology Conference, vol. 47, pp. 1415–1419 (1997)
42. Stapleton, S., Carbo, X., McKeen, T.: Tracking and diversity for a mobile communications base stationarray antenna. In: IEEE 46th Vehicular Technology Conference, 1996.'Mobile Technology for the Human Race'., vol. 3
43. Stapleton, S., Carbo, X., McKeen, T.: Spatial channel simulator for phased arrays. In: IEEE 44th Vehicular Technology Conference, pp. 1789–1792 (1994)
44. Tanaka, T., Kozono, S., Akeyama, A.: Urban multipath propagation delay characteristics in mobile communications. Electronics and Communications in Japan (Part I: Communications) **74**(8), 80–88 (1991)
45. Trump, T., Ottersten, B.: Estimation of nominal direction of arrival and angular spread using an array of senso. Signal Processing **50**(1), 57–69 (1996)
46. Valenzuela, R.: A ray tracing approach to predicting indoor wireless transmission. In: IEEE 43rd Vehicular Technology Conference, pp. 214–218 (1993)

47. Van Rees, J.: Measurements of the wide-band radio channel characteristics for rural, residential, and suburban areas. IEEE Transactions on Vehicular Technology **36**(1), 2–6 (1987)
48. de Weck, J., Merki, P., Lorenz, R.: Power delay profiles measured in mountainous terrain [radiowave propagation]. In: IEEE 38th Vehicular Technology Conference, pp. 105–112 (1988)
49. Wepman, J., Hoffman, J., Loew, L.: Characterization of macrocellular PCS propagation channels in the 1850–1990 MHz band. Proc. IEEE ICUPC94 pp. 165–170
50. Williams, J., Jakes, W.: Microwave mobile communications. Wiley-IEEE Press (1974)
51. Y. Okumura E. Ohmori, T.K., Fukua, K.: Field strength and its variability in uhf and vhf land-mobile radio service. Review Electronics Communication Lab **16**(9) (1968)
52. Zetterberg, P., Ottersten, B.: The spectrum efficiency of a base station antenna array system forspatially selective transmission. IEEE Transactions on Vehicular Technology **44**(3), 651–660 (1995)

Chapter 4
An Introduction to Integer and Large-Scale Linear Optimization

J. Cole Smith and Sibel B. Sonuc

Abstract This chapter provides an introductory analysis of linear programming foundations and large-scale methods. The chapter begins by discussing the basics of linear programming modeling and solution properties, duality principles for linear programming problems, and extensions to integer programming methods. We then develop Benders decomposition, Dantzig-Wolfe decomposition, and Lagrangian optimization procedures in the context of network design and routing problems that arise in telecommunications operations research studies. The chapter closes with a brief discussion and list of basic references for other large-scale optimization algorithms that are commonly used to optimize telecommunications systems, including basis partitioning, interior point, and heuristic methods.

4.1 Introduction

Complex system-wide design and operations problems arise in many contemporary applications, and particularly in several facets of communications networks. The last several decades have seen an explosion in the application breadth and sophistication of optimization algorithms to address these problems, due to both the substantial strides made in developing computing technology and the optimization theory that serves as a foundation for these algorithms.

Two particularly relevant areas of research applied to telecommunication optimization involve *integer programming* and *large-scale linear programming* theory. Algorithms developed for large-scale linear programs attempt to exploit special

J. Cole Smith
University of Florida, Department of Industrial and Systems Engineering, Gainesville, FL 32611, USA, e-mail: cole@ise.ufl.edu

Sibel B. Sonuc
University of Florida, Department of Industrial and Systems Engineering, Gainesville, FL 32611, USA, e-mail: sibel.bilge@ufl.edu

J. Kennington et al. (eds.), *Wireless Network Design: Optimization Models and Solution Procedures*, International Series in Operations Research and Management Science 158,
DOI 10.1007/978-1-4419-6111-2_4,

structures that may be present in certain formulations in order to improve computational efficiency over standard approaches. For instance, these algorithms may attempt to leverage the existence of parallel processors, or may convert the given problem into a related problem that can be solved by efficient algorithms. Understanding large-scale algorithms is of fundamental importance in addressing emerging problems in the field of wireless network optimization. These techniques have been used in a wide array of practical telecommunications studies, such as in [4, 24, 29, 30].

This chapter provides an accessible introduction to basic integer programming and large-scale linear optimization theory, along with references to related literature that explores these rich topics in more detail. We begin in Section 4.2 by discussing fundamental concepts related to optimization modeling, linear programming, and extensions to discrete and nonlinear models, and end the section with a description of branch-and-bound techniques for solving integer programming problems. Section 4.3 examines three common large-scale optimization techniques – Benders decomposition, Dantzig-Wolfe decomposition, and Lagrangian optimization – in the context of network optimization challenges. We then close that section (and this chapter) by briefly introducing some of the ongoing large-scale techniques whose discussion is beyond the scope of this chapter.

4.2 Basics of Modeling and Optimization

We begin this chapter by introducing the reader to basic modeling concepts in Section 4.2.1. We then consider the specific case of linear programming problems in Section 4.2.2, and discuss the key concept of duality in Section 4.2.3 related to linear programming. Next, we provide a brief discussion on nonlinear and discrete versions of optimization problems in Section 4.2.4. Finally, we discuss basic techniques for solving integer programming problems in Section 4.2.5.

4.2.1 Foundations of Optimization Models

Many complex optimization problems arising in real-world settings cannot reasonably be solved by hand, or by a team of subject-matter experts. The limiting factors may include the size of the problem (e.g., data too large for people to optimize over), its inherent complexities (e.g., having far too many options to consider), or a very small time-frame in which the problem must be solved. Optimization algorithms that are used to solve these problems must first represent the problem at hand as a *mathematical model.*

The first step in modeling a problem involves choosing *decision variables* for the problem, whose values represent the elements of the problem that should be calculated as the output. For example, in manufacturing systems, one may want to determine optimal production levels, perhaps along with optimal selling prices of these products (if demand is price elastic). For transportation networks, finding a best route for travelers is a common problem. Similarly, in communication networks, one may be interested in finding a collection of routes for data transferred through the system. However, it may be unclear how to best define decision variables in some problems. For instance, while it may be straightforward to define some variable x_a to represent the amount of production, and x_p to denote its price, defining variables corresponding to data routes can be done in several different ways. The choice of variable definition plays a large role in how the optimization problem will ultimately be solved.

In optimization models, one seeks a "best" set of decision variable values, according to some specified metric value. A function that maps a set of variable values to this metric is called an *objective function*, which is either minimized or maximized depending on the context of the problem. Some of the most common objectives minimize cost for an operation or maximize a profit function. A solution that achieves the best objective function value is called an *optimal solution*.

Decision variables are usually bounded by some restrictions, called *constraints*, related to the nature of the problem. (If not, the problem is called *unconstrained*; however, problems considered in this chapter are virtually all subject to constraints.) Constraints can be used for many purposes, and are most commonly employed to state resource limitations on a problem, enforce logical restrictions, or assist in constructing components of an objective function. For the purposes of this chapter, we will assume that constraints are either of the form $g(x) \leq 0$, $g(x) = 0$, or $g(x) \geq 0$, where x is a vector of decision variables. For example, if multiple data streams are to be sent on a route, the bandwidth of links on the network may limit the maximum sum of traffic that passes through the link at any one time, and needs to be accounted for in the optimization process via constraints. We may also state logical constraints which stipulate that each communication pair must send and receive its desired traffic. Another common set of logical constraints are *nonnegativity constraints* on the variables, which prevent negative flows, production values, etc., when they do not make sense for a given problem. An example of using constraints to assist in constructing an objective function arises when one wishes to minimize the maximum of several functions, say, $f_1(x), \ldots, f_k(x)$, where x is a vector of decision variables. One can define a decision variable z that represents $\max_{i=1,\ldots,k}\{f_i(x)\}$, and enforce constraints $z \geq f_i(x)$ for each $i = 1, \ldots, k$. Collectively, these constraints actually state that $z \geq \max_{i=1,\ldots,k}\{f_i(x)\}$, but when an objective of minimizing z is given, z takes its smallest possible value, and hence $z = \max_{i=1,\ldots,k}\{f_i(x)\}$ at optimality. A mathematical optimization model that seeks to optimize an objective function, possibly over a set of constraints is called a *mathematical program*.

Any solution that satisfies all constraints is called a *feasible solution*; those that violate at least one constraint are *infeasible*. The set of all feasible solutions defines the *feasible region* of the problem. Hence, optimization problems seek a feasible solution that has the best objective function value. When there exist multiple feasible solutions that achieve the same optimal objective function value, we say that *alternative optimal solutions* exist.

4.2.2 Linear Programming Problems

A mathematical program is called a *linear program* (LP) if no nonlinear terms appear in the objective function or in the constraints, and if no variables are restricted to take on a discrete set of values (i.e., all variables are *continuous* and allowed to take on fractions). LPs carry special attributes that enable them to be solved more easily than general mathematical programs. For instance, the following problem is an LP:

$$\max \quad x_1 + x_2 \tag{4.1a}$$
$$\text{s.t.} \quad 3x_1 + x_2 \leq 9 \tag{4.1b}$$
$$x_1 + 2x_2 \leq 10 \tag{4.1c}$$
$$x_1 - x_2 \leq 1 \tag{4.1d}$$
$$x_1, x_2 \geq 0. \tag{4.1e}$$

The feasible region of this LP is given in Figure 4.1.

The dotted line closest to the origin in Figure 4.2 represents the set of solutions that have an objective function value $x_1 + x_2 = 1$, and the next dotted line represents the set of solutions with objective function value 2. These *objective contours* are parallel, and increase in the direction of the objective function gradient (which is (1,1) in this case). To optimize a problem in two dimensions, one can slide the objective contours as far as possible in the direction of the objective gradient (for a max problem), which yields the optimal solution for this problem as depicted by Figure 4.2. Note that because this solution lies at the intersection of the first two constraints, the solution can be obtained by setting those two constraints as equalities, and then by solving the simultaneous system of equations. This yields $x_1 = 8/5$ and $x_2 = 21/5$, with objective function value $29/5$.

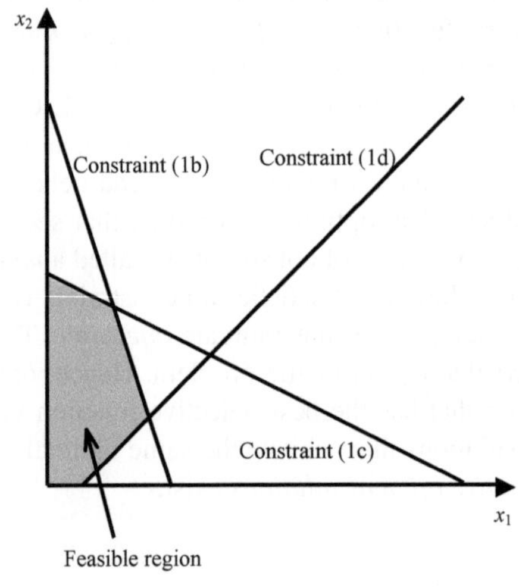

Fig. 4.1 Feasible region of the problem (4.1)

In general, we can formulate any linear programming problem in *standard form* as

$$\max \; c^T x \tag{4.2a}$$
$$\text{s.t.} \; Ax = b \tag{4.2b}$$
$$x \geq 0, \tag{4.2c}$$

where x is an n-dimensional (column) vector of decision variables, c is an n-dimensional vector of objective coefficient data, A is an $m \times n$ constraint coefficient matrix (where without loss of generality, rank(A) = m), and b is an m-dimensional right-hand-side vector that often states resource or requirement values of the optimization model. Note that inequalities are readily accommodated by adding a nonnegative *slack variable* or subtracting a nonnegative *surplus variable* from the left-hand-side of less-than-or-equal-to or greater-than-or-equal-to constraints, respectively. (The term "slack" is often used in conjunction with both less-than-or-equal-to and greater-than-or-equal to constraints.) For instance, the constraint $7x_1 + 2x_2 + 8x_3 \leq 4$ is equivalent to $7x_1 + 2x_2 + 8x_3 + s = 4$, along with the restriction $s \geq 0$ (where s serves as the slack variable). Minimization problems are converted to maximization problems by multiplying the objective function by -1 and changing the optimization to maximization. Nonpositive variables x_i are substituted with nonnegative variables (replace x_i with $-\bar{x}_i$), and unrestricted variables x_j are substituted with the difference of two nonnegative variables (replace x_j with $x'_j - x''_j$, with $x'_j \geq 0$ and $x''_j \geq 0$).

Now, observe that the optimal solution for (4.2) is given at a "corner," i.e., *extreme point* of the feasible region. In fact, regardless of the orientation of the objective function, it is easy to see that there will always exist an optimal solution at one of these extreme points. More precisely, let X be the feasible region contained

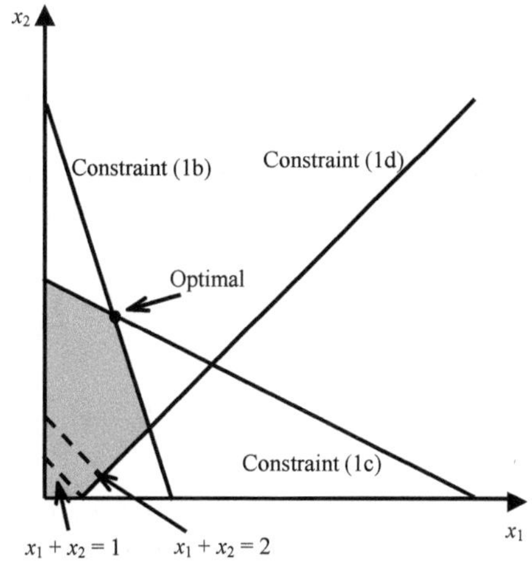

Fig. 4.2 Optimization process for solving the linear program (4.1)

within a *polyhedron* (intersection of halfspaces formed by linear constraints) for a linear program. A point $\bar{x}$ is an extreme point of X if and only if it cannot be written as a strict convex combination of any two distinct points in X, i.e., if there is no $x' \in X$, $x'' \in X$, and $0 < \lambda < 1$ such that $\bar{x} = \lambda x' + (1-\lambda)x''$. Then the following critical theorem holds true.

Theorem 4.1. *If there exists an optimal solution to a linear program, at least one extreme point must be optimal.*

Note that this theorem does not (incorrectly) assert that all optimal solutions exist at extreme points. When alternative optimal solutions exist, then an infinite number of non-extreme-point solutions are optimal. For instance, examining (4.1), if the objective function is changed to "max $x_1 + 2x_2$," then all solutions on the line segment between (x_1, x_2) = (0, 5) and (8/5, 21/5) are optimal.

The caveat in Theorem 4.1 is necessary because LPs need not have optimal solutions. If there are no feasible solutions, the problem is termed *infeasible*. If there is no limit to how good the objective can become (i.e., if the objective can be made infinitely large for maximization problems, or infinitely negative for minimization problems), then it is *unbounded*. If a problem is unbounded, then there exists a *direction d*, such that if $\hat{x}$ is feasible, then $\hat{x} + \lambda d$ is feasible for all $\lambda \geq 0$, and the objective improves in the direction d (i.e., $c^T d > 0$ for a maximization problem). Note that the set of all directions to a feasible region, unioned with the origin, is itself a polyhedron that lies in the nonnegative orthant of the feasible region (assuming $x \geq 0$). If this polyhedron is normalized with a constraint of the form $e^T x \leq 1$ (where e is a vector of 1's), then the nontrivial extreme points are *extreme directions* of the feasible region. It follows that all directions can be represented as nonnegative linear combinations of the extreme directions. Indeed, if a linear program is unbounded, it must be unbounded in at least one extreme direction. For the remaining discussion, we assume that extreme directions d have been normalized so that $e^T d = 1$.

The following *Representation Theorem* gives an important alternative characterization of polyhedra, which we use in the following section.

Theorem 4.2. *Consider a polyhedron defining a feasible region X, with extreme points x^i for $i = 1, \ldots, E$ and extreme directions d^j for $j = 1, \ldots, F$. Then any point $x \in X$ can be represented by a convex combination of extreme points, plus a nonnegative linear combination of extreme directions, i.e.,*

$$x = \sum_{i=1}^{E} \lambda_i x^i + \sum_{j=1}^{F} \mu_j d^j, \text{ where} \tag{4.3}$$

$$\lambda, \mu \geq 0, \text{ and } \sum_{i=1}^{E} \lambda_i = 1. \tag{4.4}$$

Full details of LP solution techniques are beyond the scope of this chapter, but we present a sketch of the *simplex method* for linear programs here. Observe that each extreme point can be associated with at least one set of m columns of A that form a *basis* of $\mathbb{R}^m$. The method begins by choosing a basis of A that corresponds to an extreme-point solution, and moves to an *adjacent basis* (i.e. a set of columns that differs by only one column from the current basis) at each iteration. Each move should improve some extended metric of the objective function (see the *lexicographic rule*), which then guarantees that the algorithm terminates either with an optimal solution, or in a direction of unboundedness. This method can easily accommodate the case in which a starting basis corresponding to a basic feasible solution is not readily available (using the *two-phase* or *big-M* methods [5]), and identifies when a problem is infeasible as well. There are also alternative *interior point* approaches to solving LPs, inspired by nonlinear optimization ideas generated in the 1960s. These algorithms are discussed very briefly in Section 4.3.4.

4.2.3 Duality

Duality is an important complement to linear programming theory. Consider the general linear programming form given by (4.2). A *relaxation* of the problem is given as:

$$z(\pi) = \max \ c^T x + \pi^T(b - Ax) \tag{4.5a}$$

$$\text{s.t.} \ x \geq 0, \tag{4.5b}$$

where $\pi \in \mathbb{R}^m$ is a set of *dual values* that are specified as inputs for now, and where $z(\pi) = \infty$ if (4.5) is unbounded. Assume for the following discussion that an optimal solution exists to (4.2), and let $z^\star$ be the optimal objective function value to (4.2). Then $z(\pi) \geq z^\star$. To see this, note that any feasible solution to problem (4.2) is also feasible to (4.5), and this solution yields the same objective function value in both formulations. However, the reverse is not necessarily true. Hence, for each choice of π, we obtain an upper bound on $z^\star$. To achieve the tightest (i.e., *lowest*) bound possible, we wish to optimize the following problem:

$$\min_{\pi} z(\pi), \text{ where } z(\pi) = \max_{x \geq 0}(c^T - \pi^T A)x + \pi^T b. \tag{4.6}$$

Because x is bounded only by nonnegativity restrictions, $z(\pi)$ will be infinite if any component of $(c^T - \pi^T A)$ is positive. On the other hand, if $(c^T - \pi^T A) \leq 0$, then choosing $x = 0$ solves problem (4.5), and $z(\pi) = \pi^T b$. Therefore, (4.6) is equivalent to

$$\min \ \pi^T b \tag{4.7a}$$

$$\text{s.t.} \ A^T \pi \geq c. \tag{4.7b}$$

The formulation (4.7) is called the *dual* of (4.2), which we refer to as the *primal* problem. (Note that "primal" and "dual" are relative terms, and either can refer to a minimization or a maximization problem.) Each primal variable is associated with a dual constraint, and each dual variable is associated with a primal constraint. If some constraints (4.2b) are of the $\leq$ sense (or $\geq$ sense), then their associated dual (π) variables are restricted to be nonnegative (or nonpositive). Also, if (4.2c) does not enforce nonnegativity of the x-variables (or enforces nonpositivity), then their associated constraints (4.7b) are of the $=$ sense (or of the $\leq$ sense).

There are vital relationships that exist between the primal and dual. First, suppose that there exists a feasible primal solution x' and a feasible dual solution π'. Then $(\pi')^T Ax' = (\pi')^T b$ due to primal feasibility, and $(\pi')^T Ax' \geq c^T x'$ due to dual feasibility and nonnegativity of x. Therefore, combining these expressions, $(\pi')^T b \geq c^T x'$, i.e., the dual objective function value for any feasible solution π' is greater than or equal to the primal objective function value for any primal feasible solution x'. A more general statement is given by the following *Weak Duality Theorem.*

Theorem 4.3. *Consider a set of primal and dual linear programming problems, where the primal is a maximization problem and the dual is a minimization problem. Let $z_D(\pi')$ be the dual objective function value corresponding to any dual feasible solution π', and let $z_P(x')$ be the primal objective function value corresponding to any primal feasible solution x'. Then $z_D(\pi') \geq x_P(x')$.*

Corollary 4.1. *If the primal (dual) is unbounded, then the dual (primal) is infeasible.*

Theorem 4.3 states that any feasible solution to a maximization problem yields a lower bound on the optimal objective function value for its dual minimization problem, and any feasible solution to a minimization problem yields an upper bound on the optimal objective function value for its dual maximization problem. It thus follows that if one problem is unbounded, there cannot exist a feasible solution to its associated dual as stated by Corollary 4.1. (It is possible for both the primal and dual to be infeasible.)

The *Strong Duality Theorem* below is vital in large-scale optimization.

Theorem 4.4. *Consider a set of primal and dual linear programming problems, and suppose that both have feasible solutions. Then both problems have optimal solutions, and letting $z_P^\star$ and $z_D^\star$ be the primal and dual optimal objective function values, respectively, we have that $z_P^\star = z_D^\star$.*

For instance, the dual of problem (4.1) is given by

$$
\begin{aligned}
\min\ & 9\pi_1 + 10\pi_2 + \pi_3 && (4.8a)\\
\text{s.t. } & 3\pi_1 + \pi_2 + \pi_3 \geq 1 && (4.8b)\\
& \pi_1 + 2\pi_2 - \pi_3 \geq 1 && (4.8c)\\
& \pi_1, \pi_2, \pi_3 \geq 0. && (4.8d)
\end{aligned}
$$

Recalling that the primal problem (4.1) has an optimal objective function value of 29/5, observe that, as stated by the Weak Duality Theorem, all feasible solutions to problem (4.8) have objective function value at least 29/5. For instance, $\pi^1 = (1,0,0)$ and $\pi^2 = (1/3, 1/2, 0)$ are both feasible, and have objective function values 9 and

8, respectively. By the Strong Duality Theorem, (4.8) must have an optimal solution with objective function value 29/5; note that this is achieved at $\pi^\star = (1/5, 2/5, 0)$.

Dual variables are often called *shadow prices* for the primal constraints. Recall that each dual variable π_i is associated with a primal constraint i. The value of the dual variable states a "price" for the constraint's right-hand-side value. That is, if the right-hand-side for constraint i is changed from b_i to $b_i + \varepsilon_i$, the optimal objective function value changes by $\pi_i \varepsilon_i$ (i.e., $\partial z_P^\star / \partial b_i = \pi_i$), provided that the change ε is not too great. (It is straightforward to compute the allowable range of values that ε_i can take for which $\partial z_P^\star / \partial b_i = \pi_i$.) Understanding the interpretation of these values leads to insights that are useful in large-scale optimization algorithms.

Remark 4.1. Given a primal optimal solution x, there exists a corresponding dual optimal solution π that is *complementary slack* to the primal. Mathematically, we have $\text{diag}(A^T\pi - c)x = 0$ at optimality for the primal/dual formulations given as above (see [5] for a deep and complete discussion of primal/dual relationships in linear programming). That is, for every dual constraint, either its slack or its associated primal variable equals zero. (If the primal constraints were inequalities, we would also need to state that for every primal constraint, either the primal slack variable or its associated dual variable equals zero.) This corresponding dual solution allows us to introduce the *reduced cost* concept for primal variables. The reduced cost of a variable x_j gives the rate of change in the objective function value when x_j is increased from its current value (i.e., it gives the value $\partial z_P^\star / \partial x_j$). Given an optimal solution to (4.2), let π be the corresponding complementary slack (and thus optimal) dual variables associated with constraints $Ax = b$ given in (4.2b). Define a_j to be the jth column of A. Then the reduced cost of variable x_j is given by $c_j - \pi^T a_j$. The intuition behind this formula is given as follows: If x_j is increased by Δ, the objective function value changes by $c_j\Delta$. Also, the left-hand-side of constraint i would change by $a_{ij}\Delta$ (where a_{ij} is the coefficient of variable j in row i) due to the increase in x_j by Δ. Thus, the right-hand-side of constraint i changes by $-a_{ij}\Delta$. Given the interpretation of shadow prices above, this implies that the objective function value will change by $-\pi_j a_{ij}\Delta$. Taking all of these terms together, the net change in the objective is $\Delta(c_j - \pi^T a_j)$, and hence, the rate of objective change as x_j increases is $c_j - \pi^T a_j$. For a minimization problem, we require that

$$c_j - \pi^T a_j \geq 0 \quad \forall j = 1, \ldots, n$$

at optimality, or else, the solution could be improved by increasing x_j. Note that these restrictions correspond exactly to the dual feasibility conditions given by (4.7b). In addition, the concept of reduced costs is vital in understanding the Dantzig-Wolfe decomposition algorithm given in Section 4.3.2.

4.2.4 Modeling with Nonlinear and Discrete Variables

Naturally, many real-world challenges involve nonlinear objective functions and/or constraints, and/or variables that must take on a discrete set of values. These problems are no longer considered to be linear programs, and cannot directly be solved using the framework developed above. We briefly discuss in this subsection when these problems arise, and what challenges we face in solving them.

For instance, consider a situation in which two wireless routers must be arranged over a Euclidean space, and must be no more than d meters from each other. Letting (x_i, y_i) be decision variables that specify where router $i = 1, 2$ will be placed (on a grid with one-meter spacing), we would have the constraint

$$\sqrt{(x_1 - x_2)^2 + (y_1 - y_2)^2} \leq d.$$

Since this constraint is nonlinear, we say that the problem is a nonlinear program. Nonlinearities may also arise in the objective function, for example, when minimizing a nonlinear function of communication latency, which may be represented by a nonlinear function of data transmission over a particular communication link. Nonlinear programs can be extremely difficult to solve, even if there are few nonlinear terms in the problem (see, e.g., [22]). The difficulty of, and methods used for, solving these problems depend on the convexity or concavity of the objective function, and whether or not the feasible region forms a convex set.

Several other applications may call for variables that take on only discrete values; usually some subset of the integer values (and often just zero or one). These problems are called *integer programs* (IP). An integer programming problem is often called a *binary* or *0-1* program if all variables are restricted to equal 0 or 1. Problems having a mixture of discrete and continuous variables are called *mixed-integer programs* (MIP).

There are a vast array of uses for integer variables in modeling optimization problems. The most obvious arises when decision variable values refer to quantities that must logically be integer-valued, such as personnel or non-splittable units of travel (e.g., cars on a transportation network). Logical entities may also be integer-valued (especially with 0-1 variables). For instance, variables may represent whether or not to establish a link between two nodes in a network, or whether or not to assign routing coverage to some point of access. Integer variables are also very useful in constructing certain cost functions, when these functions do not satisfy certain well-behaved properties that would allow linear programming formulations to be used (e.g., for minimization problems, piecewise-linear convex functions can be modeled as LPs, while piecewise-linear nonconvex functions generally require the use of binary variables).

4.2.5 Integer Programming Methods

We now discuss the most common methods of solving (mixed-) integer programming problems. (Although there exist differences in the literature regarding solution techniques for IPs and MIPs, the foundations in this chapter are valid for either class of problems. Hence, we do not distinguish between IPs and MIPs in this discussion.) Throughout this section, we suppose that all problems are bounded for the sake of simplicity.

To begin, consider the IP (4.9), which has the same objective function and constraints as LP (4.1), but with variables x_1 and x_2 now restricted to be integers by (4.9f).

$$\max \quad x_1 + x_2 \tag{4.9a}$$
$$\text{s.t.} \quad 3x_1 + x_2 \le 9 \tag{4.9b}$$
$$x_1 + 2x_2 \le 10 \tag{4.9c}$$
$$x_1 - x_2 \le 1 \tag{4.9d}$$
$$x_1, x_2 \ge 0 \tag{4.9e}$$
$$x_1, x_2 \in \mathbb{Z}. \tag{4.9f}$$

Clearly, the optimal LP solution given by $x = (8/5, 21/5)$ is no longer a feasible solution to the IP.

The solution methods used to solve LPs are not directly applicable when solving IPs, because IP feasible regions are generally nonconvex, while LP feasible regions are always convex. Figure 4.3 shows the feasible points for IP (4.9). The dashed line depicts an objective function contour passing through the optimal solutions $(0,5)$, $(1,4)$, and $(2,3)$, all with an objective function value of 5.

The *LP relaxation* of an IP formulation is simply the IP formulation with the variable integrality constraints omitted. Because the relaxation formulation has the same objective as the original IP formulation, but with only a subset of the constraints, the optimal objective function value to the LP relaxation is always at least

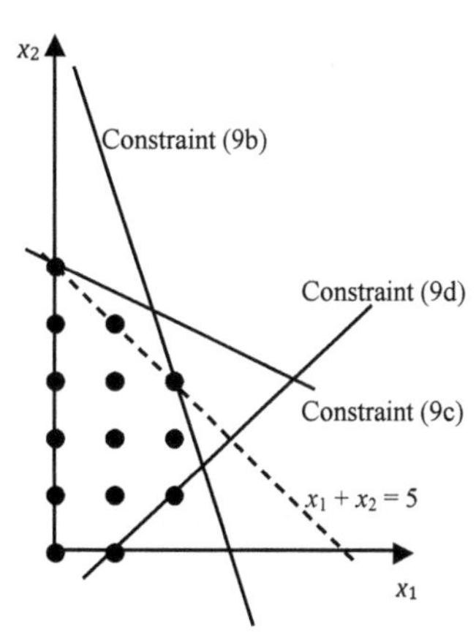

Fig. 4.3 Feasible points for problem (4.9)

as good as that of the original IP. That is, for a maximization problem, the optimal objective function value of the LP relaxation is at least as large as that of the IP; and for a minimization problem, it is at least as small as that of the IP. Therefore, for a maximization problem, the optimal LP relaxation objective function value is an *upper bound* on the optimal IP objective function value. A *lower bound* on the optimal (maximization) IP objective function value can be found by computing the objective function value of any feasible solution. This is evident because an optimal solution cannot be worse than any candidate solutions thus identified. In the subsequent discussion, we refer to the *incumbent* solution as the best solution identified for the IP thus far. (Note that for a minimization problem, the LP relaxation solution yields a lower bound, and feasible solutions yield upper bounds, on the optimal IP objective function value.)

Returning to our example, suppose that we add constraints $x_1 + x_2 \leq 5$ and $x_1 \leq 2$ to the LP (4.9), as in Figure 4.4, so that all extreme points to the LP relaxation are integer-valued. In general, we refer to a formulation that yields the convex hull of integer-feasible solutions as an *ideal* formulation. Then solving this ideal linear program (and identifying an extreme-point optimal solution) is equivalent to solving the IP. This is evident because the optimal objective function value of the LP relaxation yields an upper bound on the optimal IP objective function value, but in this case, one also obtains a lower bound from this relaxation due to the fact that an extreme point optimal LP solution is integer-valued as well. Noting that the lower and upper bounds match, the LP optimal solution must be IP optimal as well.

However, obtaining an ideal formulation is extremely difficult when dealing with large-scale problems. Not only is it computationally difficult to obtain each of the required constraints, but there may be an exponential number of these constraints. Instead, we can improve a given mathematical formulation to contain fewer fractional points, while retaining all of the feasible integer points. For example, consider the addition of constraint $2x_1 + 3x_2 \leq 15$ to (4.9), as depicted in Figure 4.5. This constraint does not eliminate any of the integer feasible points, but improves

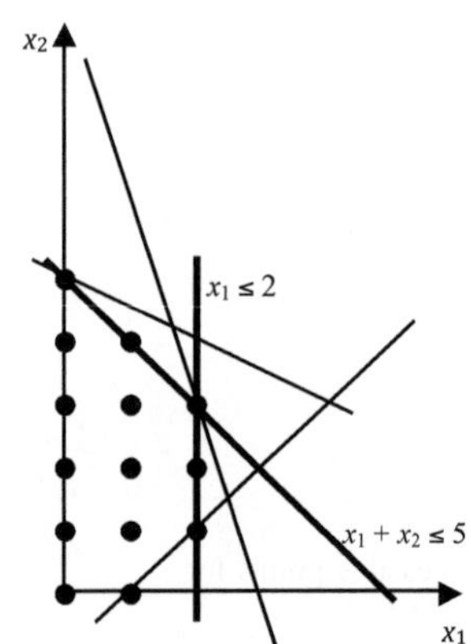

Fig. 4.4 Ideal formulation of problem (4.9)

the formulation by "cutting off" a set of fractional solutions. The new formulation is written as:

$$\max \quad x_1 + x_2 \tag{4.10a}$$
$$\text{s.t.} \quad (4.9b) - (4.9f) \tag{4.10b}$$
$$2x_1 + 3x_2 \leq 15 \tag{4.10c}$$

A constraint that does not remove any integer-feasible solution (i.e., of the form $\alpha^T x \leq \beta$ such that $\alpha^T x^i \leq \beta \; \forall x^i \in X$, where X is the set of all feasible solutions), such as (4.10c), is called a *valid inequality*. We say that the formulation (4.10) is *stronger* than formulation (4.9), because the set of fractional feasible solutions to the LP relaxation of (4.10) is a strict subset of those to the LP relaxation of (4.9). The classic work of Nemhauser and Wolsey [20] discusses common techniques used to generate valid inequalities for integer programming problems, often based on specific problem structures.

One generic approach for solving an IP is to solve its LP relaxation and obtain an optimal solution $\hat{x}$. (If no solution exists to the LP relaxation, then the IP is also infeasible.) If $\hat{x}$ is integer-valued, then it is optimal. Otherwise, add a valid inequality that cuts off $\hat{x}$, i.e., find a valid inequality $\alpha^T x \leq \beta$ such that $\alpha^T \hat{x} > \beta$. This type of valid inequality is called a *cutting plane* (or just a "cut"), and the given technique is called a *cutting-plane algorithm.* Practically speaking, though, pure cutting-plane algorithms are not practical, due to the fact that many cuts must usually be generated before termination, and the cutting-plane process is potentially not numerically stable.

Instead, most common algorithms to solve integer programs involve the *branch-and-bound* process. The branch-and-bound algorithm starts in the same fashion as the cutting-plane algorithm above, obtaining optimal LP relaxation solution $\hat{x}$ and terminating if $\hat{x}$ is integer-valued (or if no solution exists to the LP relaxation). Otherwise, at least some variable x_i equals to a value f that is fractional-valued (non-integer) in the optimal LP solution $\hat{x}$. Clearly, all feasible solutions must obey either the condition that $x_i \leq \lfloor f \rfloor$ or $x_i \geq \lceil f \rceil$. We thus *branch* into two subproblems, one

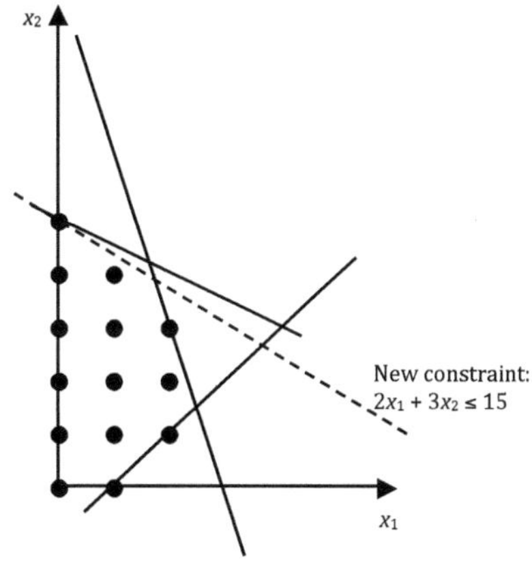

Fig. 4.5 Feasible region for problem (4.10)

containing the condition that $x_i \leq \lfloor f \rfloor$, and the other with $x_i \geq \lceil f \rceil$. The subproblems are solved recursively, and the best solution obtained from the two subproblems is optimal.

Note that the recursive application of branch-and-bound implies that the branching operations may need to be applied to subproblems at multiple levels of the recursive algorithm. The branches created throughout this algorithm yield a *branch-and-bound tree.*

Before presenting the remainder of the branch-and-bound algorithm, we illustrate the first stages of the branching process on formulation (4.10). First, observe that the optimal solution to the LP relaxation of (4.10) is $x^0 = (12/7, 27/7)$. Of course, x^0 is infeasible to (4.10), because x_1^0 and x_2^0 are both fractional. Therefore, the next step in our algorithm is to branch and either eliminate the portion of the feasible region in which $1 < x_1 < 2$, or the portion in which $3 < x_2 < 4$. We (arbitrarily) choose to branch on x_2. As a result, we create two subproblems:

$$\begin{aligned} \text{Subproblem 1: max} \quad & x_1 + x_2 && (4.11a)\\ \text{s.t.} \quad & \text{constraints (4.9b)} - \text{(4.9e), (4.10c)} && (4.11b)\\ & x_2 \leq 3 && (4.11c) \end{aligned}$$

$$\begin{aligned} \text{Subproblem 2: max} \quad & x_1 + x_2 && (4.12a)\\ \text{s.t.} \quad & \text{constraints (4.9b)} - \text{(4.9e), (4.10c)} && (4.12b)\\ & x_2 \geq 4. && (4.12c) \end{aligned}$$

Figure 4.6 illustrates these two new subproblems, which we call R_1 and R_2, respectively associated with subproblems 1 and 2.

Now, both regions R_1 and R_2 are *active*, meaning that they must be searched by recursively solving both of the associated subproblem LPs. This process is depicted in the branch-and-bound tree shown in Figure 4.7. We must solve subproblems over

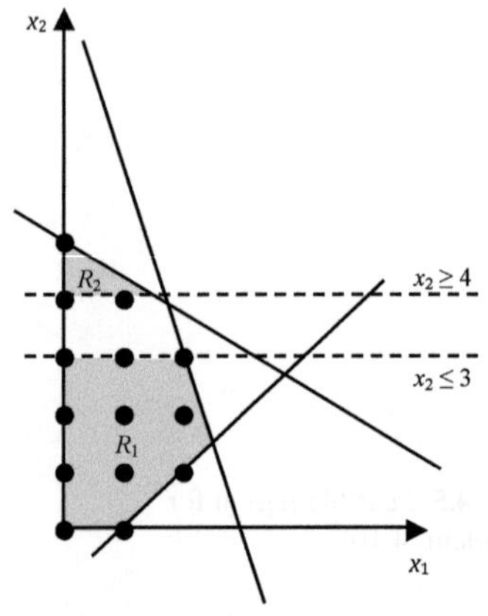

Fig. 4.6 Feasible regions for branch-and-bound algorithm for IP (4.10)

regions R_1 and R_2 after the initial branching step. The optimal solution to subproblem 1 is $x^1 = (2,3)$ with objective function value of 5, and the optimal solution to subproblem 2 is $x^2 = (3/2,4)$ with objective function value 11/2. A critical observation is that all feasible solutions must lie either in R_1 or in R_2. Because the relaxed optimal objective function values for subproblems 1 and 2 are 5 and 11/2, respectively, it is impossible to find an integer-feasible solution having an objective function value larger than 5 in region 1, or larger than 11/2 in region 2. Moreover, we have actually identified an integer feasible solution with objective 5 in region 1. The implication is that (a) we have identified an optimal solution to region 1 and do not need to search it further, (b) the optimal objective function value to the overall IP is no less than 5 (noting that x^1 serves as an incumbent), and (c) the optimal objective function value to the overall IP is no greater than 11/2.

We employ these concepts to generate simple rules for terminating search over these subproblems. In general, when a subproblem is solved within the branch-and-bound algorithm, four scenarios can arise.

1. If the subproblem is infeasible, then there is clearly no integer-feasible solution within the region represented by the subproblem. Therefore, we need not search further in this region. We say that we *fathom* this subproblem, or equivalently, that we *prune* search in this area of the branch-and-bound tree.

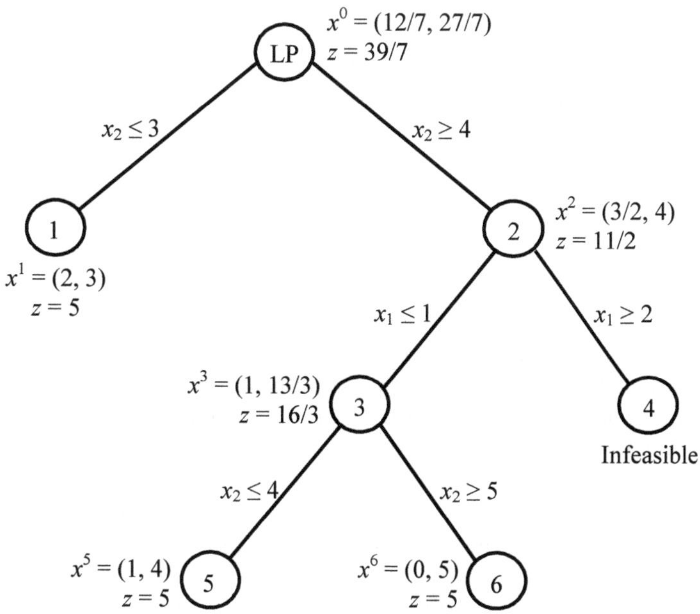

Fig. 4.7 Branching tree for branch-and-bound algorithm applied to IP (4.10)

2. If the optimal objective function value is not strictly better than the incumbent objective function value, then no integer-feasible solution in the subproblem can improve upon the best solution identified thus far. (It is possible that no incumbent solution yet exists; if not, then this particular case cannot arise.) We would thus fathom search in this region.
3. If the subproblem solution is integer-feasible, then it represents an optimal solution to the integer program defined over the subproblem region. Therefore, we fathom search in this region. If this solution is better than the incumbent, replace the incumbent with the subproblem solution.
4. If none of the previous three cases occurs, then the subproblem's solution is fractional, and has a strictly better objective function value than the incumbent solution. In this case, we must choose a variable x_i that takes on a fractional value f in this subproblem solution, and branch into two new subproblems (one by appending $x_i \leq \lfloor f \rfloor$ into the constraint set, and the other by appending $x_i \geq \lceil f \rceil$, as before).

In cases 1–3 above, the regions are fathomed, and are made *inactive* since search no longer needs to take place. In case 4, the region is also made inactive, but the two new subproblems created via the branching process both need to be explored, and are thus active. Note that an upper bound for maximization problems is given by the largest objective function value at any active subproblem. (For a minimization problem, we find a lower bound by computing the least objective function value at all active subproblems.) The algorithm terminates when no nodes are active, implying that search has concluded across the entire feasible region. The incumbent solution in memory when the algorithm terminates is an optimal solution; if no incumbent has been found, then the original IP is infeasible.

Continuing with our example, subproblem 1 is fathomed due to case 3 (having an integer-feasible solution), with x^1 becoming the incumbent solution with objective 5 as mentioned above. The only active subregion is now R_2 with optimal solution $x^2 = (3/2, 4)$. Since x_1^2 is the only fractional value, we branch on variable x_1 in R_2, creating two new active branches as follows. Let R_3 be the intersection of R_2 with the condition $x_1 \leq 1$, and let R_4 be the intersection of R_2 with $x_1 \geq 2$. After we branch on x_1, R_2 becomes inactive, and the new set of active regions is $\{R_3, R_4\}$.

The optimal solution over R_3 is $x^3 = (1, 13/3)$ with objective function value of $16/3$. Note that $16/3 > 5$ and x_2^3 is fractional; hence, we branch on x_2 as: $x_2 \leq 4$ (call this new region R_5) and $x_2 \geq 5$ (call this new region R_6). Note that these new regions retain the previous requirements $x_2 \geq 4$ and $x_1 \leq 1$. For subproblem 4, we have the new branch restriction that $x_1 \geq 2$ in addition to the restriction that $x_2 \geq 4$. Thus, constraint (4.9b) (which states that $3x_1 + x_2 \leq 9$) cannot be satisfied because the left-hand-side of (4.9b) must be at least 10. This problem is infeasible, and R_4 is fathomed. The active regions are now R_5 and R_6.

Solving over R_5 gives an optimal solution $x^5 = (1, 4)$ with objective function value of 5, which is integer-valued but fails to improve the incumbent solution. (It turns out that this is an alternative optimal solution for (4.10), but the goal of branch-and-bound is to find one optimal solution rather than all.) This region is *fathomed by bound* in case 2 because its objective is not strictly greater than 5. (We could also say that it is fathomed by integrality in case 3 since it is an integer solution.) Subproblem 6 yields another integer solution $x^6 = (0,5)$ which gets fathomed both by bound and by integrality as well. Because there are no more active branches left, the algorithm terminates, and we conclude that the incumbent solution $x^\star = (2,3)$ is an optimal solution for our problem.

It is important to note that the number of subproblems generated within branch-and-bound can grow exponentially in terms of the number of integer variables in the corresponding IP. Therefore, a practical concern is that the algorithm will not terminate within a reasonable amount of time. The execution time for branch-and-bound can be reduced by adding certain valid inequalities that cut off fractional solutions, which would otherwise need to be explored and eliminated by branch-and-bound. Algorithmic considerations may attempt to improve computational time by prescribing "node selection" rules (i.e., the order in which subproblems are branched on and explored) and "variable selection" rules (i.e., the choice of variables to branch on when case 4 above is encountered). The field is far too large and complex to be discussed in any detail here, and so we refer to [20] for a comprehensive treatment of the subject.

However, some problems appear to be too difficult to solve, either because of the problem complexity or size, regardless of the algorithmic tricks used to solve them. If the algorithm must be terminated early, then we can still attempt to assess the quality of the incumbent solution. Recall that, for a maximization problem, the incumbent solution's objective function value is a lower bound (*LB*) on the optimal IP solution. The upper bound (*UB*) is the largest objective function value present at any active node of the branch-and-bound tree. We say that the *optimality gap* is the difference *UB* − *LB*, and the *relative optimality gap* is $(UB - LB)/LB \times 100\%$. When the relative optimality gap is sufficiently small, we may terminate the branch-and-bound procedure with the knowledge that the incumbent solution is within a certain percentage of optimality. (The relative optimality gap is given with respect to the incumbent solution's objective function value. Thus, for minimization problems, the incumbent solution's objective function value yields an upper bound, and the relative optimality gap replaces *LB* with *UB* in the denominator.) In the example above, after subproblems 2 and 3 are solved, we have that *LB* = 5 and *UB* = 11/2. Thus, the relative optimality gap is (0.5/5)×100% = 10%.

4.3 Large Scale Optimization

Many real-world optimization applications require very large mathematical programming models for which classical optimization techniques require too much time or memory to solve to optimality, due to a prohibitively large number of variables and constraints. In some formulations, though, the problem would be separable (solvable as a series of smaller optimization problems) if not for the presence of variables that are present in multiple sets of constraints, or constraints that jointly constrain several sets of variables. In this situation, we say that the problem contains *complicating variables* or *complicating constraints*, respectively. (Examples of these situations are shown in Sections 4.3.1 and 4.3.2, respectively.)

In this case, we can attempt to exploit the special structure of these mathematical programming problems by using *large-scale optimization* methods. In fact, "large-scale optimization" is a broad term and refers to many different types of advanced techniques for solving large optimization problems. For this chapter, we focus primarily on decomposition techniques due to their common use in solving telecommunication-related problems.

The rest of this section is organized as follows. We discuss the Benders decomposition technique in Section 4.3.1, and illustrate it using a capacitated evacuation network example. Next, we discuss the Dantzig-Wolfe decomposition method in Section 4.3.2, illustrated using a multicommodity flow example. We then describe in Section 4.3.3 key principles behind Lagrangian optimization. Finally, we briefly discuss several other large-scale optimization strategies in Section 4.3.4.

4.3.1 Benders Decomposition

To motivate the use of Benders decomposition [7], we begin by introducing a capacitated network design problem in which arcs can fail. Consider a network $G = (V, A)$ with node set V and candidate arc set A, and let d_i denote the supply/demand value of node $i \in V$ (where $d_i > 0$ indicates that there is a supply of d_i units at node i, and $d_i < 0$ indicates that a demand of $|d_i|$ is requested at node i). Define the *forward star* of node $j \in V$ (i.e., the nodes adjacent from node j) as $FS(j) = \{i \in V : (j, i) \in A\}$, and the *reverse star* of node $j \in V$ (those nodes adjacent to node j) as $RS(j) = \{i \in V : (i, j) \in A\}$.

Arc (i, j) can be constructed at a cost of $r_{ij} \geq 0$. If constructed, the arc will have a capacity of $q_{ij} > 0$, and a per-unit flow cost of $c_{ij} \geq 0$. (Note that all flows are assumed to be simultaneous, meaning that the total flow across arc (i, j) is no more than q_{ij}.)

A network design problem would hence seek to minimize the cost of building arcs plus the cost of routing flows across the established arcs. However, we also consider in this problem the situation that some sets of arcs can possibly fail. For ease of exposition, we consider here the case in which any single arc can fail, although the discussion that follows easily extends to the case in which combinations of arcs can fail. In fact, we adopt the terminology *failure scenario* from [11], where scenario $s \in A$ refers to the event that some arc $s \in A$ fails. Let $p^{ij} \geq 0$ be the probability that (only) arc (i, j) fails. Another interpretation here is that p^{ij} represents the proportion of time that arc (i, j) will fail over a long-term network flow situation. (Note that we use a single index for arcs where convenient, e.g., p^s and p^{ij} are equivalently used. Also, for simplicity, we assume that $\sum_{(i,j)\in A} p^{ij} = 1$, which is without loss of generality by adding a dummy arc if necessary.)

Now, the objective becomes the minimization of arc construction costs, plus the *expected* flow costs that will take place after any arc fails. Note that the cost for transmitting network flows is only assessed after an arc failure (if any arc fails). Define binary variable x_{ij} equal to 1 if x_{ij} is built, and 0 otherwise. Also, let variables f^s_{ij} represent the flow value on arc $(i,j) \in A$ given that arc $s \in A$ would fail if built. For each scenario $s \in A$, we define $q^s_{ij} = q_{ij}$ if $s \neq (i,j)$, and $q^{ij}_{ij} = 0$ to model the remaining potential capacity on network arcs after a failure occurs. Then this network design problem can be formulated as follows.

$$\min \sum_{(i,j)\in A} r_{ij}x_{ij} + \sum_{s\in A} p^s \sum_{(i,j)\in A} c_{ij} f^s_{ij} \tag{4.13a}$$

$$\text{s.t. } f^s_{ij} \le q^s_{ij} x_{ij} \qquad \forall (i,j) \in A,\ s \in A \tag{4.13b}$$

$$\sum_{i\in FS(j)} f^s_{ji} - \sum_{i\in RS(j)} f^s_{ij} = d_j \qquad \forall j \in V,\ s \in A \tag{4.13c}$$

$$f^s_{ij} \ge 0 \qquad \forall (i,j) \in A,\ s \in A \tag{4.13d}$$

$$x_{ij} \in \{0,1\} \qquad \forall (i,j) \in A \tag{4.13e}$$

The first term of (4.13a) gives the fixed costs for constructing the arcs, while the second term gives the expected total flow cost. Constraints (4.13b) restrict each arc-flow value if the corresponding arc has not been built, or was built but failed in the given scenario. Otherwise, the flow variable is bounded by the corresponding arc capacity. Constraints (4.13c) state the flow balance constraints. Constraints (4.13d) and (4.13e) enforce non-negativity and logical conditions on the variables, respectively.

Note that if the x-variables are fixed values, the problem would be split into $|A|$ separate network flow problems, each one taking place over the set of variables $\{f^s_{ij}, \forall (i,j) \in A\}$, for each $s \in A$. For each s, we could then minimize $\sum_{(i,j)\in A} c_{ij} f^s_{ij}$ subject to constraints (4.13b), (4.13c), and (4.13d) corresponding only to failure scenario s, which is just a minimum cost flow problem [1]. The optimal solution to the overall problem would then be the collection of these individual flow problem solutions.

In this sense, we say that the problem has *complicating variables* x_{ij}. It is possible to solve (4.13) directly by mixed-integer programming techniques, but this technique is often very inefficient for problems such as these (especially when the number of scenarios is not limited by $|A|$, but may involve combinations of arcs that simultaneously fail). However, problems such as (4.13) that contain these complicating variables are amenable to Benders decomposition.

We begin by stating a generic model on which we will study Benders decomposition. Here, S can be thought of as a set of "scenarios," and if a set of complicating variables x exists, the problem can be separated into $|S|$ different optimization problems in terms of variables y^s, for each $s \in S$.

$$\min \ c^T x + \sum_{s \in S} (h^s)^T y^s \tag{4.14a}$$
$$\text{s.t.} \ Ax = b \tag{4.14b}$$
$$T^s x + G^s y^s = g^s \quad \forall s \in S \tag{4.14c}$$
$$x \geq 0, \ y^s \geq 0, \quad \forall s \in S, \tag{4.14d}$$

where $A \in \mathbb{R}^{m^0 \times n}$ and $T^s \in \mathbb{R}^{m^s \times n}$, $\forall s \in S$, $G^s \in \mathbb{R}^{m^s \times k}$, $\forall s \in S$, and all over vectors have conforming dimensions. For the sake of simplicity, we suppose here that problem (4.14) is not unbounded; the case in which (4.14) is unbounded requires only slight modifications.

In Benders decomposition, we create a *relaxed master problem*, in which we determine candidate x-solutions, and anticipate how those solutions affect the objective function contribution terms corresponding to variables y^s, $\forall s \in S$. Given the values for x-variables, we can then formulate $|S|$ *subproblems*, where subproblem $s \in S$ corresponds to the optimization over variables y^s. We begin by developing the subproblem formulation below, given a fixed set of values $\hat{x}$ corresponding to x.

$$\mathbf{SP}^s(\hat{x}) : \min \ (h^s)^T y^s \tag{4.15a}$$
$$\text{s.t.} \ G^s y^s = g^s - T^s \hat{x} \tag{4.15b}$$
$$y^s \geq 0 \tag{4.15c}$$

The dual of this problem is given as follows.

$$\mathbf{DSP}^s(\hat{x}) : \max \ (g^s - T^s \hat{x})^T \pi \tag{4.16a}$$
$$\text{s.t.} \ (G^s)^T \pi \leq h^s \tag{4.16b}$$

First, let us assume that there exists an optimal solution to $\text{SP}^s(\hat{x})$, with objective value z^s. Then by the strong duality theorem, there is an optimal solution to $\text{DSP}^s(\hat{x})$ with the same objective function value. Furthermore, let Ω^s represent the set of all extreme points to the dual feasible region for scenario s, and note that this dual feasible region is invariant with respect to the given value of $\hat{x}$. We therefore have that

$$z^s = \max_{\pi \in \Omega^s} (g^s - T^s \hat{x})^T \pi. \tag{4.17}$$

Next, suppose that $\text{SP}^s(\hat{x})$ is infeasible, and therefore, that $\text{DSP}^s(\hat{x})$ is either infeasible or unbounded. In this case, it is convenient to place certain "regularization" restrictions on the subproblems to ensure that the dual is always feasible. (An easy way of doing this is to add upper bounds on variables y^s by adding the constraints $y^s \leq u$, where $u = (M, \ldots, M)^T$ and M is an arbitrarily large number. In this case, each $y^s < M$ at optimality by assumption that the problem is bounded. Reversing the constraints as $-y^s \geq -u$, and associating duals β with these primal constraints, our dual constraints become $(G^s)^T \pi - \beta \leq h^s$, with $\beta \geq 0$, and a feasible solution exists in which β is set to be sufficiently large.) Now, if $\text{SP}^s(\hat{x})$ is infeasible, then $\text{DSP}^s(\hat{x})$ must unbounded. There would therefore exist an extreme direction d to the dual feasible solution such that $(g^s - T^s \hat{x})^T d > 0$. Letting Γ^s denote the set of all extreme rays to the dual feasible region for scenario s, we have that $\mathrm{SP}^s(\hat{x})$ is feasible if and only if

$$(g^s - T^s\hat{x})^T d \le 0 \qquad \forall d \in \Gamma^s. \tag{4.18}$$

Based on (4.17) and (4.18), we can reformulate (4.14) as the following equivalent problem, where now z^s is a decision variable that takes on the optimal objective function value of $\mathrm{SP}^s(x)$.

$$\min \; c^T x + \sum_{s \in S} z^s \tag{4.19a}$$

$$\text{s.t. } Ax = b \tag{4.19b}$$

$$z^s \ge (g^s - T^s x)^T \overline{\pi} \qquad \forall s \in S,\, \overline{\pi} \in \Omega^s \tag{4.19c}$$

$$(g^s - T^s x)^T \overline{d} \le 0 \qquad \forall s \in S,\, \overline{d} \in \Gamma^s \tag{4.19d}$$

$$x \ge 0. \tag{4.19e}$$

Note that (4.19c) is equivalent to (4.17) since optimality forces z^s to the maximum value of $(g^s - T^s x)^T \overline{\pi}$ over all extreme points $\overline{\pi}$ in Ω^s. Inequalities (4.19c) are sometimes called "optimality inequalities" (or "optimality cuts" if they are used to remove a targeted solution that reports an infeasible candidate solution with respect to the relationship between the x- and z-variables), while (4.19d) are sometimes called "feasibility inequalities" (or "feasibility cuts" if they are used to remove a solution x to which no feasible solution corresponds in (4.15) for some $s \in S$).

Observe, however, that problem (4.19) is likely to be far more difficult to solve than (4.14) itself because the sets Ω^s and Γ^s may both contain an exponential number of elements. Hence, we adapt a cutting-plane strategy in which inequalities of the form (4.19c) and (4.19d) are dynamically added to the problem only as needed. Initially, we start with only a subset (usually empty) $\overline{\Omega}^s \subseteq \Omega^s$ of extreme points to each dual feasible region, and also an (empty) subset $\overline{\Gamma}^s \subseteq \Gamma^s$ of extreme rays to each dual feasible solution. Then, the *relaxed master problem* is the same as (4.19), but with Ω^s replaced by $\overline{\Omega}^s$, and with Γ^s replaced by $\overline{\Gamma}^s$. (The problem is called "relaxed" because many or all of the constraints (4.19c) and (4.19d) are initially missing from the formulation. Most studies refer simply to a *master problem* instead, with "relaxed" being implied.) The Benders decomposition algorithm then proceeds as follows.

Step 0. Initialize by setting $\overline{\Omega}^s = \overline{\Gamma}^s = \emptyset$ for each $s \in S$, and continue to Step 1.

Step 1. Solve the relaxed master problem. If the relaxed master problem is infeasible, then the original problem is infeasible as well, and the process terminates. Otherwise, obtain optimal solution $\hat{x}$, $\hat{z}^s$, for each $s \in S$. Observe that the problem may be unbounded, with some $\hat{z}^s$ values taking on values of $-\infty$. Continue to Step 2.

Step 2. Solve dual subproblem $\mathrm{DSP}^s(\hat{x})$, $\forall s \in S$, given the value of $\hat{x}$ computed in Step 1. If a dual subproblem is unbounded for some $s \in S$, then an extreme direction $\hat{d} \in \Gamma^s$ exists for which $(g^s - T^s\hat{x})^T \hat{d} > 0$. Therefore, we add $\hat{d}$ to $\overline{\Gamma}^s$ (thus creating a new "feasibility cut" of the form (4.19d)) to avoid regenerating an $\hat{x}$ that makes

$\text{DSP}^s(\hat{x})$ unbounded in the direction $\hat{d}$. Add these constraints to the relaxed master problem for any $s \in S$ such that $\text{DSP}^s(\hat{x})$ is unbounded. If any such constraints are added, return to Step 1. Otherwise, continue to Step 3.
Step 3. For each $s \in S$, let $\hat{\pi}^s$ be an optimal solution to $\text{DSP}^s(\hat{x})$. If $\hat{z}^s < (g^s - T^s x)^T \hat{\pi}^s$, then a constraint of the form (4.19c) is being violated. We therefore need to add $\hat{\pi}$ to $\overline{\Omega}^s$, thus generating an additional constraint (optimality cut) of the form (4.19c). Otherwise, we will have that $\hat{z}^s = (g^s - T^s x)^T \hat{\pi}^s$, and no constraint needs to be generated for s. If no such constraints need to be generated, then the algorithm terminates with $\hat{x}$ being optimal, and the corresponding solutions to $\text{SP}^s(\hat{x})$ being the optimal solutions to y^s. Otherwise, add all necessary optimality cuts to the restricted master problem, and return to Step 1.

Remark 4.2. One potential difficulty with the algorithm above is that, at each iteration of this process, a maximum of $|S|$ inequalities can potentially be added to the restricted master problem. In this case, it is possible that too many inequalities will be added to the restricted master problem, thus impairing the overall computational performance of the solver. Therefore, some implementations of this algorithm will add at most one inequality at each iteration (i.e., one feasibility cut, or one optimality cut). Alternatively, the optimality cuts in particular are sometimes aggregated by reformulating the restricted master problem to have an objective function of $c^T x + z$, summing $z^s \geq (g^s - T^s x)^T \hat{\pi}^s$ over all $s \in S$ (even those that do not require an optimality cut at the current iteration), and replacing $\sum_{s \in S} z^s$ with z. These are referred to as "single-cut" Benders implementations, whereas the procedure given above is a "multi-cut" Benders implementation.

Remark 4.3. Note that model (4.14) assumes that x is continuous. The Benders decomposition algorithm can be run without modification if x is also restricted to be integer-valued. In this case, one might use the optimality and feasibility cuts within a *branch-and-bound* scheme to solve the mixed-integer programming relaxed master problem, which is called a *branch-and-cut* algorithm [20]. However, if any of the y-values are restricted to be integer-valued, then in general we cannot use the Benders decomposition algorithm as stated above. The reason that the algorithm fails in this case is that no dual formulation is easily attainable for subproblems containing integer-restricted variables.

Finally, let us return to the application on capacitated network design presented earlier. The relaxed master problem would be given by:

$$\min \sum_{(i,j)\in A} r_{ij} x_{ij} + \sum_{s\in S} z^s \tag{4.20a}$$

$$\text{s.t. [Dynamically generated feasibility and optimality cuts]} \tag{4.20b}$$

$$x_{ij} \in \{0,1\} \qquad \forall (i,j) \in A. \tag{4.20c}$$

Note that initially, the relaxed master problem contains no constraints at all on the x-variables, aside from the binary restrictions on those variables. There are no constraints on the z-variables either, although it would be valid to state nonnegativity restrictions for each z^s (which would prevent unboundedness of the problem). The (primal) subproblem for scenario s, given $\hat{x}$, would be:

$$\min \quad p^s \sum_{(i,j)\in A} c_{ij} f_{ij}^s \tag{4.21a}$$

$$\text{s.t.} \quad f_{ij}^s \leq q_{ij}^s \hat{x}_{ij} \qquad \forall (i,j) \in A \tag{4.21b}$$

$$\sum_{i\in FS(j)} f_{ji}^s - \sum_{i\in RS(j)} f_{ij}^s = d_j \qquad \forall j \in V \tag{4.21c}$$

$$f_{ij}^s \geq 0 \qquad \forall (i,j) \in A. \tag{4.21d}$$

Let α_{ij} be the dual variables associated with (4.21b) and let β_j be the dual variables associated with (4.21c). Then if (4.21) is infeasible, we can show that the dual would be unbounded (because the trivial solution in which all α and β variable values equal zero is necessarily feasible, due to the assumption that all cost data is nonnegative). Therefore, there exists an extreme direction to the dual feasible region that makes the dual unbounded. Letting the components of this unbounded direction be given by $\overline{\alpha}$, $\overline{\beta}$, our feasibility cuts have the form:

$$\sum_{(i,j)\in A} \overline{\alpha}_{ij} q_{ij}^s x_{ij} + \sum_{j\in V} \overline{\beta}_j d_j \leq 0. \tag{4.22}$$

(Note that the α-values are nonpositive, and so this constraint effectively requires that additional capacity be built in the network as a necessary condition for feasibility.) Similarly, if the dual subproblem has an optimal solution, then the optimality cuts are of the form

$$z^s \geq \sum_{(i,j)\in A} \hat{\alpha}_{ij} q_{ij}^s x_{ij} + \sum_{j\in V} \hat{\beta}_j d_j, \tag{4.23}$$

where $\hat{\alpha}$ and $\hat{\beta}$ are components of an optimal solution to the dual subproblem.

4.3.2 Dantzig-Wolfe Decomposition

In contrast to Benders decomposition, Dantzig-Wolfe decomposition [10] is designed to exploit linear programming structures that have complicating constraints, rather than complicating variables. We illustrate such a scenario in the context of *multicommodity flows*, which are prevalent in several telecommunication routing scenarios. As before, we have a network $G = (V,A)$ with node set V and (established) arc set A, where each arc (i,j) has a capacity of $q_{ij} > 0$. We use the notion of forward stars and reverse stars as in Section 4.3.1.

In our multicommodity flow problem, we have a set K of commodities that must be transported across the network. We let d_i^k denote the supply/demand value for commodity k at node $i \in V$ (where again, a positive value represents a supply and a negative value represents a demand). The per-unit cost for shipping commodity $k \in K$ over arc $(i,j) \in A$ is given by c_{ij}^k, and each unit of commodity k shipped over

(i,j) consumes s^k_{ij} units of its total capacity. (For instance, if capacity is given by the bandwidth of some particular communication link, video communication will consume bandwidth at a different rate than voice communication.) The objective of the problem is to satisfy each commodity's demands and exhaust each commodity's supplies at a minimum cost, while obeying capacity limits on the arcs.

Define flow variables x^k_{ij} as the flow of commodity $k \in K$ over arc $(i,j) \in A$. Then the multicommodity network flow problem is formulated as follows.

$$\min \sum_{k \in K} \sum_{(i,j) \in A} c^k_{ij} x^k_{ij} \tag{4.24a}$$

$$\text{s.t.} \sum_{k \in K} s^k_{ij} x^k_{ij} \leq q_{ij} \qquad \forall (i,j) \in A \tag{4.24b}$$

$$\sum_{i \in FS(j)} x^k_{ji} - \sum_{i \in RS(j)} x^k_{ij} = d^k_j \qquad \forall j \in V,\ k \in K \tag{4.24c}$$

$$x^k_{ij} \geq 0 \qquad \forall (i,j) \in A,\ k \in K. \tag{4.24d}$$

The objective (4.24a) minimizes the commodity flow costs over the network. Constraints (4.24b) enforce the arc capacity restrictions. Constraints (4.24c) enforce the flow balance constraints (at each node, and for each commodity), and (4.24d) enforce nonnegativity restrictions on the flow variables.

Here, there is no set of complicating variables. However, if the capacity constraints (4.24b) were removed from the problem, then we could solve (4.24) as $|K|$ separable network flow problems. Hence, we say that (4.24b) are *complicating constraints*. One common way of solving problems like this is via Dantzig-Wolfe decomposition, which employs the Representation Theorem (Theorem 4.2) as a key component. An alternative method for solving these problems is via Lagrangian optimization, which we discuss in Section 4.3.3.

Our generic model for this subsection contains $|K|$ sets of variables, x^k for $k \in K$, which would be separable if the complicating constraints were removed from the problem.

$$\min \sum_{k \in K} (c^k)^T x^k \tag{4.25a}$$

$$\text{s.t.} \sum_{k \in K} A^k x^k = b \tag{4.25b}$$

$$H^k x^k = g^k \qquad \forall k \in K \tag{4.25c}$$

$$x^k \geq 0 \quad \forall k \in K, \tag{4.25d}$$

where $A^k \in \mathbb{R}^{m^0 \times n^k}$ and $H^k \in \mathbb{R}^{m^k \times n^k}$, $\forall k \in K$, and all other vectors have conforming dimensions.

In the Benders decomposition method, we decomposed the dual feasible region into its extreme points and extreme directions, and included a constraint in a relaxed master problem corresponding to each extreme point and extreme direction, added to the formulation as necessary. In the Dantzig-Wolfe decomposition method, we will decompose the feasible regions given by $X^k = \{x^k \geq 0 : H^k x^k = g^k\}$ into its extreme points and directions. We will then have a very large optimization problem that contains variables (instead of constraints) corresponding to each of these extreme points and directions.

Let Ω^k and Γ^k denote the set of extreme points and extreme directions to X^k, respectively. We replace constraints $H^k x^k = g^k$, $x^k \geq 0$ by insisting that x^k is equal to a convex combination of points in Ω^k plus a nonnegative linear combination of extreme directions in Γ^k as in Theorem 4.2. Here, it is convenient to index Ω^k by $1,\ldots,E^k$, so that $\Omega^k = \{x^{1k},\ldots,x^{E^k k}\}$. Also, we index Γ^k by $1,\ldots,F^k$, i.e., $\Gamma^k = \{d^{1k},\ldots,d^{F^k k}\}$. Then we can substitute x^k by

$$x^k = \sum_{i=1}^{E^k} \lambda_{ik} x^{ik} + \sum_{j=1}^{F^k} \mu_{jk} d^{jk} \quad \forall k \in K, \tag{4.26}$$

and include constraints $\lambda_{ik} \geq 0, \forall k \in K,\ i = 1,\ldots,E^k$, $\mu_{jk} \geq 0, \forall k \in K,\ j = 1,\ldots,F^k$, and $\sum_{i=1}^{E^k} \lambda_{ik} = 1, \forall k \in K$. By doing so, we can eliminate (4.25c) and (4.25d), which yields the following exponential-size problem.

$$\min \sum_{k\in K} (c^k)^T \left(\sum_{i=1}^{E^k} x^{ik} \lambda_{ik} + \sum_{j=1}^{F^k} d^{jk} \mu_{jk} \right) \tag{4.27a}$$

$$\text{s.t.} \sum_{k\in K} A^k \left(\sum_{i=1}^{E^k} x^{ik} \lambda_{ik} + \sum_{j=1}^{F^k} d^{jk} \mu_{jk} \right) = b \tag{4.27b}$$

$$\sum_{i=1}^{E^k} \lambda_{ik} = 1 \qquad \forall k \in K \tag{4.27c}$$

$$\lambda, \mu \geq 0. \tag{4.27d}$$

(Recall that the x- and d-vectors are parameters of this formulation; the only variables are the λ- and μ-variables.) As in the Benders decomposition technique, we note that the enumeration of extreme points and extreme directions for X^k is computationally prohibitive, and dynamically generate those extreme points and directions that are necessary at optimality. Therefore, only a subset of extreme points (say, $\overline{\Omega}^k$) and extreme directions ($\overline{\Gamma}^k$) of X^k are present for each $k \in K$. Because there are missing columns instead of rows in this case, the Dantzig-Wolfe procedure considers a *restricted* master problem rather than a relaxed one.

Suppose that the restricted master problem is solved to optimality, and has an optimal dual solution with complementary slack dual values π associated with (4.27b) and γ associated with (4.27c). In this case, we compute the reduced cost of a variable λ_{ik} as $(c^k)^T x^{ik} - \pi^T(A^k x^{ik}) - \gamma_k$, and of μ_{jk} as $(c^k)^T d^{jk} - \pi^T(A^k d^{jk})$. The column-generation strategy is essentially the following: If there exists an ungenerated extreme point (or extreme direction) whose reduced cost is negative, generate the column, add it to the restricted master problem, and re-solve the restricted master problem. Otherwise, the current restricted master problem solution is optimal.

Fortunately, we need not enumerate each possible extreme point and extreme direction to find out if there exists a negative reduced-cost column. Given dual values $\hat{\pi}$, consider the *Danzig-Wolfe subproblem* of the form:

$$\mathbf{SP}^k(\hat{\pi}) : \min \quad ((c^k)^T - \hat{\pi}^T A^k)x \tag{4.28a}$$
$$\text{s.t.} \quad H^k x = g^k \tag{4.28b}$$
$$x \geq 0. \tag{4.28c}$$

Note that the feasible region is given by X^k. If problem $\mathrm{SP}^k(\hat{\pi})$ is unbounded, it is unbounded in some extreme direction $\hat{d}$, and therefore $((c^k)^T - \hat{\pi}^T A^k)^T d < 0$. This represents an extreme direction whose corresponding μ-variable in the restricted master problem must have a negative reduced cost. The Dantzig-Wolfe process would add $\hat{d}$ to $\overline{\Gamma}^k$ and would re-solve the restricted master problem, obtaining a new primal/dual optimal pair of solutions to (4.27).

Now, suppose instead that $\mathrm{SP}^k(\hat{\pi})$ has an optimal solution (if it were infeasible, we would conclude that the overall problem must also be infeasible), and consider an extreme-point optimal solution $\hat{x}$. If the optimal objective function value of $\mathrm{SP}^k(\hat{\pi})$ is less than $\hat{\gamma}_k$, then we would have $(c^k)^T\hat{x} - \hat{\pi}^T(A^k\hat{x}) - \hat{\gamma}_k < 0$. This means that we have found a negative-reduced-cost extreme point; we therefore add $\hat{x}$ to $\overline{\Omega}^k$ and re-solve the restricted master problem.

If no new columns are added to the restricted master problem after examining all subproblems k (i.e., no negative reduced cost extreme points or extreme directions are found), then the current solution must be optimal to (4.25). The only remaining issue is how to initialize the process in order to find an initial feasible solution to the restricted master problem. (Note that starting with empty $\overline{\Omega}^k$ and $\overline{\Gamma}^k$ sets for each $k \in K$ would imply that the restricted master problem is infeasible due to (4.27c).) One simple way of initializing the extreme point sets for the Dantzig-Wolfe algorithm is to insert artificial variables into each set Ω^k, which are present to guarantee an initial feasible solution, but are penalized by a prohibitively large value in the objective function so that they will become equal to zero at optimality (e.g., as in the big-M method [5]). If any such artificial variable is still positive at optimality, then there is no solution to the original formulation (4.25). The overall Dantzig-Wolfe decomposition algorithm can now be given as follows.

Step 0. Initialize by setting $\overline{\Omega}^k$ to any (nonempty) set of extreme points to X^k, and set $\overline{\Gamma}^k = \emptyset$ for each $k \in K$. Add any necessary artificial variables to the restricted master problem as described above. Continue to Step 1.

Step 1. Solve the restricted master problem. If the restricted master problem is unbounded, then the original problem is unbounded as well, and the process terminates. Otherwise, obtain the complementary slack (and optimal) dual solution $\hat{\pi}$ and $\hat{\gamma}$ to the restricted master problem. Continue to Step 2.

Step 2. Solve each subproblem $\mathrm{SP}^k(\hat{\pi})$. If the subproblem is unbounded, then an extreme direction $\hat{d} \in \Gamma^k$ exists with a negative reduced cost. Therefore, we add $\hat{d}$ to $\overline{\Gamma}^k$. Otherwise, let $\hat{x}$ be an extreme point solution that optimizes $\mathrm{SP}^k(\hat{\pi})$. If $(c^k)^T\hat{x} -$

$\hat{\pi}^T(A^k\hat{x}) - \hat{\gamma}_k < 0$, then add $\hat{x}$ to $\overline{\Omega}^k$. If any new columns are added (corresponding to extreme points or extreme directions) for any $k \in K$, then return to Step 1. Otherwise, the current solution to the restricted master problem is optimal, and the procedure stops.

For the multicommodity flow problem, we have a restricted master problem of the form (4.27), and a subproblem that takes the form of a simple network flow problem:

$$\min \sum_{(i,j)\in A} (c_{ij}^k - \hat{\pi}_{ij} s_{ij}^k) x_{ij} \tag{4.29a}$$

$$\text{s.t.} \sum_{i\in FS(j)} x_{ji} - \sum_{i\in RS(j)} x_{ij} = d_j^k \quad \forall j \in V,\ k \in K \tag{4.29b}$$

$$x \geq 0. \tag{4.29c}$$

Note that the cost for pushing flow across arc (i, j) is given by the per-unit flow cost c_{ij}, plus the shadow price $\hat{\pi}_{ij}$ on the capacity of arc (i, j) multiplied by the rate of capacity usage $s_{ij}^k x_{ij}$ of commodity k on arc (i, j). That is, the costs include the direct cost of sending a unit of flow on (i, j), plus the costs associated with utilizing capacity on arc (i, j). (Note that the π-values here will be nonpositive, because those duals are associated with $\leq$ constraints in a minimization problem; therefore, all coefficients of the x-variables in the objective function are nonnegative.) Furthermore, because there are no negative coefficients in the objective function, we will never generate any extreme directions for the restricted master problem in this application. (We refer interested readers to classical surveys [2, 15] of multicommodity flow algorithms as a starting point for a more thorough treatment of these problems, and to [17] for more recent work in the area.)

Remark 4.4. The structure of (4.29) suggests another important facet of these decomposition procedures. Note that not only are the subproblems (4.29) separable, but they are also solvable by much more efficient means than linear programming. (See [1] for a discussion of low-polynomial time algorithms for network flow problems.) Hence, even if the number of separable problems is small, one may still opt to use decomposition in order to exploit specially-structured subproblems that are quickly solvable.

Remark 4.5. In this subsection, we have compared in some detail how Benders decomposition compares to Dantzig-Wolfe decomposition. In fact, taking the dual of a problem (4.25) (after some simple adjustments) yields a problem of the form (4.14). It is thus not surprising that applying Benders decomposition to the dual of (4.25) is equivalent to solving (4.25) directly by Dantzig-Wolfe decomposition.

4.3.3 Lagrangian Optimization

Another common (and more general) technique for addressing problems having complicating constraints is by *Lagrangian optimization*. The idea here is to penalize violations to the complicating constraints within the objective function, and omit those complicating constraints from the constraint set. Indeed, we have already seen a basic application of this concept in Section 4.2.3, where constraints $Ax = b$ were omitted from the constraint set, and a penalty term $\pi^T(b - Ax)$ was added to the objective function. (This strategy led us to develop the dual formulation for linear programs.) The multipliers that are used to penalize violations to these constraints are called *Lagrangian duals*, and the process of moving constraints out of a constraint set and into the objective function is often referred to as "dualizing" the constraints.

Whereas all constraints were dualized in Section 4.2.3, it is more common to dualize only the complicating constraints, leaving behind a separable problem (or otherwise easily solvable problem, as mentioned in Remark 4.4). In the context of the multicommodity flow problem, we would dualize the capacity constraints (4.24b).

We again consider the generic formulation (4.25). Assume for simplicity that $X^k = \{x \geq 0 : H^k x^k = g^k\}$ is nonempty and bounded. (If X^k is empty, then the problem is infeasible. If X^k is unbounded, we can bound it by placing a constraint $e^T x^k \leq M$ for a very large value M, and note that any solution lying on this constraint indicates a direction of unboundedness.) A relaxation of (4.25) is given as follows (to which we will later refer as an "inner problem," and often called the "Lagrangian function" or "Lagrangian subproblem"), for any given vector π:

$$\textbf{IP}(\pi): \min \sum_{k \in K} (c^k)^T x^k + \pi^T (b - \sum_{k \in K} A^k x^k) \tag{4.30a}$$

$$\text{s.t.} \quad H^k x^k = g^k \quad \forall k \in K \tag{4.30b}$$

$$x^k \geq 0 \quad \forall k \in K. \tag{4.30c}$$

Problem IP(π) is a relaxation of (4.25) because any solution to (4.25) is a feasible solution to IP(π) with the same objective function value (regardless of the value of π). That is, let z_P be the optimal objective function value to (4.25) (assuming that (4.25) has an optimal solution), and let $z_{IP(\pi)}$ be the optimal objective function value to IP(π), where $z_{IP(\pi)} = \infty$ if IP(π) is unbounded. Then $z_{IP(\pi)} \leq z_P$ for any vector π. We can therefore use problem IP(π) to yield lower bounds to the optimal objective function value of (4.25).

Note here that if constraint (4.25b) were written as the inequality $\sum_{k \in K} A^k x_k \geq b$, then we would formulate (4.30) in the same way, but with the stipulation that π should be nonnegative. This will ensure that the objective penalty term, $\pi^T (b - \sum_{k \in K} A^k x^k)$, remains nonpositive when x is feasible to (4.25b), and that $z_{IP(\pi)} \leq z_P$ as desired. If (4.25b) were written in the $\leq$ sense, we would insist that $\pi \leq 0$.

In order to provide the best possible lower bound for z_P, we can seek a value of π that maximizes $z_{IP(\pi)}$. Thus, the Lagrangian dual problem is:

$$\max_{\pi} \; z_{IP(\pi)}. \tag{4.31}$$

Observe that because $z_{IP(\pi)}$ is determined by means of the inner minimization problem (4.30), problem (4.31) is a maximin problem (maximization of a minimum objective). For linear programming problems, such as (4.25), there will exist a $\pi = \pi^\star$ such that $z_{IP(\pi^\star)} = z_P$ (assuming again that z_P has an optimal solution). We say that there is no *duality gap* for these problems. For integer and/or nonlinear programming problems, duality gaps may indeed exist.

In some situations, (4.31) can be solved directly. When direct solution of this Lagrangian dual problem is not possible, a very common technique for solving (4.31) is via *subgradient optimization*, which exploits the fact that IP(π) is a piecewise-linear concave function of π. (In fact, when maximizing a concave function, one correctly refers to "supergradients" rather than subgradients, although both terms are used in the literature in this case.) We refer the reader to [6, 8] for excellent discussions of this technique, along with the classical text [28]. However, we provide a sketch of the subgradient optimization method here for completeness.

Step 0. Set iteration counter $i = 0$, and choose an initial estimate, π^0, of the dual values π. (One often starts with $\pi^0 \equiv 0$.) Select a positive integer R to be a maximum number of iterations, and $\varepsilon > 0$ to be an acceptable subgradient norm for algorithm termination. Initialize the best lower bound computed thus far as $LB = -\infty$. Continue to Step 1.

Step 1. Solve IP(π^i), and obtain an optimal solution $(\hat{x}^{1,i},\ldots,\hat{x}^{|K|,i})$. (Note that in (4.30), the x^k-vectors can be optimized in $|K|$ separable problems.) If $z_{IP(\pi^i)} > LB$, then set $LB = z_{IP(\pi^i)}$. (Note that because subgradients are being used in this procedure, we are not guaranteed that $z_{IP(\pi^0)}$, $z_{IP(\pi^1)}, \ldots$ will be a nondecreasing sequence of values.) Continue to Step 2.

Step 2. One subgradient of IP(π^i) is given by $s^i = (b - \sum_{k \in K} A^k \hat{x}^{k,i})$. If the (Euclidean) norm of s^i is sufficiently small, then the current dual estimate is sufficiently close to optimal, and we stop. Thus, we terminate if $||s^i|| < \varepsilon$. Also, to control the maximum computational effort that we exert, we can also terminate if $i = R$. Otherwise, continue to Step 3.

Step 3. Based on the subgradient computed in Step 2, we adjust our dual estimate by moving in the (normalized) direction of the subgradient multiplied by a step length parameter α^i determined based on the iteration. That is,

$$\pi^{i+1} = \pi_i + \alpha^i \frac{s^i}{||s^i||}. \tag{4.32}$$

Note that the step-length parameter α^i is a critical consideration in guaranteeing key convergence properties of the subgradient optimization algorithm; see [6, 8, 28]. Set $i = i + 1$, and return to Step 1.

Remark 4.6. There are many computational and theoretical issues that arise in subgradient optimization that are beyond the scope of an introductory chapter such as this. Some of the finer points that are particularly important regard the dual updating function (4.32), the convergence of the algorithm, and the recovery of primal solutions.

First, there are several techniques that exist in the literature that offer computational guidance regarding the choice of step sizes. A "block-halving" scheme appears to be particularly effective, in which step sizes α are constant through several iterations, and are then halved every w iterations, for some positive integer w. Also, the direction of dual movement need not be directly in the subgradient direction s; instead, a deflected (or average) subgradient strategy may choose the dual updating direction as the bisection of the previous direction and the computed subgradient. See [13, 27], along with classical Lagrangian works [8, 28], for details. However, one must use these step sizes and direction strategies in the context of convergence results developed in the foregoing papers for Lagrangian optimization procedures. These results provide sufficient conditions regarding step sizes and direction strategies that guarantee the optimization of (4.31).

Finally, note that this method is designed to generate dual optimal solutions π, along with the optimal objective function value to (4.31) if it exists (which coincides with z_P for linear programming problems). To compute an optimal primal solution, we refer the reader to primal recovery methods within Lagrangian optimization as developed in [3, 26].

4.3.4 Other Methods

The field of large-scale optimization is expanding rapidly, and many promising optimization techniques have been omitted in this chapter for the sake of brevity. We briefly remark here on three areas of large-scale optimization studies that play significant roles in telecommunciations optimization.

4.3.4.1 Basis Partitioning

In solving a linear program with constraints $Ax = b$ and $x \geq 0$, where $A \in \mathbb{R}^{m \times n}$ and the rows of A are linearly independent (i.e., rank(A) = m), we have that $n > m$ in any interesting optimization problem. Although we have skipped in this chapter the details of the simplex method for solving linear programs, a key step in the simplex method represents an extreme point as a *basic feasible solution* (BFS). A BFS is obtained by partitioning the columns of A into m linearly independent basic columns B and $n-m$ nonbasic columns N, and the variable set into basic variables x_B and nonbasic variables x_N. Hence, the constraints can now be written as $Bx_B + Nx_N = b$. Setting $x_N \equiv 0$, we can solve for the basic variables as $x_B = B^{-1}b$, where the inverse of B exists due to the assumption that B contains linearly independent columns. If $x_B \geq 0$, then it is an extreme point to the feasible region.

Because B^{-1} is a critical component in identifying the current extreme point, and also in determining how to improve the solution or prove its optimality, computing

B^{-1} is a bottleneck operation in the simplex method. (It is worth noting that modern simplex implementations do not compute B^{-1} directly, but instead use numerical linear algebra methods that improve numerical stability and reduce computational effort by exploiting the sparsity of the A matrix.) Therefore, if special structures exist in the constraint matrix, it may be possible to compute B^{-1} by specialized algorithms rather than by using general methods from linear algebra. Basis partitioning algorithms compute B^{-1} by decomposing B into specially structured blocks, and then employing a specialized algorithm to find corresponding blocks to its inverse. Only a small submatrix of B^{-1} needs to be computed under these schemes, which reduces both computational and memory requirements for solving problems by the simplex method. Foundational work in this area was performed by Schrage [25], and has been applied to network flow problems in several studies; see [18, 19] for multicommodity flow applications.

4.3.4.2 Interior Point Methods

As mentioned previously, interior-point methods were originally developed to solve nonlinear optimization problems. These algorithms had a dramatic implication in the field of linear programming in the early 1980s by yielding the first polynomial-time algorithm for linear programming problems. Interior point algorithms are more sophisticated than simplex-based algorithms. Due to worst-case time-complexity functions that are polynomial in the input size of the linear program, they often outperform simplex approaches for very-large-scale problems (roughly 10,000 variables or more), depending on problem-specific details such as the presence of special constraint structures. See, e.g., [12, 21, 31] for discussion on these methods on linear and nonlinear optimization problems. There is currently much development in computation and theory for interior-point algorithms, some of which is specifically focused on interior-point approaches for large-scale problems (e.g., [9]).

4.3.4.3 Heuristics

Exact optimization techniques such as the ones described above can become impractical because of the size and/or complexity of real-world instances. Rather than eschewing optimization as an option, one can turn to *heuristic* methods for solving problems. Heuristics are designed to quickly provide good-quality feasible solutions, but without a guarantee of optimality, or in some cases, without a guarantee of feasibility.

Naturally, heuristic techniques are as diverse in complexity and effectiveness as are exact optimization techniques. Many heuristic algorithms are tailored specifically for the problem at hand, employing specific properties of the problem to obtain fast and effective heuristic techniques. There are too many such approaches to name, but one of the most famous of these is the Lin-Kernighan algorithm [16] for the Euclidean traveling salesman problem. A common trend over the last few decades in heuristic development has sought algorithms that provide a performance guarantee. These *approximation algorithms* work by guaranteeing that the heuristic solution is within some $\varepsilon\%$ of optimality (see, e.g., [14]).

Alternatively, *metaheuristic* optimization algorithms are designed to provide a general framework for optimizing difficult problems. The key concept behind these algorithms is that the basic heuristic algorithm need not change depending on the application, aside from the specialization of a subset of its component routines. As a result, these algorithms have garnered much attention in the literature due to their practical ease of implementation and their effectiveness in providing good solutions to real-world problems (see [23] for a collection of recent metaheuristic applications to industrial problems). While there is some disagreement over what properly constitutes a metaheuristic, some of the most common metaheuristics include local search, simulated annealing, tabu search, genetic algorithms, ant colony optimization, greedy randomized adaptive search procedure, and variable neighborhood search algorithms.

References

1. Ahuja, R.K., Magnanti, T.L., Orlin, J.B.: Network Flows: Theory, Algorithms, and Applications. Prentice Hall, Upper Saddle River, NJ (1993)
2. Assad, A.A.: Multicommodity network flows – a survey. Networks **8**(1), 37–91 (1978)
3. Barahona, F., Anbil, R.: The volume algorithm: Producing primal solutions with a subgradient method. Mathematical Programming **87**, 385–399 (2000)
4. Barnhart, C., Hane, C.A., Vance, P.H.: Using branch-and-price-and-cut to solve origin-destination integer multicommodity flow problems. Operations Research **48**(2), 318–326 (2000)
5. Bazaraa, M.S., Jarvis, J.J., Sherali, H.D.: Linear Programming and Network Flows, second edn. John Wiley & Sons, New York (1990)
6. Bazaraa, M.S., Sherali, H.D., Shetty, C.M.: Nonlinear Programming: Theory and Algorithms, second edn. John Wiley & Sons, New York (1993)
7. Benders, J.F.: Partitioning procedures for solving mixed variables programming problems. Numerische Mathematik **4**, 238–252 (1962)
8. Bertsekas, D.P.: Nonlinear Programming. Athena Scientific, Cambridge, MA (1999)
9. Byrd, R.H., Hribar, M.E., Nocedal, J.: An interior point algorithm for large scale nonlinear programming. SIAM Journal on Optimization **9**(4), 877–900 (1999)
10. Dantzig, G.B., Wolfe, P.: Decomposition principle for linear programs. Operations Research **8**(1), 101–111 (1960)
11. Garg, M., Smith, J.C.: Models and algorithms for the design of survivable networks with general failure scenarios. Omega **36**(6), 1057–1071 (2008)
12. Hager, W.W., Hearn, D.W., Pardalos, P.M. (eds.): Large Scale Optimization: State of the Art. Springer, New York, NY (1994)
13. Held, M., Wolfe, P., Crowder, H.: Validation of subgradient optimization. Mathematical Programming **6**, 62–88 (1974)
14. Hochbaum, D.S. (ed.): Approximation Algorithms for NP-Hard Problems. PWS Publishing Company, Boston, MA (1996)
15. Kennington, J.L.: A survey of linear cost multicommodity network flows. Operations Research **26**(2), 209–236 (1978)
16. Lin, S., Kernighan, B.W.: An effective heuristic algorithm for the traveling-salesman problem. Operations Research **21**(2), 498–516 (1973)

17. Mamer, J.W., McBride, R.D.: A decomposition-based pricing procedure for large-scale linear programs: An application to the linear multicommodity flow problem. Management Science **46**(5), 603–709 (2000)
18. McBride, R.D.: Solving embedded generalized network flow problems. European Journal of Operational Research **21**, 82–92 (1985)
19. McBride, R.D., Mamer, J.W.: Implementing an LU factorization for the embedded network simplex algorithm. INFORMS Journal on Computing **16**(2), 109–119 (2004)
20. Nemhauser, G.L., Wolsey, L.A.: Integer and Combinatorial Optimization. John Wiley & Sons, New York (1999)
21. Nesterov, Y., Nemirovskii, A.: Interior-Point Polynomial Algorithms in Convex Programming. Society for Industrial and Applied Mathematics, Philadelphia, PA (1994)
22. Pardalos, P.M., Vavasis, S.A.: Quadratic programming with one negative eigenvalue is NP-hard. Journal of Global Optimization **1**(1), 15–23 (1991)
23. Resende, M.G.C., de Sousa, J.P., Viana, A. (eds.): Metaheuristics: Computer decision-making. Kluwer Academic Publishers, Norwell, MA (2004)
24. Rosenberger, J.M., Olinick, E.V.: Robust tower location for code division multiple access networks. Naval Research Logistics **54**(2), 151–161 (2007)
25. Schrage, L.: Implicit representation of generalized variable upper bounds in linear programming. Mathematical Programming **14**(1), 11–20 (1978)
26. Sherali, H.D., Choi, G.: Recovery of primal solutions when using subgradient optimization methods to solve Lagrangian duals of linear programs. Operations Research Letters **19**(3), 105–113 (1996)
27. Sherali, H.D., Ulular, O.: A primal-dual conjugate subgradient algorithm for specially structured linear and convex programming problems. Applied Mathematics and Optimization **20**, 193–221 (1989)
28. Shor, N.Z.: Minimization Methods for Nondifferentiable Functions. Springer, Berlin (1985)
29. Smith, J.C., Schaefer, A.J., Yen, J.W.: A stochastic integer programming approach to solving a synchronous optical network ring design problem. Networks **44**(1), 12–26 (2004)
30. Sutter, A., Vanderbeck, F., Wolsey, L.A.: Optimal placement of add/drop multiplexers: Heurisic and exact algorithms. Operations Research **46**(5), 719–728 (1998)
31. Wright, S.J.: Primal-Dual Interior-Point Methods. Society for Industrial and Applied Mathematics, Philadelphia, PA (1997)

Part II
Optimization Problems for Networks with Infrastructure

Chapter 5
Mathematical Programming Models for Third Generation Wireless Network Design

Eli Olinick

Abstract This chapter develops a series of optimization models for third-generation (3G) cellular network design leading up to a comprehensive model including the selection of base stations and mobile telephone switching office (MTSO) locations, and the assignment of mobiles to base stations. The models also consider the design of a backbone network connecting the base stations, MTSOs, and public switched telephone network (PSTN) gateways. These models take as input demand for cellular service in a given planning area, a set of candidate tower and MTSO locations with corresponding costs, costs for provisioning links between the candidate tower and MTSO locations as well as costs for linking the MTSOs to PSTN gateway nodes. Based on these inputs and propagation data for the planning area, the comprehensive model can be used to determine the optimal selection of radio towers, MTSO locations, backbone network topology, and the service capacity of the resulting radio network. For problem instances of practical interest, the model leads to large, difficult integer programs that require special solution techniques. Nevertheless, researchers have reported finding high-quality solutions, in reasonable amounts of CPU time, to problems with hundreds of candidate tower locations supporting tens of thousands of simultaneous cellular phone sessions. This chapter surveys strategies that have been developed to achieve these results.

5.1 Introduction

This chapter presents fundamental mathematical programming models for designing third-generation (3G) cellular networks. Previous generations of cellular networks employed *frequency division multiple access* (FDMA) or *time division multiple access* (TDMA) schemes to allow multiple users to share the system's traffic-

Eli Olinick
Department of Engineering Management, Information, and Systems, Bobby B. Lyle School of Engineering, Southern Methodist University, Dallas, TX 75275-0123, USA, e-mail: olinick@smu.edu

J. Kennington et al. (eds.), *Wireless Network Design: Optimization Models and Solution Procedures*, International Series in Operations Research and Management Science 158, DOI 10.1007/978-1-4419-6111-2_5,

carrying capacity. The planning process for FDMA/TDMA systems was typically addressed in two phases. The first phase consisted of solving a *tower location problem*, optimizing the location of radio towers (also called *base stations* or *Node Bs*) to provide coverage for the area being served, and the second involved assigning frequencies to the selected tower locations (cells) [3]. Sherali et al. [53] and Mathar and Niessen [42] apply mathematical programming techniques to solve tower location problems, and Murphey et al. [46] give an extensive survey of the graph coloring problems that arise during the frequency assignment phase. A model that solves both problems jointly is given in Kalvenes et al. [33]. However, as noted in the seminal paper by Amaldi et al. [7], the use of *code division multiple access* (CDMA) schemes in the current generation (3G) networks gives rise to a fundamentally different network design problem.

CDMA schemes enable service providers to use all of their licensed frequencies in each cell, and so the models described in this chapter do not have a frequency assignment component. Instead, capacity in a CDMA-based system is determined by a *signal-to-interference ratio* (SIR). Every active connection between a mobile device and a tower in a CDMA system can potentially interfere with every other active mobile-tower connection. Thus, CDMA systems use *admission control policies* to ensure that new connections are only allowed if they will not cause too much interference with currently active connections. Since the assignment of mobiles to towers (*subscriber assignment*) determines how much interference each connection causes with every other connection, the most efficient CDMA network designs are found by simultaneously solving the tower location and subscriber assignment problems [7].

Calls from mobiles in cellular networks are typically relayed via radio links to cell sites (towers). Each tower is connected to a *mobile telephone switching office* (MTSO). MTSOs are switching offices in cellular networks that perform various operational functions such as *handoffs* (determining when to switch a mobile device's connection from one tower to another to improve reception) [56] and keeping track of billing information [44]. The MTSOs are connected (usually by fiber-optic cable) to form a *backbone network* for the CDMA system which allows for routing traffic between cells in the network as well as to and from public switched telephone network (PSTN) gateways to provide cellular users with access to long distance networks and other services. This chapter develops a comprehensive optimization model for CDMA network design including the selection of base stations and MTSO locations, the assignment of mobiles to base stations, and the design of a backbone network connecting the base stations, MTSOs, and the PSTN gateways. This model takes as input a set of candidate tower and MTSO locations with corresponding costs, costs for provisioning links between the candidate tower and MTSO locations as well as costs for linking the MTSOs to PSTN gateway nodes. Demand for traffic is modeled by *test points* in the service area. Test points are described in [7] as centroids of demand where given amounts of traffic must be satisfied at a given minimum SIR, $SIR_{\min}$ [7, 39]. Based on these data, the model can be used to determine the selection of radio towers, MTSO locations, backbone network topology, and the service capacity of the resulting radio network. A typical problem instance and solution can be can be represented graphically as in Figures 5.1 and 5.2, respectively.

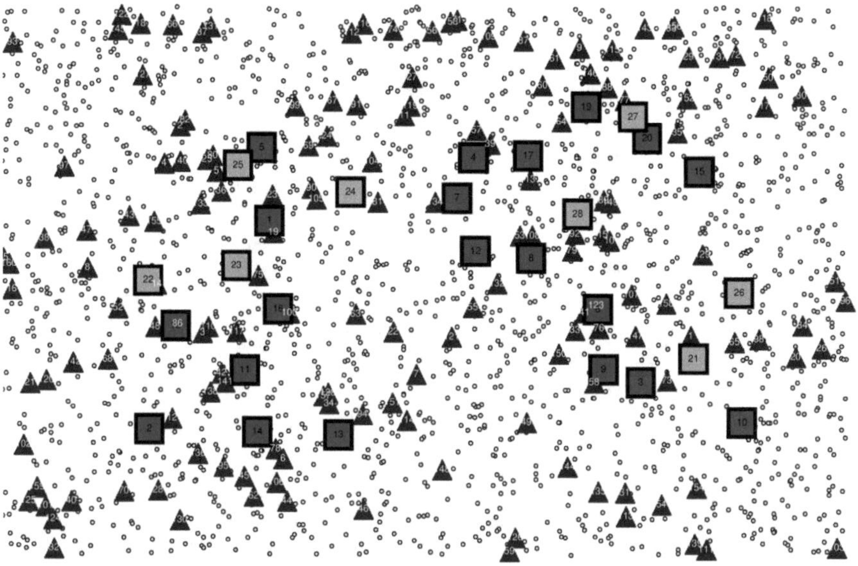

Fig. 5.1 Graphical representation of a CDMA planning problem with test points uniformly distributed over the service area (circles are test points, triangles are candidate tower locations, dark squares are candidate MTSO locations, and light squares are candidate PSTN gateway locations).

This chapter is structured as follows. In Section 5.2 we review the core model for jointly optimizing tower location and subscriber assignment in CDMA networks proposed in a series of influential papers by Amaldi et al. [4, 5, 7]. A search on Google Scholar for [7] at the time of this writing returned over 140 citations. Rather than attempting a comprehensive review of this considerable literature, this chapter provides a tutorial on fundamental mathematical programming models for CDMA network design. For recent surveys see Amzallag [9], Mannino et al. [41], and D'Andreagiovanni [18]. Section 5.3 extends the core model of Amaldi et al. to include the elements necessary for the comprehensive design described above and depicted in Figures 5.1 and 5.2. The core and extended models are large, difficult integer programs that require special techniques for practical solution. These techniques are discussed in Section 5.4.

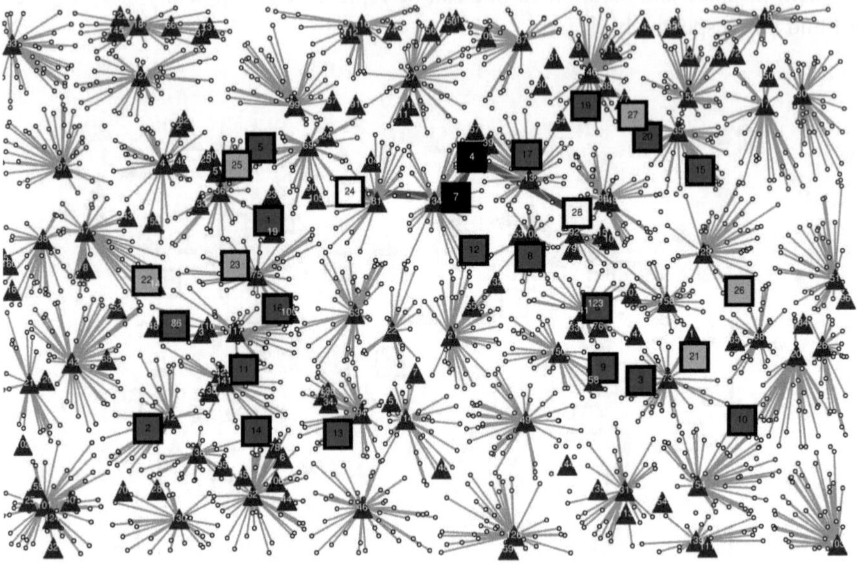

Fig. 5.2 Graphical representation of the solution to the problem depicted in Figure 5.1 (not including the lines from the selected towers to the two selected MTSO locations). Thick lines indicate the fiber backbone. Assignments of subscribers to base stations are indicated by thin lines from circles to triangles.

5.2 Tower Location and Subscriber Assignment in CDMA Networks

In this section, we review the core tower location and subscriber assignment model for CDMA networks proposed by Amaldi et al. [7]. We develop the formulation of the core model in Section 5.2.1 and discuss some of the challenges involved in solving it in Section 5.2.2. Techniques for overcoming these challenges are discussed in Section 5.4.

5.2.1 The Core Model

To describe the core model of Amaldi et al. [7], let L and M denote sets of candidate tower locations and test points, respectively, in the service area of interest. The model has three key parameters associated with these two sets: (1) the cost a_ℓ for using a tower at location ℓ, (2) the forecasted peak demand for service d_m at test point $m \in M$, and (3) the *attenuation factor* $g_{m\ell}$ of the radio signal between test point m and tower location ℓ. The demand for service at a test point is measured in terms of the number of mobile users who simultaneously request service at that location. The attenuation factors are assumed to be measured empirically or derived from an appropriate propagation model for the planning region (e.g., the Hata model [31]). Note that throughout this presentation the terms "subscriber" and "mobile user" are used synonymously to refer to users of the CDMA network (i.e., customers who subscribe to the service provided by the network owner). We also use the terms "subscriber assignment" and "test point assignment" interchangeably since demand is modeled using test points.

The tower location and subscriber assignment decisions are represented in the model by binary indicator variables. The variable y_ℓ is equal to one if tower ℓ is selected for the design and zero, otherwise. Likewise $x_{m\ell}$ is equal to one if, and only if, subscribers at test point m are assigned to tower ℓ.

The core model in [7] assumes a *power-control mechanism* by which connections from mobile handsets to towers must be received at a guaranteed power level of P_{target}. Thus, a handset at test point m will transmit with power level $P_{\text{target}}/g_{m\ell}$ in order to send a signal to tower ℓ that is received at power level $g_{m\ell}\frac{P_{\text{target}}}{g_{m\ell}} = P_{\text{target}}$. The model further assumes that the maximum transmission power of a mobile handset is $P_{\max}$, and so test point m may only be assigned to tower ℓ if $P_{\text{target}}/g_{m\ell} \leq P_{\max}$. The subset of towers to which test point m may be assigned is denoted by $L_m = \{\ell \in L : P_{\text{target}}/g_{m\ell} \leq P_{\max}\}$, and the set of test points that could be assigned to (covered by) tower ℓ is denoted by $M_\ell = \{m \in M : P_{\text{target}}/g_{m\ell} \leq P_{\max}\}$. In a CDMA-based network every mobile-to-tower connection can, in theory, interfere with every other mobile-to-tower connection. Consequently, the total strength of all signals received at tower location ℓ from all mobile users in the system is $P_{\text{target}} \sum_{m\in M} \sum_{j\in L_m} d_m \frac{g_{m\ell}}{g_{mj}} x_{mj}$ and the interference with respect to a given connection assigned to the tower is $P_{\text{target}} \sum_{m\in M} \sum_{j\in L_m} d_m \frac{g_{m\ell}}{g_{mj}} x_{mj} - P_{\text{target}}$ [7]. The SIR for a connection to tower ℓ, denoted by SIR_ℓ, is thus given by

$$SIR_\ell = \frac{P_{\text{target}}}{P_{\text{target}} \sum_{m\in M} \sum_{j\in L_m} d_m \frac{g_{m\ell}}{g_{mj}} x_{mj} - P_{\text{target}}} = \frac{1}{\sum_{m\in M} \sum_{j\in L_m} d_m \frac{g_{m\ell}}{g_{mj}} x_{mj} - 1}. \tag{5.1}$$

To ensure a minimum acceptable SIR of $SIR_{\min}$ for each session assigned to tower ℓ, test points must be assigned to towers in such a way that $SIR_\ell \geq SIR_{min}$. That is, a feasible solution that uses tower ℓ must satisfy the constraint

$$\frac{1}{\sum_{m\in M} \sum_{j\in L_m} d_m \frac{g_{m\ell}}{g_{mj}} x_{mj} - 1} \geq SIR_{\min}, \tag{5.2}$$

which is equivalent to the linear constraint

$$\sum_{m\in M} \sum_{j\in L_m} d_m \frac{g_{m\ell}}{g_{mj}} x_{mj} \leq s. \tag{5.3}$$

where s denotes $1 + \frac{1}{SIR_{\min}}$.

Amaldi et al. [7] propose the following integer programming model, ACMIP, to minimize the cost of providing service to all test points in the coverage area:

$$\min \sum_{\ell \in L} a_\ell y_\ell + \lambda \sum_{m \in M} \sum_{\ell \in L_m} \frac{d_m}{g_{m\ell}} x_{m\ell} \tag{5.4}$$

$$\sum_{\ell \in L_m} x_{m\ell} = 1 \qquad \forall m \in M, \tag{5.5}$$

$$x_{m\ell} \le y_\ell \qquad \forall m \in M, \ell \in L_m, \tag{5.6}$$

$$\sum_{m \in M} \sum_{j \in L_m} d_m \frac{g_{m\ell}}{g_{mj}} x_{mj} \le s + (1 - y_\ell)\beta_\ell \qquad \forall \ell \in L, \tag{5.7}$$

$$x_{m\ell} \in \{0,1\} \qquad \forall m \in M, \ell \in L_m, \tag{5.8}$$

$$y_\ell \in \{0,1\} \qquad \forall \ell \in L. \tag{5.9}$$

The first term in the objective function (5.4) is the total tower usage cost and the second term is the total handset transmission power when all connections are active. The user-specified parameter $\lambda \ge 0$ determines the trade off between infrastructure cost and power usage (energy). This trade off is discussed in Section 5.3.4. Constraint set (5.5) forces each test point to be assigned to exactly one tower location, and constraint set (5.6) links the tower location and subscriber assignment decision variables so that test points can only be assigned to selected towers. A minimum acceptable SIR is ensured for all active connections by the *quality-of-service constraints* (QoS) (5.7). If tower location ℓ is selected, then $y_\ell = 1$ and (5.7) enforces the signal-to-interference requirement (5.3). However, if tower location ℓ is *not* selected then we do not need to enforce (5.3) at that location. Indeed, doing so may artificially constrain the solution. Therefore, the constant β_ℓ is selected to be large enough so that the constraint is satisfied automatically by any set of binary $x_{m\ell}$ values when y_ℓ is zero. A suitable value is

$$\beta_\ell = \sum_{m \in M} d_m \left\{ \max_{j \in L_m \setminus \{\ell\}} \left(\frac{g_{m\ell}}{g_{mj}} \right) \right\} \tag{5.10}$$

where

$$\max_{j \in L_m \setminus \{\ell\}} \left(\frac{g_{m\ell}}{g_{mj}} \right) = 0 \text{ if } L_m \setminus \{\ell\} = \emptyset. \tag{5.11}$$

Constraint sets (5.8) and (5.9) define the domains of the decision variables. Given a solution to ACMIP, we refer to the set of selected towers as $\bar{L} \subseteq L$.

5.2.2 Challenges in Solving the Core Model

The optimization problem modeled by ACMIP belongs to the class of theoretically intractable NP-hard problems meaning that it is unlikely that there is an efficient algorithm for finding provably optimal solutions to every problem instance [25]. While this fact alone does not necessarily imply that practical instances of the problem are hard to solve, the values of the constants in the QoS constraints (5.7) make it very challenging to solve ACMIP with the branch-and-bound procedure described in Chapter 4. Formulations with relatively large constants such as β_ℓ are avoided if possible in integer programming since they tend to produce weak linear programming relaxations that can take considerably longer to solve with branch-and-bound than equivalent formulations that do not use these constants [47]. Generally speaking, the larger the β_ℓ values, the longer it takes for branch-and-bound to converge to a provably optimal solution. On the other hand, values that are too low enforce the QoS constraints even on towers that aren't selected and thus artificially restrict the feasible region. Another difficulty with the formulation is that it gives rise to ill-conditioned problems as the largest and smallest attenuation factor ratios ($\frac{g_{m\ell}}{g_{mj}}$) in a given QoS constraint can differ by several orders of magnitude [34]. We discuss the implications of this issue in Section 5.4.4.

As noted in [7], some instances of ACMIP may be infeasible. That is, it might not be possible to cover all test points in a particular area at a given $SIR_{\min}$ with a given set of candidate tower locations. In such situations Amaldi et al. propose maximizing the amount of traffic carried by the network [7]. To achieve this, they relax (5.5) to

$$\sum_{\ell \in L_m} x_{m\ell} \leq 1 \qquad \forall m \in M \tag{5.12}$$

so that test points are allowed to be uncovered. They then maximize an objective function of the form

$$\sum_{m\in M} \sum_{\ell \in L_m} d_m x_{m\ell} - \omega \sum_{\ell \in L} a_\ell y_\ell \tag{5.13}$$

where ω is a user-selected value that favors increasing demand coverage over reducing cost. Note that if 100% coverage is possible, then solving this coverage-maximizing variant of ACMIP with an appropriate value for ω minimizes the cost of covering all test points.

5.3 Extensions to the Core Model

In this section we discuss important variations of, and extensions to, the core model. In Section 5.3.1 we discuss an alternative power control model proposed by Amaldi et al. [7] that allows transmissions to be received at the towers with varying power levels (possibly less than P_{target}) as long as the SIR for each connection is at least $SIR_{\min}$. Allowing mobiles to reduce transmission power can lead to more efficient use of system resources, however the resulting mathematical programming model is a nonlinear integer program that is extremely difficult to solve to provable optimality. In Sections 5.3.2 and 5.3.3 we build up to the comprehensive planning model depicted in Figures 5.1 and 5.2. The model presented in Section 5.3.2 extends ACMIP to maximize profit subject to government-imposed minimum-service requirements. This model is then extended in Section 5.3.3 to include network infrastructure (i.e, MTSOs, PSTN gateways, and the backbone links). We discuss the trade-off between profit and power in Section 5.3.4 and conclude by outlining other design considerations that researchers have incorporated into the these models in Section 5.3.5.

5.3.1 SIR-Based Power Control

In their computational study Amaldi et al. observed that SIR values often exceed $SIR_{\min}$ in solutions to ACMIP [7]. That is, after the tower location and subscriber assignment decisions have been made it may turn out that an acceptable SIR can be achieved even if signals from certain test points are received at their assigned towers at power levels less than P_{target}. This means that system resources can be used more efficiently by allowing mobiles at appropriate test points to transmit at reduced power levels. This improves the total transmission power term in the objective function (5.4) and reduces interference systemwide thereby increasing capacity. For a given solution to ACMIP, an iterative power-control algorithm given by Yates [57] can determine transmission powers at the test points to minimize total transmit power under minimum SIR constraints. While this approach cannot be directly embedded in ACMIP, it can be approximated by using a more sophisticated *SIR-Based Power Control* model. As an alternative to assuming that each connection is received at P_{target}, Amaldi et al. [7] propose a model in which the transmission power of a mobile at test point m is a decision variable, p_m. In this model, the signal from test point m is received by tower ℓ at power level $p_m g_{m\ell}$, and the total strength of all signals received at tower ℓ (assuming all test points are covered) is $\sum_{i \in M} d_i p_i g_{i\ell}$. The model assumes that there is a *thermal* or background noise of η (typically -130 dB) at each tower location. If test point m is assigned to tower ℓ, then the SIR for the connection is

$$\frac{p_m g_{m\ell}}{\sum_{i \in M} d_i p_i g_{i\ell} + \eta - p_m g_{m\ell}}. \tag{5.14}$$

Thus, the model imposes the following set of constraints for each allowable test point assignment:

$$x_{m\ell}\Big(\sum_{i \in M} d_i p_i g_{i\ell} + \eta - p_m g_{m\ell}\Big) \le \frac{p_m g_{m\ell}}{SIR_{\min}} \qquad \forall m \in M, \ell \in L_m. \tag{5.15}$$

Observe that if test point m is assigned to tower ℓ, (5.15) ensures that the SIR (5.14) is at least $SIR_{\min}$. However, if test point m is *not* assigned to tower ℓ the constraint reduces to $p_m \geq 0$ which is trivially satisfied. The ACMIP with SIR-based power control is the following mathematical program:

$$\min \sum_{\ell \in L} a_\ell y_\ell + \lambda \sum_{m \in M} \sum_{\ell \in L_m} d_m p_m x_{m\ell}, \tag{5.16}$$

subject to (5.5), (5.6), (5.15), (5.8), (5.9) and

$$0 \leq p_m \leq P_{\max} \qquad \forall m \in M. \tag{5.17}$$

The potential advantage of using SIR-based power control is that mobiles can emit lower powers and therefore generate less interference with towers they are not assigned to. Indeed, in their computational study Amaldi et al. found that solutions to the SIR-based power control model used at least 20% fewer towers than solutions to their power-based power control model ACMIP [7]. The disadvantage of using this formulation is that (5.15) is a non-linear constraint which makes the problem extremely difficult to solve — even more difficult than ACMIP. Consequently Amaldi et al. did not attempt to find provably optimal solutions to this model. Instead, they solved it with a heuristic procedure described in Section 5.4.2 to compare it with the *power-based model* described in Section 5.2.1.

5.3.2 Profit Maximization with Minimum-Service Restrictions

Building on the work of Amaldi et al. [4, 5, 7], Kalvenes et al. [34] propose a profit-maximization variant of ACMIP, called KKOIP, in which the objective function is

$$r \sum_{m \in M} \sum_{\ell \in L_m} x_{m\ell} - \sum_{\ell \in L} a_\ell y_\ell \tag{5.18}$$

where r is the revenue generated for each unit of demand served in the planning area. Unlike ACMIP, KKOIP allows for partial demand satisfaction at the test points. Thus, in this model $x_{m\ell}$ is defined as the maximum number of users at test point m that can be simultaneously assigned to tower ℓ. Consequently, KKOIP replaces (5.5), (5.6) and (5.8) with

$$\sum_{\ell \in L_m} x_{m\ell} \leq d_m \qquad \forall m \in M, \tag{5.19}$$

$$x_{m\ell} \leq d_m y_\ell \qquad \forall m \in M, \ell \in L_m, \tag{5.20}$$

$$x_{m\ell} \in \mathbb{N} \qquad \forall m \in M, \ell \in L_m, \tag{5.21}$$

and the QoS constraints (5.7) with

$$\sum_{m \in M} \sum_{j \in L_m} \frac{g_{m\ell}}{g_{mj}} x_{mj} \leq s + (1 - y_\ell)\beta_\ell \qquad \forall \ell \in L. \tag{5.22}$$

KKOIP also includes a set of constraints that model a minimum-service requirement that the United States Federal Communications Commission (FCC) imposes on cellular service providers as part of its licensing regulations. After receiving a license from the FCC, the service provider has a five-year time window during which it must install enough network capacity so that it could provide service to at least a certain fraction ρ of the population in the market or else forfeit its license. The value of ρ depends on the frequency band of the license [23, 34]. The constraints enforcing the FCC requirement are

$$q_m \leq \sum_{\ell \in L_m} y_\ell \qquad \forall m \in M, \tag{5.23}$$

$$q_m \geq y_\ell \qquad \forall m \in M, \ell \in L_m, \tag{5.24}$$

$$\sum_{m \in M} d_m q_m \geq \rho \sum_{m \in M} d_m, \tag{5.25}$$

$$q_m \in \{0, 1\} \qquad \forall m \in M, \tag{5.26}$$

where q_m is one if, and only if, test point m can be serviced by at least one of the selected towers. Constraint set (5.23) ensures that a test point is only considered capable of receiving service if there is at least one selected tower within its reach as determined by $P_{\max}$. Conversely, constraint set (5.24) forces the indicator variable q_m equal to one if there is at least one selected tower in range of test point m. Constraint (5.25) ensures that the proportion of the demand at the test points that could be assigned to the selected towers is at least ρ, and (5.26) defines the q_m variables as binary. Hence, KKOIP maximizes (5.18) subject to (5.19)-(5.26) and (5.9).

5.3.3 Infrastructure

As described earlier, a comprehensive CDMA planning model should consider other elements of the network infrastructure (e.g., MTSOs) in addition to the tower locations. A cost minimization model that includes tower-to-MTSO wiring cost and handoff cost for a given traffic volume at the base stations was developed by Merchant and Sengupta [43] and refined by Li et al. [40]. Cai et al. [16, 17] build on this work and on the KKOIP model to develop an integrated profit maximization model considering the selection of tower, MTSO and PSTN gateway locations, tower-to-MTSO connections, subscriber assignment, and the backbone network as depicted in Figures 5.1 and 5.2. In addition to the input parameters described previously, the model in [17] includes a set of candidate MTSO locations denoted by K. The set K_0 denotes the union of K and the PSTN gateway (i.e., location zero). The binary variable z_k is equal to one if, and only if, the design uses an MTSO at location k, and binary variable $s_{\ell k}$ indicates whether or not tower ℓ is connected to the MTSO at location k. The model imposes the following constraint sets to ensure that each tower is connected to a selected MTSO

$$\sum_{k\in K} s_{\ell k} = y_\ell \qquad \forall \ell \in L, \tag{5.27}$$

$$s_{\ell k} \leq z_k \qquad \forall \ell \in L, \forall k \in K. \tag{5.28}$$

The model assumes that each MTSO has capacity to connect to at most α towers and enforces this capacity limit with the constraint set

$$\sum_{\ell\in L} s_{\ell k} \leq \alpha z_k \qquad \forall k \in K. \tag{5.29}$$

To model the construction of the backbone network, CKKOIP uses a set of network flow constraints that describe a spanning tree structure that is rooted at the PSTN gateway and connects all of the selected MTSO locations[1]. The basic idea is that a unit of flow is generated at MTSO location k if $z_k = 1$, and that unit of flow either goes directly to the PSTN gateway or to a neighboring MTSO on the backbone. The model uses the binary variable w_{jk} to indicate whether or not there is a backbone link between MTSO location $j \in K$ and MTSO/PSTN gateway location $k \in K_0$, and integer variable u_{jk} to represent the amount of flow on that link. Note that u_{jk} does not represent the amount of telecommunications traffic on link (j,k), but rather the number of MTSO's that are connected to the PSTN gateway via a path using the link between locations j and k. The constraints that construct the backbone network are

$$z_0 = 1, \tag{5.30}$$

$$\sum_{j\in K\setminus\{k\}} u_{kj} + u_{k0} = \sum_{j\in K\setminus\{k\}} u_{jk} + z_k \qquad \forall k \in K, \tag{5.31}$$

$$\sum_{k\in K} u_{k0} = \sum_{k\in K} z_k, \tag{5.32}$$

$$u_{jk} \leq |K| z_k \qquad \forall j \in K, k \in K_0 \setminus \{j\}, \tag{5.33}$$

$$u_{jk} \leq |K| w_{jk} \qquad \forall j \in K, k \in K_0 \setminus \{j\}. \tag{5.34}$$

Constraint (5.30) ensures that the backbone network includes the PSTN gateway. The second set of constraints, (5.31), are for flow conservation. They state, for MTSO k, that the out-flow is equal to the in-flow plus the generated flow, z_k. Constraint set (5.32) ensures that the total amount of flow reaching the PSTN gateway is equal to the number of selected MTSOs. Constraint sets (5.33) and (5.34) ensure that there is flow on the link between MTSO k and MTSO j (or the PSTN gateway) only if both locations are selected for the design and connected by a direct link in the backbone network.

[1] Editor's note: Network flow constraints are also used to model spanning tree structures in Chapter 10.

Using b_k to denote the cost of using an MTSO at location k, $c_{\ell k}$ to denote the cost of connecting tower location ℓ to MTSO location k, and h_{jk} to represent the cost of providing a backbone link between locations j and k, CKKOIP is stated as

$$\max \underbrace{r \sum_{m\in M}\sum_{\ell\in L_m} x_{m\ell}}_{\text{Subscriber revenue}} - \underbrace{\sum_{\ell\in L} a_\ell y_\ell}_{\text{Tower cost}} - \underbrace{\sum_{k\in K} b_k z_k}_{\text{MTSO cost}} - \underbrace{\sum_{\ell\in L}\sum_{k\in K} c_{\ell k} s_{\ell k}}_{\text{Connection cost}} - \underbrace{\sum_{j\in K}\sum_{k\in K_0\setminus\{j\}} h_{jk} w_{jk}}_{\text{Backbone cost}}. \tag{5.35}$$

subject to (5.9), (5.19)-(5.22), (5.27)-(5.34), and the following sets of constraints on the domains of the decision variables

$$s_{\ell k} \in \{0,1\} \qquad \forall \ell \in L, k \in K, \tag{5.36}$$

$$u_{jk} \in \mathbb{N} \qquad \forall j \in K, k \in K_0 \setminus \{j\}, \tag{5.37}$$

$$w_{jk} \in \{0,1\} \qquad \forall j \in K, k \in K_0 \setminus \{j\}, \tag{5.38}$$

$$z_k \in \{0,1\} \qquad \forall k \in K. \tag{5.39}$$

St-Hilaire et al. [55] propose a generalization of CKKOIP that includes differentiated service classes. Their model has separate demand values for voice and data service at each test point. Consequently, the backbone must connect the MTSOs to a gateway to a public data network for data services in addition to the PSTN gateway. They use their model to demonstrate the trade-off between solving an integrated model and using a decomposition approach that first solves the tower location and subscriber assignment subproblem, then determines which MTSOs to select, and finally selects the gateway nodes and finds an optimal backbone topology. They point out that the second approach is faster, but not necessarily optimal, and demonstrate this with a case study.

Cai et al. [16] extend the CKKOIP model so that it selects at least two PSTN gateways from a set of candidate locations, and makes the network more reliable by using a *biconnected* backbone network for the selected MTSOs and PSTN gateways. Thus, the backbone in [16] is required to be reliable enough that it will not be disconnected by any single-node failure. This is the *biconnectivity augmentation* problem [25] which is known to belong to the class of NP-hard problems [22] and has been extensively studied as a fundamental problem in network design. Khuller [36] presents approximation results for this and a series of related network design problems. Surveys of the integer programming literature related to the biconnectivity augmentation problem may be found in Fortz et al. [24] and Grötschel et al. [28, 29]. In addition to making the backbone biconnected, the model in [16] requires each tower to be connected to two different MTSOs so that if one MTSO fails the tower is still connected to the backbone.

5.3.4 The Power-Revenue Trade-Off

In this section we briefly illustrate how the models described above can be used to investigate the tradeoff between profit and power consumption. The objective here is neither minimization of total transmit power in the network nor maximization of network operator revenue. Instead we find the optimal tradeoff between total revenue and total transmit power. Network operators could use this type of analysis to trade a portion of their net revenue for a substantial reduction in transmit power for their subscribers. Such lowered transmit powers would lead to larger talk and stand-by times for the mobile devices and could be an effective and new dimension of product differentiation.

The objective function in this illustration is

$$\max \underbrace{\sum_{m\in M}\sum_{\ell\in L_m} r_m x_{m\ell}}_{\text{Subscriber revenue}} - \underbrace{\sum_{\ell\in L} a_\ell y_\ell}_{\text{Tower cost}} - \underbrace{\lambda \sum_{m\in M}\sum_{\ell\in L_m} \frac{x_{m\ell}}{g_{m\ell}}}_{\text{Power cost}} \tag{5.40}$$

where r_m is the annual revenue (in \$) generated from each subscriber serviced in area m and λ is a penalty term balancing the dollar costs with power. In this example we seek a solution that optimizes (5.40) subject to the KKOIP constraints related to tower location and subscriber assignment, (5.19)-(5.22) and (5.9).

Table 1: The Power-Revenue Trade-Off.

λ	Tot. Power	Demand	Tot. Rev.	Tower Cost	Towers	Profit
2.00E-12	1.70822E+12	100.00%	$4,067,900	$729,725	5	$3,338,172
3.00E-12	1.70822E+12	100.00%	$4,067,900	$729,725	5	$3,338,170
4.00E-12	1.70822E+12	100.00%	$4,067,900	$729,725	5	$3,338,168
5.00E-12	1.70822E+12	100.00%	$4,067,900	$729,725	5	$3,338,166
6.00E-12	1.70822E+12	100.00%	$4,067,900	$729,725	5	$3,338,165
7.00E-12	1.17521E+12	100.00%	$4,067,900	$875,670	6	$3,192,222
8.00E-12	7.45158E+11	100.00%	$4,067,900	$1,021,615	7	$3,046,279
9.00E-12	7.45158E+11	100.00%	$4,067,900	$1,021,615	7	$3,046,278
1.00E-11	7.45158E+11	100.00%	$4,067,900	$1,021,615	7	$3,046,278
1.00E-10	1.60E+11	82.11%	$3,339,960	$1,313,505	9	$2,026,439

The results given in Table 1 illustrate this tradeoff quantitatively. It can be seen that all the demand can be satisfied using different amounts of power and at different costs. Thus, a network operator may decide to provide service at lower net revenue since it could reduce power consumption for users in the network. For example, a reduction in net profit from $3,338,172 to $3,046,278 results in 3.6dB savings in power. This substantially lower power consumption in the mobiles would lead to increased battery, talk, and standby times and indirectly result in higher customer satisfaction and increases in revenue. The reduction in power is due to the fact that more towers are used in service, thus reducing the distance over which each mobile needs to transmit.

5.3.5 Additional Design Considerations

The models presented thus far consider the key aspects of 3G network design. In this section we briefly describe some of the other decisions that network designers face and how the they may be incorporated within the framework developed in this chapter. The ACMIP model and its extensions assume symmetric requirements for *uplink* (mobile-to-tower) and *downlink* (tower-to-mobile) communication. However, future cellular systems are expected to have larger data rate requirements on the downlink than on the uplink. The common belief is that asymmetric applications such as web browsing will dominate traffic leading to larger data throughputs on the downlink. The models described above are readily modified to achieve desired downlink performance as described below, and in Amaldi et al. [2, 6, 8] and Rajan et al. [51].

Let $SIR^{\downarrow}_{\min}$ be the minimum acceptable signal-to-interference ratio for the downlink. For simplicity, we assume a single path channel and perfect orthogonality of the spreading codes assigned to users connected to the same base station. Thus, at each mobile, the interference is only from base station signals that do not service that particular mobile. Moreover, we assume that the base station transmits at constant power for each mobile that it serves. The downlink from tower ℓ to subscribers at test point m is subject to ambient noise at location m, N_m. Thus, the SIR requirement for the downlink from tower ℓ to subscribers at test point m is

$$\frac{P_{\text{target}}}{\sum_{i\in M\setminus\{m\}}\sum_{j\in L_i\setminus\{\ell\}} P_{\text{target}}\frac{g_{mj}}{g_{ij}}x_{ij} + N_m} \geq SIR^{\downarrow}_{\min}. \tag{5.41}$$

To add this to the models described above, let $v_{m\ell}$ be a binary variable such that $v_{m\ell}$ is one if and only there is at least one subscriber at a test point m who is served by tower ℓ and add the following downlink SIR constraints to the model:

$$\sum_{i\in M\setminus\{m\}}\sum_{j\in L_i\setminus\{\ell\}} \frac{g_{mj}}{g_{ij}}x_{ij} \leq \frac{1}{SIR^{\downarrow}_{\min}} - \frac{N_m}{P_{\text{target}}} + (1-v_{m\ell})\Gamma_{m\ell} \qquad \forall m\in M, \ell\in L_m, \tag{5.42}$$

where $\Gamma_{m\ell}$ is a suitably large constant such that (5.42) is non-binding (i.e., automatically satisfied) when $v_{m\ell}$ is zero. Thus, the Γ's play the same role in the downlink SIR constraints as the β's play in the uplink QoS constraints. The following constraints must also be added to ensure a logical connection between the x's and v's, and to define the v's as binary:

$$x_{m\ell} \leq d_m v_{m\ell} \qquad \forall m\in M, \ell\in L_m, \tag{5.43}$$

$$v_{m\ell} \in \{0,1\} \qquad \forall m\in M, \ell\in L_m. \tag{5.44}$$

Interference between users assigned to the same tower in a CDMA system can be reduced by a technique known as *sectorization*. Sectorization uses directional antennas to focus energy in particular directions or sectors of the cell, and signals from any particular test point are only received in one sector of each cell. Cai [15] extends the CKKOIP model to include a choice of an omnidirectional antenna (as in the models described earlier in this chapter) or a directional antenna with either three 120° sectors or six 60° sectors.

The models described above are driven by demand forecasts. In telecommunications (and many other industries) it is exceedingly difficult to accurately forecast demand [38]. Optimizing a CDMA network design for a forecast that turns out to be inaccurate can be very costly. If the forecast overestimates the demand, then the provider will incur unnecessary costs in building an over-capacitated network. On the other hand, if the forecast underestimates the demand, then the provider pays an opportunity cost for missing potential revenue and risks losing market share in a highly competitive business (at least until they add sufficient capacity to capture all the demand). Rosenberger and Olinick [48, 52] develop a *stochastic programming model with simple recourse* [50] to optimize tower location in CDMA networks under demand uncertainty. Specifically, the model chooses a set of tower locations that maximizes the expected profit in the KKOIP model (without the FCC constraints) given a probability distribution for the demand at each test point.

Robust optimization is another popular strategy for handling uncertainty in mathematical programming models. The idea behind robust optimization is to create a design that will be fairly good (robust) no matter which of a given set of scenarios occurs. There are several different concepts of *robustness* that appear in the literature (e.g., [10, 11, 12, 37, 45, 49]) and applications of robust optimization models for designing fiber-optic telecommunications networks under demand uncertainty may be found in [14, 30, 35, 38, 54]. Bienstock and D'Andreagiovanni [13] and D'Andreagiovanni [18] use a robust optimization framework to address another source of uncertainty in planning wireless networks, the attenuation factors. Even when these factors are measured empirically, the actual attenuation experienced when the network is deployed may differ from what was measured due to weather conditions, new construction, and other factors. This difference can lead to less coverage and/or lower quality of service than was planned. The model proposed in [13, 18] produces designs that are "protected" against changes in the attenuation factors. Eisenblätter et al. [20, 21] describe a very detailed, robust optimization-based 3G planning system developed for the Models and Simulations for Network Planning and Control of UMTS (MOMENTUM) project conducted at the Zuse Institute Berlin. The model developed for MOMENTUM is a robust model that considers such design elements as sectorization, specific antenna configurations (e.g., type, height, and tilt), and transmission power model (power-based vs. SIR-based) in addition to tower location and subscriber assignment.

The models described in this chapter have been presented as so-called "green field" problems in which a network is being designed for an area that currently has no infrastructure in place. In many cases, however, providers seek to leverage their existing infrastructure when deploying 3G networks. Kalvenes et al. [34] describe several variations on the theme of using the type of planning models discussed in this chapter to optimize the expansion of existing infrastructure to increase network capacity, provide new services, etc. Typically these situations are modeled by fixing variables representing infrastructure that is already in place to one while setting their costs to zero, and there is often a budget constraint on the investment in new towers, MTSOs, etc. Thus, the models described above are easily adapted to so-called *infrastructure expansion problems*.

5.4 Solving the Models

In Section 5.2.2 we described aspects of the CDMA network design problem that make it computationally intractable. Nevertheless, researchers have reported finding high-quality solutions to problems with hundreds of candidate tower locations and thousands of test points in reasonable amounts of CPU time [16]. This section surveys strategies that have been developed to achieve these results. Section 5.4.1 discusses a key insight into the structure of optimal subscriber assignments that Amaldi et. al [7] exploit to develop fast heuristics for finding high-quality solutions to ACMIP. These heuristics are described in Section 5.4.2. Enhancements to the branch-and-bound procedure for finding exact solutions to KKOIP (which can also be applied to ACMIP and CKKOIP) proposed by Kalvenes et al. [34] are presented in Section 5.4.3. Section 5.4.4 describes techniques for dealing with the numerical issues mentioned in Section 5.2.2 that are inherent in these models. The size of the integer programs corresponding to the models described above can be reduced, and solution times improved, by deciding a priori that certain subscriber assignments will not be made. The trade-offs in this approach are discussed in Section 5.4.5.

5.4.1 The Nearest Available Tower Principle

Due to the inherent difficulties of solving ACMIP with branch-and-bound, Amaldi et al. propose randomized greedy search procedures for finding high-quality solutions to the coverage-maximizing variant of the problem within reasonable CPU time limits [7]. These algorithms exploit an important property of ACMIP to quickly hone in on high-quality solutions.

Suppose that test point m is assigned to tower i in a feasible solution to (5.5)-(5.9). If there is another selected tower j such that $g_{mj} > g_{mi}$, then we can obtain another feasible solution by reassigning test point m from tower i to tower j. This reassignment allows mobiles at test point m to reduce their transmission power from P_{target}/g_{mi} to P_{target}/g_{mj} thereby reducing interference and increasing SIR throughout the system. Thus, in solving either variant of ACMIP it is sufficient to only consider solutions where each covered test point is assigned to a "nearest" available tower (i.e., one that requires the least transmission power to reach with a signal strength of P_{target}).

5.4.2 Randomized Greedy Search

Given a set of selected towers $\bar{L}$, the algorithms in [7] assign test point m to the tower $\ell \in \bar{L}$ with the largest attenuation factor $g_{m\ell}$. If it turns out that this subscriber assignment is infeasible, then test points assigned to towers where the SIR is less than $SIR_{\min}$ are removed (i.e., the corresponding x variables are set to zero) one at a time, in non-increasing order of $\frac{d_m}{g_{m\ell}}$ until the QoS constraints (5.7) are satisfied for every tower in $\bar{L}$. Thus, the subscriber assignment sub-problem can be solved very quickly for any given tower subset $\bar{L}$.

Given $\bar{L}$, an iteration of the first procedure proposed in [7], called Add, takes each unselected tower $\ell \in L \setminus \bar{L}$ and evaluates the change in the objective function (5.13) resulting from adding ℓ to $\bar{L}$. The procedure then randomly selects one tower from the subset of unselected towers whose addition to $\bar{L}$ leads to the greatest improvement in (5.13). The procedure starts with no selected towers (i.e., $\bar{L} = \emptyset$) and stops when it reaches a point where the addition of any of the unselected towers to $\bar{L}$ leads to a decrease in (5.13). That is, it stops when it reaches a *local optimum*. The second procedure, called Remove, works in essentially the opposite manner; it starts with $\bar{L} = L$ and removes towers in a randomized greedy fashion until no further improvement in (5.13) is possible.

In [7] the authors compare these procedures against CPLEX—the "gold standard" for commercial branch-and-bound codes—on a set of 10 small test problems in which 22 tower locations and 95 test points are randomly placed in a 400 m by 400 m service area. In these problems each test point has a demand of $d_m = 1$ and the attenuation factors in these problems are such that it is possible to cover all 95 test points. Since the tower locations all have the same usage cost (i.e., $a_\ell = a$ for all $\ell \in L$), solving these problems amounts to minimizing the number of towers needed to cover all the test points. Using CPLEX to obtain optimal solutions, it was determined that in each instance 4 towers were sufficient to cover all test points. Add and Remove were each run 50 times on each problem instance on a Pentium III/700-MHz processor. Taking the best solution of the 50 found by each procedure, Add and Remove found optimal solutions to 5 and 8 out of the 10 instances, respectively. In the other cases, the best solution found used 5 towers. The average CPU times for the 50 runs of Add and Remove were 10 and 30 seconds, respectively. By contrast, CPLEX 7.0 took 5 to 20 minutes to find optimal solutions on a faster machine [7].

At the cost of increased CPU time, the authors were able to improve solution quality (in terms of the average and standard deviation of the number of towers used over a set of 50 runs) using the solutions found by Add and Remove as starting points for a *tabu search* algorithm. Developed in the seminal papers by Glover [26, 27], tabu search (TS) is a *meta-heuristic* for improving the performance of *local search* (LS) heuristics. In a LS heuristic one starts with a feasible solution and considers the "neighborhood" of feasible solutions formed by applying various problem-specific "moves" to generate new solutions. The LS then moves to the best solution in the neighborhood of the current solution, or stops if no move produces an improved solution. LS has been shown to be an effective approach to solving difficult optimization problems, however it has the potential drawback of converging to a local optimum. TS is designed to overcome this drawback by allowing non-improving moves, making recently used moves "tabu" (i.e., not allowing them) for a certain number of iterations to avoid cycling back to previously visited neighborhoods, employing "diversification" strategies to encourage the search to explore new areas of the feasible region, and using "intensification" strategies to hone in on high-quality solutions. TS has been successfully applied to a wide range of optimization problems [32]. In this case, the moves in the LS are adding a tower to $\bar{L}$, removing a tower from $\bar{L}$, or doing both simultaneously.

The heuristics presented in [7] were also tested on 10 medium ($|L| = 120$ and $|M| = 400$) and 5 large ($|L| = 200$ and $|M| = 750$) problem instances. The service areas for the medium and large instances are 1 km by 1 km, and 1.5 km by 1.5 km, respectively. The demands in the medium size instances are randomly selected from $\{1,2,3\}$ and the demands in the larger instances are randomly selected from $\{1,2\}$. Using tabu search, the authors found solutions that satisfy all the demand in the medium instances using 40 to 48 towers in an average CPU time on the order of 2 hours and 40 minutes. For the larger instances, the tabu search took on the order of 12 hours and the best solutions found covered at least 99% of the demand using 54 to 64 towers. Note that the results described above are for *multistart TSs*. That is, the tabu search was run until certain stopping criteria were met and then restarted with another Add or Remove solution. This process was repeated 50 times for each problem instance. The authors report that solutions of comparable quality were found in about half the CPU time by making a single run of Remove followed by a longer tabu search.

For a given set of selected towers Amaldi et al. [7] determine optimal power levels in the SIR-based model by assigning each test point to the nearest available tower and solving the resulting system of linear equations implied by constraint set (5.15) [7]. This results in considerably longer solution times compared to using the power-based model, but with a considerable savings in the number of towers needed to satisfy all the demand. For example, on the medium sized problem instances the tabu search algorithms took approximately 6 times longer with the SIR-based model but found solutions that used at at least 20% fewer towers than those found with the power-based model. Due to the relatively long running time on the small problem instances, the authors did not attempt to solve the medium or large instances of ACMIP to provable optimality with CPLEX. In the next section we discuss ways to improve the running time for CPLEX on power-based models.

5.4.3 Improving Branch-and-Bound Performance

The observation made by Amaldi et al. [7] that optimal subscriber assignments can be found by only considering assignment of test points to their nearest selected towers can be exploited in two different ways to improve the performance of the branch-and-bound process when applied to these types of network design problems.

At any particular node in the branch-and-bound tree, there can be multiple variables that could be selected to branch on, and in the generic branch-and-bound algorithm described in Chapter 4 the choice is made arbitrarily. In fact, this happens at the root node of the example branch-and-bound tree shown in Figure 4.7. For tower-location/subscriber-assignment problems, however, adopting a *branching rule* that gives priority to branching on fractional y variables over branching on non-integer x variables has been shown to speed up the branch-and-bound process [34]. Intuitively, this branching rule implements a strategy for searching the feasible region of first determining the tower locations and then solving for the best subscriber assignment.

Limiting test point assignments to the nearest selected tower can also be built into the integer programming model itself via the following *cuts* proposed by Kalvenes et al. [34] for KKOIP:

$$x_{m\ell} \leq d_m(1-y_j) \qquad \forall m \in M, j \in L_m \text{ such that } g_{m\ell} < g_{mj}. \tag{5.45}$$

If tower j is selected, then (5.45) forces $x_{m\ell} = 0$ for all any selected tower ℓ that receives signals from test point m at lower power. That is, it only allows solutions where every test point served is assigned to a nearest selected tower. The advantage of adding these cuts to the model is that they can eliminate a large number of sub-optimal points from the feasible region. However, there can be a relatively large number of these cuts for any given problem instance, and so adding them increases the number of constraints in the model and this in turn can lead to significantly longer solution times. Kalvenes et al. [34] studied this trade-off in a series of computational experiments which we summarize in Section 5.4.4. Note that when applied to ACMIP, where $x_{m\ell}$ is a binary variable, the right-hand side of (5.45) is simply 1 - y_j.

By focusing only on the interference from test points assigned to tower ℓ, Amaldi et al. [7] observe that the QoS constraint (5.7) implies

$$\sum_{m\in M_\ell} d_m x_{m\ell} \leq s \Rightarrow \sum_{m\in M_\ell} x_{m\ell} \leq s. \tag{5.46}$$

Hence, the QoS constraints implicitly impose a capacity limit on the number of test points that can be assigned to a tower, and the second set of cuts proposed in [34] are the valid inequalities

$$\sum_{m\in M_\ell} x_{m\ell} \leq s \qquad \ell \in L. \tag{5.47}$$

Although these valid inequalities are implied by (5.7) and (5.22), Kalvenes et al. [34] demonstrate that including them explicitly in the KKOIP formulation improves solution quality.

Chapter 4 describes how the branch-and-bound process can be stopped before it finds a provably optimal solution, and how in such cases the relative optimality gap is used to measure the quality of the best (incumbent) solution found at that point. The relative optimality gap can be used as a stopping criterion for branch-

and-bound so that the process is stopped when the relative optimality gap of the incumbent solution is within a user-defined threshold. This strategy is widely used in practice to make a trade-off between solution quality and running time; the process terminates more quickly with larger thresholds but at the cost of leaving portions of the feasible region unexplored and/or finding a good (tight) dual bound on the objective function (i.e., the quality of the incumbent solution may in fact be better than that implied by the relative optimality gap). It is up to the user to determine an acceptable threshold for the relative optimality gap; typical values used in the literature are 1%, 5%, and 10% [34].

5.4.4 Dealing with Numerical Instability

As noted earlier, the wide range of coefficients on the left-hand side of the QoS constraints (5.7) or (5.22) can lead to numerical instability of CPLEX's branch-and-bound algorithm. CPLEX, and presumably other commercially available branch-and-bound codes, scale the constraint matrix of the user-supplied problem so that the absolute value of all the coefficients in any given constraint are between zero and one. It then solves the scaled problem and applies the reverse scaling to the result to obtain the solution returned to the user. Kalvenes et al. [34] observed that although the solutions CPLEX found to the scaled problems in their test sets were feasible (within a given tolerance), the unscaled solutions often violated some of the QoS constraints. To correct this, they apply a post-processing procedure that minimizes the number of subscribers that must be dropped from the scaled solution to ensure that the unscaled solution satisfies the QoS constraints. The procedure involves solving the following integer programming problem

$$\min \sum_{(m,\ell)\in A} \delta_{m\ell} \tag{5.48}$$

$$\text{s.t.} \sum_{(m,j)\in A} \frac{g_{m\ell}}{g_{mj}} (\bar{x}_{mj} - \delta_{mj}) \le s \qquad \forall \ell \in LV \tag{5.49}$$

$$\delta_{m\ell} \in \mathbb{N} \qquad \forall (m,\ell) \in A, \tag{5.50}$$

where in the unscaled solution constant $\bar{x}_{m\ell}$ is the value of $x_{m\ell}$, $A = \{m \in M, \ell \in L_m : \bar{x}_{m\ell} > 0\}$, and LV is set of selected towers, and $\delta_{m\ell}$ is the number of subscribers at test point m that are dropped from tower ℓ in the scaled solution. After solving the integer program above, the post-processing procedure sets $x_{m\ell} = \bar{x}_{m\ell} - \delta_{m\ell}$ for all $m \in M$ and $\ell \in L_m$. Thus, Kalvenes et al. [34] propose a two-phase solution: Phase I applies branch-and-bound to the KKIOP model and if the solution violates any of the QoS constraints, Phase II applies the post-processing procedure to the Phase I solution.

Kalvenes et al. [34] present a computational study using this post-processing procedure, the branching rule and cuts, (5.45) and (5.47), described in Section 5.4.3

on a series of 300 KKOIP problem instances. The largest problem instances in the study have 40 candidate tower locations and 250 test points with demands generated from the discrete uniform distribution on the range $[1, 32]$. Using the default CPLEX settings and KKOIP without the branching rule or cuts, they solved 10 of the largest problem instances in an average CPU time of over 7 hours on a Compaq AlphaServer DS20E with dual EV 6.7(21264A) 667MHz processors and 4,096MB of RAM. The average and maximum optimality gap of the Phase II solutions were 15.41% and 49.69%, respectively. Using a 1% relative optimality gap as a stopping criterion for Phase I, the branching rule, and both sets of cuts, they solved a set of 100 of these problems (including the 10 mentioned above) in an average CPU time of 13 minutes. In these runs the average and maximum relative optimality gap of the Phase II solutions were reduced to 1.18% and 4.27%, respectively.

Overall, the results in [34] indicate that using the branching rule significantly reduces CPU time and that both cuts (5.45) and (5.47) improve CPLEX's performance; as a general rule, the authors recommend using all three for solving CDMA network design problems. However, in instances where there are a relatively large number of tower locations within range of each test point (i.e., $|L_m|$ is close to $|L|$ for most test points) adding cut set (5.45) to the model can actually increase solution time rather significantly. Section 5.4.5 describes an additional strategy that has been shown to be effective in these cases.

The core model presented in this chapter is arguably the most straight-forward, or "natural", framework to formulate 3G network design problems, however it is by no means the only practical approach. Indeed, D'Andreagiovanni [18], D'Andreagiovanni et al. [19], and Mannino et al. [41] report encouraging computational results with recently proposed alternative formulations. The details of these formulations are beyond the scope of this chapter, but in general the idea is to enforce the QoS requirements without using constraints of the form of (5.7) and thereby avoid the computational difficulties arising from the use of large β_ℓ-type constants and from the potentially wide range of attenuation factors in a given problem instance.

5.4.5 Reducing Problem Size

In their computational study of KKOIP, Kalvenes et al. [34] found that problem difficulty increases with *subscriber density*. Subscriber density can be measured in terms of the number of allowable subscriber-to-tower assignments. In an extreme case where $L_m = M$ for every test point there are $|L||M|$ subscriber assignment variables (x's). This happens, for example, in problem instances generated using the parameters given in [7]. This leads to integer programs with more decision variables and denser constraint matrices, both of which contribute to longer solution times. Thus, a practical strategy for finding high-quality solutions to dense problem instances is to limit L_m to a relatively small number of the closest tower locations to test point m. As with using the relative optimality gap as a stopping criterion for branch-and-bound, restricting the size of L_m also trades solution quality for faster solution times. Cai et al. [17] found this to be a very effective strategy in their computational study of the CKKOIP model. For the largest instances in the study this strategy reduced solution time by an order of magnitude with only a modest loss in revenue (typically 5%) compared to the best solution found using the full L_m for each test point. Thus, they were able to find near-optimal solutions to problem instances with up to 160 potential tower locations, 2,000 test points and 5 candidate MSTO locations in less than an hour of CPU time on the Compaq AlphaServer described above.

Rosenberger and Olinick [52] propose a heuristic based on a geographical decomposition approach whereby a large planning region is subdivided into several smaller regions. After solving their model for one of the smaller problems they fix the y variables for all towers that are not near the boundary of the region. They repeat this for each of the sub-regions and then solve a problem for the entire large region in which many of the tower location decisions have already been made.

As computing resources continue to become less expensive and more efficient, the techniques described in this section should enable network planners to "scale up" the models described in Sections 5.2 and 5.3 to tackle even larger design problems. For example, it takes only a matter of seconds to solve small AMCIP instances with newer hardware and a more recent version of CPLEX than the combination used in [7]. The fact remains, however, that the optimization problem represented by the core model, and its variations and extensions, is computationally intractable, and so as wireless technology advances and planners attempt to design larger and more complex networks the "combinatorial explosion" of the models seems likely to eventually outstrip the performance gains from advances in computing technology. Thus, we expect that the development of good, fast heuristics and improvements to the branch-and-bound process for solving cellular network design problems will continue to be active areas of research.

References

1. Ahuja, R.K., Magnanti, T.L., Orlin, J.B.: Network Flows: Theory, Algorithms, and Applications. Prentice Hall, Upper Saddle River, NJ (1993)
2. Amaldi, E., Capone, A., Malucelli, F.: Base station configuration and location problems in UMTS networks. In: Proceedings of the 9th International Conference on Telecommunication Systems - Modeling and Analysis, pp. 341–348 (2001)
3. Amaldi, E., Capone, A., Malucelli, F.: Discrete models and algorithms for the capacitated location problems arising in UMTS network planning. In: Proceedings of the 5th International Workshop on Discrete Algorithms and Methods for Mobile Computing and Communications, pp. 1–8. ACM (2001)
4. Amaldi, E., Capone, A., Malucelli, F.: Improved models and algorithms for UMTS radio planning. In: IEEE 54th Vehicular Technology Conference Proceedings, vol. 2, pp. 920–924. IEEE (2001)
5. Amaldi, E., Capone, A., Malucelli, F.: Optimizing base station siting in UMTS networks. In: IEEE 53rd Vehicular Technology Conference Proceedings, vol. 4, pp. 2828–2832. IEEE (2001)

6. Amaldi, E., Capone, A., Malucelli, F.: Optimization models and algorithms for downlink UMTS radio planning. In: Proceedings of the IEEE Wireless Communications and Networking Conference, vol. 2, pp. 827–8731. IEEE (2003)
7. Amaldi, E., Capone, A., Malucelli, F.: Planning UMTS base station location: Optimization models with power control algorithms. IEEE Transactions on Wireless Communications **2**, 939–952 (2003)
8. Amaldi, E., Capone, A., Malucelli, F., Signori, F.: A mathematical programming approach for W-CDMA radio planning with uplink and downlink constraints. In: IEEE 58th Vehicular Technology Conference Proceedings, vol. 2, pp. 806–810. IEEE (2003)
9. Amzallag, D.: Approximation algorithms for optimization problems in future cellular networks. Ph.D. thesis, Technion - Israel Institute of Technology, Haifa 32000, Israel (2008)
10. Averbakh, I.: On the complexity of a class of combinatorial optimization problems with uncertainty. Mathematical Programming Series A **90**, 263–272 (2001)
11. Ben-Tal, A., Nemirovski, A.: Robust convex optimization. Mathematics of Operations Research **23**, 769–805 (1998)
12. Ben-Tal, A., Nemirovski, A.: Robust solutions of linear programming problems contaminated with uncertain data. Mathematical Programming Series A **88**, 411–424 (2000)
13. Bienstock, D., D'Andreagiovanni, F.: Robust wireless network planning. In: Proceedings of the XL Annual Conference of the Italian Operational Research Society (AIRO2009), Siena, Italy (2009)
14. Birkan, G., Kennington, J., Olinick, E., Ortynski, A., Spiride, G.: Making a case for using integer programming to design DWDM networks. Optical Networks Magazine **4**, 107–120 (2003)
15. Cai, Q.: Optimization-based design tools for W-CDMA cellular networks. Ph.D. thesis, Southern Methodist University (2006)
16. Cai, Q., Kalvenes, J., Kennington, J., Olinick, E.: Reliable W-CDMA network design. Tech. Rep., Department of Engineering Management, Information, and Systems, Bobby B. Lyle School of Engineering, Southern Methodist University, Dallas, TX (2006). http://lyle.smu.edu/ jlk/cdmarely.pdf
17. Cai, Q., Kalvenes, J., Kennington, J., Olinick, E.: W-CDMA network design. International Journal of Mobile Newtork Design and Innovation **2**(2), 92–104 (2007)
18. D'Andreagiovanni, F.: Pure 0-1 programming approaches to wireless network design. Ph.D. thesis, Sapienza Università di Roma (2009)
19. D'Andreagiovanni, F., Mannino, C., Sassano, A.: A power-indexed formulation for wireless network design. Tech. Rep. 14-2009, Department of Computer and System Sciences, Sapienza Università di Roma (2009). http://ojs.uniroma1.it/index.php/DIS_TechnicalReports/article/viewFile/8855/8826
20. Eisenblätter, A., Fügenschuh, A., Fledderus, E.R., Geerdes, H.F., Heideck, B., Junglas, D., Koch, T., Kürner, T., Martin, A.: Mathematical methods for automatic optimization of UMTS radio networks. Tech. Rep. D4.3, IST-2000-28088 MOMENTUM (2003). URL http://momentum.zib.de/paper/momentum-d43.pdf
21. Eisenblätter, A., Fügenschuh, A., Koch, T., Koster, A.M.C.A., Martin, A., Pfender, T., Wegel, O., Wessäly, R.: Modelling feasible network configurations for UMTS. In: G. Anandalingam, S. Raghavan (eds.) Telecommunications Network Design and Managment, pp. 1–23. Kluwer Academic Publishers, Dordrecht, The Netherlands (2003)
22. Eswaran, K., Trajan, R.: Augmentation problems. SIAM Journal on Computing **5**(4), 653–660 (1976)
23. FCC: Construction/coverage requirements. http://wireless.fcc.gov/licensing/index.htm?job=const_req_by_service. Accessed on June 14, 2010
24. Fortz, B., Labbé, M., Maffioli, F.: Solving the two-connected network with bounded meshes problem. Operations Research **48**, 866–877 (2000)
25. Garey, M., Johnson, D.: Computers and Intractability: A Guide to the Theory of NP-Completeness. W. H. Freeman, San Francisco, CA (1979)
26. Glover, F.: Tabu search, Part I. ORSA Journal on Computing **1**(3), 190–206 (1989)

27. Glover, F.: Tabu search, Part II. ORSA Journal on Computing **2**(1), 4–32 (1990)
28. Grötschel, M., Monma, C., Stoer, M.: Computational results with a cutting plane algorithm for designing communication networks with low-connectivity constraints. Operations Research **40**, 309–330 (1992)
29. Grötschel, M., Monma, C., Stoer, M.: Design of survivable networks. In: Handbooks in OR/MS, vol. 8, chap. 10, pp. 617–672. North-Holland, Amsterdam, The Netherlands (1995)
30. Gryseels, M., Sorbello, L., Demeester, P.: Network planning in uncertain dynamic environments. In: Networks 2000, 9th International Telecommunication Network Planning Symposium (2000).
31. Hata, M.: Empirical formula for propagation loss in land mobile radio service. IEEE Transactions on Vehicular Technology **29**, 317–325 (1980)
32. J. Blazewicz, P.L., Milostan, M.: Application of tabu search strategy for finding low energy structure of protien. Artificial Intelligence in Medicine **35**, 135–145 (2005)
33. Kalvenes, J., Kennington, J., Olinick, E.: Hierarchical cellular network design with channel allocation. European Journal of Operational Research **160**(1), 3–18 (2005).
34. Kalvenes, J., Kennington, J., Olinick, E.: Base station location and service assignments in W-CDMA networks. INFORMS Journal on Computing **18**(3), 366–376 (2006)
35. Kennington, J., Lewis, K., Olinick, E., Ortynski, A., Spiride, G.: Robust solutions for the WDM routing and provisioning problem: models and algorithms. Optical Networks Magazine **4**, 74–84 (2003)
36. Khuller, S.: Approximation algorithms for finding highly connected subgraphs. In: D. Hochbaum (ed.) Approximation Algorithms for NP-Hard Problems, pp. 236–265. PWS Publishing Company, Boston, MA (1997)
37. Kouvelis, P., Yu, G.: Robust Discrete Optimization and Its Applications. Kluwer Academic Publishers, Dordrecht, The Netherlands (1997)
38. Laguna, M.: Applying robust optimization to capacity expansion of one location in telecommunications with demand uncertainty. Management Science **44**, 5101–5110 (1998)
39. Lee, J., Miller, L.E.: CDMA Systems Engineering Handbook. Artech House Publisher, Boston, MA (1998)
40. Li, J., Kameda, H., Itoh, H.: Balanced assignment of cells in PCS networks. In: SAC '97: Proceedings of the 1997 ACM symposium on Applied computing, pp. 297–301. ACM, New York, NY, USA (1997).
41. Mannino, C., Mattia, S., Sassano, A.: Planning wireless networks by shortest path. Computational Optimization and Applications (2009). Published online at http://dx.doi.org/10.1007/s10589-009-9264-3
42. Mathar, R., Niessen, T.: Optimum positioning of base stations for cellular radio networks. Wireless Networks **6**, 421–428 (2000)
43. Merchant, A., Sengupta, B.: Assignment of cells to switches in PCS networks. ACM/IEEE Transactions on Networking **3**, 521–526 (1995)
44. Mobiledea: Mobile telephone switching office (MTSO). http://www.mobiledia.com/glossary/166.html. Accessed on June 14, 2010
45. Mulvey, J.M., Vanderbei, R.J., Zenios, S.A.: Robust optimization of large-scale systems. Operations Research **43**, 264–281 (1995)
46. Murphey, R., Pardalos, P., Resende, G.: Frequency assignment problems. In: D. Zu, P. Pardalos (eds.) Handbook of Combinatorial Optimization, pp. 1–55. Kluwer Academic Publishers (1999)
47. Nemhauser, G.L., Wolsey, L.A.: Integer and Combinatorial Optimization. John Wiley & Sons, New York (1999)
48. Olinick, E., Rosenberger, J.: Optimizing revenue in CDMA networks under demand uncertainty. European Journal of Operational Research **186**(2), 812 – 825 (2008).
49. Paraskevopoulos, D., Karakitsos, E., Rustem, R.: Robust capacity planning under uncertainty. Management Science **37**, 787–800 (1991)
50. Philpott, A.: Stochastic programming introduction. http://stoprog.org/index.html?spintroduction.html. Accessed on June 21, 2010

51. Rajan, D., Olinick, E., Venkateswaran, J.: A unified optimization framework for design of CDMA cellular networks with uplink and downlink rate guarantees. Tech. Rep. 10-EMIS-04, Department of Engineering Management, Information, and Systems, Bobby B. Lyle School of Engineering, Southern Methodist University, Dallas, TX (2010)
52. Rosenberger, J., Olinick, E.: Robust tower location for code devision multiple access networks. Naval Research Logistics **54**(2), 151–161 (2007)
53. Sherali, H.D., Pendyala, C.M., Rappaport, T.S.: Optimal location of transmitters for micro-cellular radio communication system design. IEEE Journal on Selected Areas in Communications **14**(4), 662–673 (1996)
54. Soteriou, A., Chase, R.: A robust optimization approach to improving service quality. Manufacturing and Service Operations Management **2**, 264–286 (2000)
55. St-Hilaire, M., Chamberland, S., Pierre, S.: Uplink UMTS network design–an integrated approach. Computer Networks **50**, 2747–2761 (2006)
56. TopBits.com: MTSO mobile telephone switching office. http://www.topbits.com/mtso-mobile-telephone-switching-office.html. Accessed on June 14, 2010
57. Yates, R.: A framework for uplink power control in cellular radio systems. IEEE Journal on Select Areas in Communications **13**, 1341–1348 (1995)

Chapter 6
Optimization Based WLAN Modeling and Design

Jeff Kennington, Jason Kratz, and Gheorghe Spiride

Abstract This chapter describes strategies, software tools, and optimization models that may be used to design wireless local area networks. While most small scale WLAN deployments can be designed in an ad-hoc fashion, medium to large scale networks can benefit from a more methodical design approach. In one design strategy, RF signal strength data are sampled at various locations within a site. Measurements are then used to determine transmitter locations that ensure site wide coverage. This chapter describes approaches that belong to an alternate strategy, which relies on using special purpose software tools. These tools most often attempt to solve an underlying optimization problem maximizing either coverage or capacity. Some coverage models are simple and can be solved with existing commercial optimization software. Solution procedures for sophisticated capacity models with integer variables and nonlinear constraints are not solvable using current commercial optimization software. Therefore, many industrial and academic research groups rely upon metaheuristics for problem solution. The chapter concludes with a discussion of computational results that illustrate the application of two widely known metaheuristics to the WLAN design problem.

Jeff Kennington
Department of Engineering Management, Information, and Systems, Bobby B. Lyle School of Engineering, Southern Methodist University, Dallas, TX 75275-0123, USA, e-mail: jlk@smu.edu

Jason Kratz
Department of Engineering Management, Information, and Systems, Bobby B. Lyle School of Engineering, Southern Methodist University, Dallas, TX 75275-0123, USA, e-mail: jkratz@smu.edu

Gheorghe Spiride
Department of Engineering Management, Information, and Systems, Bobby B. Lyle School of Engineering, Southern Methodist University, Dallas, TX 75275-0123, USA, e-mail: gspiride@smu.edu

J. Kennington et al. (eds.), *Wireless Network Design: Optimization Models and Solution Procedures*, International Series in Operations Research and Management Science 158, DOI 10.1007/978-1-4419-6111-2_6, © Springer Science+Business Media, LLC 2011

6.1 Introduction

Wireless local area networks (WLANs) connect two or more computers or devices using spread- spectrum or OFDM modulation technologies that allow communication over a limited area. Communication between devices on a WLAN typically requires low or limited mobility and/or roaming, and the use of unlicensed, free industrial, scientific, and medical (ISM) spectrum. This is in contrast with cellular networks, which supports high mobility subscribers and operates in licensed spectrum bands at much higher transmission power levels. Taking advantage of ISM spectrum resources at low transmission power means that WLAN technologies must exhibit robust error tolerance and tolerate interference with other devices in the same spectrum bands, e.g. cordless phones, microwave ovens, and garage door openers. WLANs are based on the IEEE 802.11 family of protocols (http://standards.ieee.org/getieee802/) that specify the set of frequencies that can be used to construct these wireless networks. The most widespread protocol is IEEE 802.11 b/g, which defines 11 channels in the 2.4 GHz portion of the spectrum as listed in Table 1. Since a signal occupies approximately 22 MHz of the spectrum, this results in 3 non-overlapping channels (1, 6, and 11).

Table 6.1 IEEE 802.11 b/g Channels

Frequency Spectrum (GHz)	Channel	Frequency Spectrum (GHz)	Channel
2.412	1	2.442	7
2.417	2	2.447	8
2.422	3	2.452	9
2.427	4	2.457	10
2.432	5	2.462	11
2/437	6	-	-

The structure of a WLAN comprises mobile devices (MDs) that communicate with access points (APs). A wireless access point bridges the communication between devices connected to the wireless network and a wired network, typically Ethernet based. The main driving factors for the growing WLAN popularity are the availability of low cost equipment, ease and speed of installation in a variety of settings, and the effective and efficient use of ISM spectrum resources. The ease and speed of installation of wireless access points is due, among other things, to the fact that most deployments use Power Over Ethernet (PoE) systems to safely transfer electrical power along with data over standard category 5 cable in an Ethernet network. Current PoE standards (IEEE 802.3af and 802.3at) allow up to 25W of power to be distributed to each access point device. This greatly simplifies installation and reduces costs since deploying a new access point requires a single type of cabling to be made available (Ethernet CAT 5) without separate electrical work and permitting. While setting up a WLAN in a residential environment requires installing one or two APs, wireless LANs that target the enterprise/university campus environment may require deploying many more APs. Recently, a number of small cities and towns across the United States deployed city wide WLAN networks that allow boradband access to Internet resources at reasonable costs for residents, local government workers, and emergency and first responders. Though initially very popular with city councils across the United States, these networks faced large operational costs that have limited their growth.

The basic WLAN design problem involves decisions concerning the number and location of radio transmitters/receivers (APs), the frequency to be assigned to each AP, and the signal power of each AP. Many organizations install WLANs as an alternative to cabled networks without spending much time on planning and design. Some simply place APs in a grid, turn them on, and add more as problems arise. However, careful planning ensures high performance at minimum cost.

6.2 Literature Survey

There are two basic strategies have been adopted by industry specialists for designing WLANs: *site surveys* and the use of *software tools*. Many of these software tools use optimization models and algorithms to solve this problem. Reseatch studies that propose the use of optimization models and algorithms generally address one or more of the following problems: *i)* determine a minumum cost coverage for a given site, *ii)* determine a channel assignment for a fixed set of APs to optimize some metric, or *iii)* determine the optimal AP placement to optimize some metric. In addition, the literature also contains descriptions of several proprietary design tools developed by industrial research groups.

6.2.1 Site Surveys

An excellent description of a site survey for the Miami International Airport may be found in [4]. Geier and his associates mounted an AP on a roll-around cart with its antenna attached to the top of a telescopic pole. Since APs are usually mounted on the ceiling, this configuration could be used to approximate the location of an installed AP. On a separate cart, they carried laptops and other equipment that could record signal-to-noise ratios. Using this equipment, they tested over sixty AP locations before developing a proposed design. This practice adopts the well known approach used in cellular radio frequency (RF) optimization, where cell site coverage is checked via extensive drive testing.

Mateus et al. [12] (see also [17]) report on their experience using a site survey to locate 3 APs to cover 2 floors of an academic building with labs and faculty offices. The authors selected 6 candidate locations for AP installation to cover 253 demand areas. Signal strength data was collected at each of the 253 demand points by slowly rotating a mobile receiver at each location. Signal strength is sensitive to mobile orientation and the largest 20% and smallest 20% of the values were discarded. The average of the remaining 60% was used in an integer programming model to determine the optimal location for their 3 APs. Using 3 APs is desirable since each may be assigned a unique frequency and thereby avoid interference.

Site surveys can be useful for developing a coverage-oriented design for low density areas, but may not produce a good capacity-oriented design for high density areas. That is, capacity measurements are not part of a site survey but can be incorporated within software tools. Some clients have found site surveys to be fairly expensive.

6.2.2 Coverage Models

A simple optimization-based coverage oriented model can be developed by superimposing two grids onto a site: one for potential AP locations and one for mobile device (MD) locations. The MD grid is usually finer than the AP grid since in practice mobile devices could occupy many more positions than APs. The idea is to select the minimum number of AP grid points so that each MD grid point is covered (within range) of at least one AP. This can be modeled as a simple binary integer program. A description of the sets and variables needed for this model may be found in Table 6.2, and the coverage-oriented model is given as follows:

$$\text{Minimize} \quad \sum_{s \in S} w_s$$

$$\text{Subject to}$$

$$\sum_{s \in CM_m} w_s \geq 1 \forall m \in M$$

$$w_s \in \{0,1\} \quad \forall s \in S$$

Table 6.2 Sets and Variables used for the Coverage Model

Name	Type	Description
S	set	set of potential APs
M	set	set of all demand points
CM_m	subset of S	set of APs within range of MD m
w_s	binary variable	$w_s = 1$ if AP s is used; 0 otherwise

This model was used to cover a square site with a dimension of 250 feet as illustrated in Figure 6.1. Each grid square has a dimension of 50 feet by 50 feet. The MDs are placed on the 36 grid points and the potential locations for APs are placed on the 16 internal grid points. For effective communication, the distance between an AP and a MD must be no more than 120 feet. The optimal solution used 4 APs located at the following grid points: $(50,50),(150,50),(50,150),(200,150)$ as illustrated in Figure 6.2. Note that all MD grid points are covered, but there is a small area near $(125,250)$ that is not covered. A finer grid could be used to eliminate this dead spot.

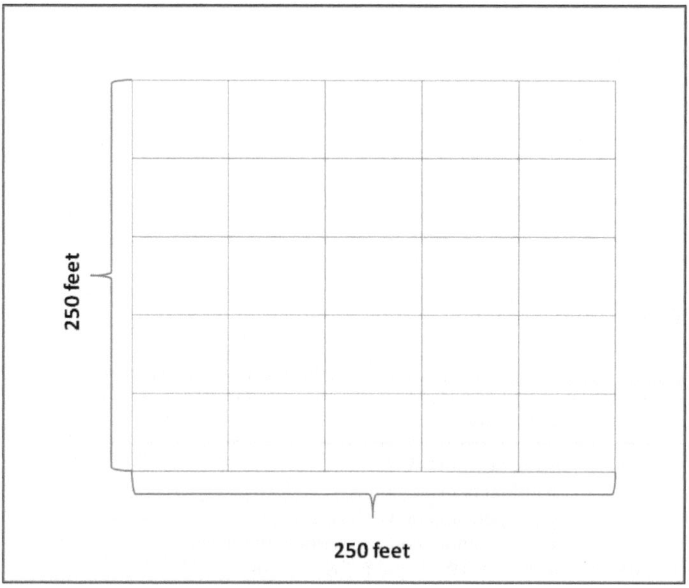

Fig. 6.1 The Site With a Superimposed Grid

A more detailed coverage-oriented model has been developed by Bahri and Chamberland [2]. They developed an optimization model for AP placement, AP transmit power, and frequency assignment. The sets and variables used to describe their model may be found in Table 6.3. The objective is to minimize the required total number of APs to be installed that guarantee coverage. Their model is a nonlinear integer mathematical program that is shown to be *NP*-hard. The authors propose using a tabu search algorithm to obtain a solution. Attenuation factors are based on an adaptation of the Hata formula ([6]). Their algorithm was empirically evaluated on a 10-floor building with several hundreds of stations. They compare their solutions to lower bounds obtained by relaxing a subset of the difficult constraints.

6.2.3 Channel Assignment

Leung and Kim [10] present a heuristic procedure for frequency assignment that is closely related to frequency assignment models for cellular networks. They demonstrate their procedure on a clover-leaf hexagon served by APs with an antenna hav-

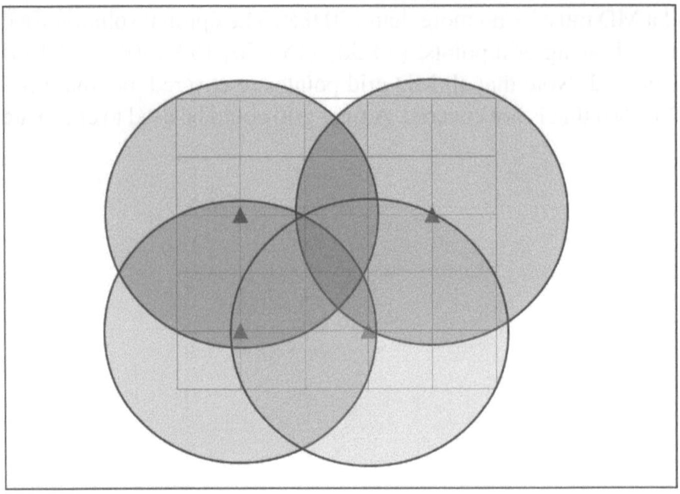

Fig. 6.2 Optimal Solution for Grid Coverage

Table 6.3 Sets and Variables Used By the Global Model of Bahri and Chamberland [2]

Name	Type	Description
S	set	set of potential APs
M	set	set of all demand points
P	set	set of potential power levels
F	set	set of potential non-overlapping frequencies
w_s	binary variable	$w_s = 1$ if AP s is used; 0 otherwise
x_{ms}	binary variable	$x_{ms} = 1$ if demand point m is assigned to AP s; 0 otherwise
y_{ps}	binary variable	$y_{ps} = 1$ if AP s is assigned power level p; 0, otherwise
z_{fs}	binary variable	$z_{fs} = 1$ if AP s is assigned frequency f; 0 otherwise

ing a beam width of 60 degrees. The type of solution obtained is illustrated in Figure 6.3.

Vanhatupa et al. [20] present a frequency assignment model for environments where multiple WLANs coexist. They consider the problem of assigning frequencies to a new AP given a set of existing, fixed AP locations, such that throughput is maximized. Their formulation is shown to be equivalent to finding a solution to the weighted graph coloring problem, which is *NP*-complete. Hence, heuristics are applied to obtain approximate solutions.

Kauffmann et al. [9] address the problem of AP frequency assignment in an environment where multiple WiFi devices are deployed in the same area. They also propose algorithms that allow wireless clients to select the best AP for communication. Distributed algorithms are recommended that do not require coordination among the various network devices. They claim that simple software modifications may be incorporated by the IEEE 802.11k task group.

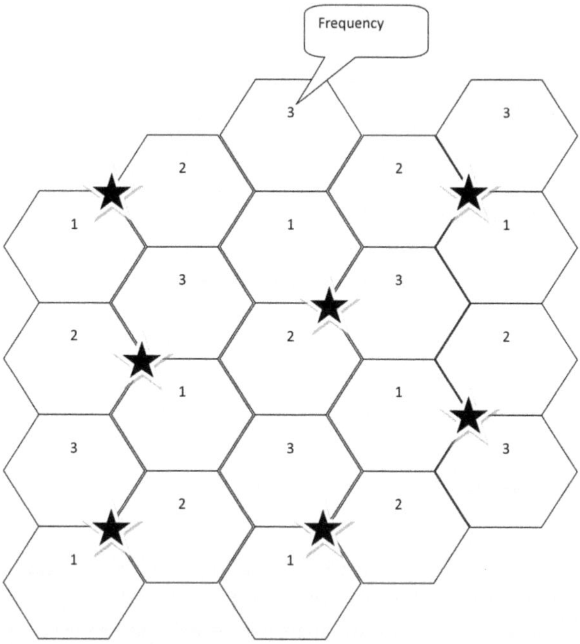

Fig. 6.3 Frequency Assignment Using Three APs with a 60 Degree Antenna at each Star cf. [10]

Mishra et al. [13] describe an environment where multiple APs are operational and mobiles scan the wireless medium and associate with the AP having the strongest signal, and the channel assignment can be changed dynamically. The APs are capable of selecting the best channel from $\{1,\ldots,k\}$ for operation. Generally, k is 11 and channel overlap is permitted.

Let $G = [V,E]$ denote a graph, where V is the set of APs and E denotes pairs of APs that are close enough to cause interference if the assigned channels overlap. Let $W_{ij}\forall(i,j) \in E$ be the weight associated with link (i,j). The authors generally use the total number of clients associated with the two APs as the edge weight. A channel assignment is a mapping $C : V \to \{1,\ldots,k\}$. Given such a mapping, I_{ij} is defined as the interference caused by the assignment of $C(i)$ to AP i and $C(j)$ to AP j. Interference is a function of the distance between the network components (MDs and APs) as well as obstacles between the devices. If there is no overlap in frequency bands, then I_{ij} is 0. Using this notation, 3 versions of the channel assignment model are defined in [13].

The first version minimizes the maximum weighted interference:

$$\text{Minimize} \quad \max(W_{ij}I_{ij}) \quad \forall(i,j) \in E$$

The second version minimizes the sum of the weighted interference:

$$\text{Minimize} \quad \sum_{(i,j)\in E} W_{ij} I_{ij}$$

The third version is identical to the second with all weights set to 1. The third model is the well-known graph coloring problem and the second is the weighted graph coloring problem. They present two distributed algorithms for problem solution. Their simulation experiments indicate that the use of overlapping channels can yield improvements over the restricted use of only 3 non-overlapping channels. Some of the simulation results indicated improvements of up to 30%. It should be noted that channel assignment is not required for some design problems. Some AP vendors (see [15]) design their equipment so that it can choose the best frequency for communication. Each AP listens for traffic on the various frequencies and selects a frequency that results in the best reception.

6.2.4 Access Point Placement

Sherali et al. [18] developed a set of nonlinear optimization models to determine optimal access point location. Mobile devices are placed at grid points and standard models are used to determine $g_i(x,y,z)$, the path loss at grid point i associated with a access point at (x,y,z). One objective function is to minimize the sum of all weighted path losses. A penalty term may be defined for mobiles at grid points where communication is not possible. A second objective function is to minimize the maximum path loss plus the aforementioned penalty. The third model is a combination of the first two. The authors evaluate three nonlinear techniques for solution (Hooke and Jeeves, a quasi-Newton method, and a conjugate gradient method) and report solving several small problems successfully.

Adickes et al. [1] use a genetic algorithm based on a circle covering heuristic to solve the transceiver placement problem. Three objective functions are developed and a Pareto ranking is used to evaluate solutions. The first objective is the fraction of covered grid points. The second uses the Shannon-Hartley law (http://www.fact-index.com/s/sh/shannon_hartley_law.html) to estimate capacity, while the third model involves average signal strength over the set of grid points. Using a population size of 30, the authors develop the Genetic Algorithm Optimizer (GAO) software tool for problem solution.

6.2.5 Proprietary Design Tools

The problem of determining the location, number, and power of a set of access points that will provide coverage for a given building at minimum cost was investigated by Fortune et al. in [3]. They developed the Wireless System Engineering (WISE) software tool to assist design personnel with this problem. WISE is designed to quickly create a good design (but not necessarily optimal) and comprises several modules: *i)* a database module to create and store a digitized version of the building, *ii)* a propagation prediction module that implements a 3D ray-tracing algorithm to determine propagation loss between two points, *iii)* an access point optimization module for transceiver location, and *iv)* a graphical user-interface for coverage display.

For the case of n base stations, where (x_j, y_j, z_j) is the location of base station j, the optimization problem is to maximize $f(x_1, y_1, z_1, \ldots, x_n, y_n, z_n)$. A grid is superimposed on the floor plan and f is defined as the ratio of the number of grid points with received power sufficient for communication to the total number of grid points. The authors use a direct heuristic search method that exploits the characteristics of this design problem and report that there are many local optima.

Panjawni et al. [14] developed a software system called Site Modeling Tool (SMT) to determine good locations for radio frequency transceivers. Their system is an extension of AutoCAD and runs on either a PC or workstation. Under the assumption that the coverage regions are connected sets without holes, the coverage regions are modeled as polygons. Their algorithm involves a search for boundary points of these polygons. The first boundary point is located by searching outward in a straight line until communication is not possible. Each additional boundary point is found in a similar manner by a sweep in a counterclockwise direction around the transceiver. The authors use 36 boundary points to define the coverage region polygon. The system supports both single and multifloor buildings and provides a visual display of the solution which allows a designer to determine the locations of the transceivers. Once transceiver locations are placed, the SMT calculates and displays the expected coverage.

Hills [8] describes the design process used to deploy the wireless infrastructure (*Wireless Andrew*) for Carnegie Mellon University. Wireless Andrew covers 65 buildings with a total floor area of approximately 3 million ft^2. A five step iterative approach is used to deploy the WLAN: *i)* initial selection of AP locations, *ii)* use signal strength measurements to adjust AP locations, *iii)* create a coverage map, *iv)* assign frequencies to APs, and *v)* document the system. The report recommends using a coverage-oriented design in low density areas and a capacity- oriented design in high density areas.

Two very sophisticated systems, one academic (S^{4}W) and one commercial (SitePlanner®) are discussed in Skidmore et al. [19]. Both systems use state-of-the-art channel modeling, global optimization procedures for transmitter placement, and a simulation system for performance evaluation. The optimization modules attempt to minimize average shortfall of penalty functions of coverage and bit error rates. These systems are defined as Problem Solving Environments (PSEs) for wireless network design and analysis.

Longo [11] and colleagues at OPNET Technologies, Inc. developed the WLAN Design Solution for the Department of the Navy as part of the Small Business Innovation Research (SBIR) program. Their system is designed to optimize AP placement for military ships or other military structures. The system is combined with a simulation module that can be used to estimate throughput for a proposed design.

6.2.6 *Current State-of-the-Art*

Based upon the experience of deploying WLANs on our own campus and the literature in the previous sections, we conclude that IT professionals use a combination of three basic strategies to design WLANs: *i)* they arbitrarily decide where to locate APs, *ii)* they conduct site-surveys, or *iii)* they use the assistance of software tools. Over the course of the last 10 years, WLAN access has evolved from a somewhat infrequent feature of an enterprise network to a "must-have". Practical WLAN design approaches have mirrored this evolution path. Initially, simple designs involving only a few APs were put together by simply select the APs' location and assigning frequencies that appear appropriate. This sort of ad-hoc design approach evolved to support more complicated designs, where site-surveys would be undertaken. Geier [4] - the principal consultant and founder of Wireless-Nets, Ltd. - illustrates the recent state-of-the-art in the practice of site surveys for WLAN design.

The proliferation of low-cost WLAN hardware "bundled" along with portable computers, smartphones, appliances, game consoles, etc. contributes to the increase in complexity of WLAN designs, especially within the scope of large enterprise deployments. Many hardware vendors provide proprietary software tools for use with their products. The tools described by Longo [11] (OPNET Technologies) and Skidmore et al. [19] (Wireless Valley Communications, Inc.) represent the current state-of-the-art of these types of tools. The initial efforts outlined in [3] were augmented by the development of the Site-Specific System Simulator for Wireless Communications (S^4W) at Virginia Tech [19]. Funded by a grant from the NSF Next Generation Software program, this PSE allows IT personnel to run simulation models that may be used to design WLANs that minimize costs, or maximize capacity or coverage. S^4W also provides support for the data management tasks that simplify the maintenance of a large WLAN design in practice.

6.3 A Maximimum Capacity-Oriented Model

A simple coverage model was presented in Section 6.2.2 and a coverage model that also involved channel assingment was developed by Bahri and Chamberland [2]. However, to our knowledge, the literature does not contain a model that attempts to determine the location and channel assignment of APs that yields a network having maximum capacity. The problem of interest may be defined as follows:

> Given a fixed number of APs that can be deployed in a given site with static MD locations, determine the location and channel assignment for each AP that results in the maximum throughput.

A new optimization-based model for this problem will be described in this section.

Solving this problem for a given building requires determining *i)* the number of access points (APs) to install, *ii)* the location of these APs, *iii)* the transmission power assigned to each AP, and *iv)* the frequency assigned to each AP. A mobile device requesting service will typically connect to the nearest AP. The MD and the AP enter into a negotiation to determine the highest transmission rate that can be safely used. This transmission rate is a function of the received signal power and the interference caused by other active MDs and other APs operating on the same frequency. Given a WLAN design that covers all floors of a building and a set of MDs with known locations requesting service, the network capacity can be calculated as the sum of the transmission rates of all MDs. We use Shannon's capacity calculation to estimate the capacity of a given design with a given set of users.

For this model, the user must provide a set of potential locations for the APs and select a set of static locations for all MDs to be serviced. Therefore, the attenuation factor between each AP and each MD can be computed and are given constants for this model. The model description shown in Table 6.4 requires four sets, two types of constants, and six types of variables.

Table 6.4 Sets, Constants, and Variables used for the Capacity Model

Name	Type	Description
S	set	set of potential APs
M	set	set of all demand points
P	set	set of potential power levels
F	set	set of potential non-overlapping frequencies
γ_{sm}	constant	attenuation between AP s and MD m
N_m	constant	noise associated with MD m
w_s	binary variable	$w_s = 1$ if AP s is used; 0 otherwise
y_{sf}	binary variable	$y_{sf} = 1$ if AP s is assigned frequency f; 0 otherwise
z_{sm}	binary variable	$z_{sm} = 1$ if MD m is affiliated with AP s; 0 otherwise
$x_{s_1 s_2 f}$	binary variable	$x_{s_1 s_2 f} = 1$ if both AP s_1 and s_2 are assigned frequency f
NS	integer variable	the number of APs to be installed in the design
P_s	variable	power setting for AP s

The first constraint ensures that exactly NS APs are selected for installation. This constraint is simply

$$\sum_{s \in S} w_s = NS \tag{6.1}$$

The next $|S|$ constraints ensure that each AP that is installed is assigned one of the possible frequencies. These constraints are as follows:

$$\sum_{f \in F} y_{sf} = w_s \quad \forall s \in S \tag{6.2}$$

The third set of constraints forbids assignment of an MD to an AP that is not installed.

$$z_{sm} \leq w_s \quad \forall s \in S, m \in M \tag{6.3}$$

The following set of constraints ensures that each mobile is assigned to the installed AP having the strongest signal.

$$z_{s_2m} \leq 1 - w_{s_1} \quad \forall m \in M, s_1, s_2 \in S | s_1 \neq s_2 \text{ and } \gamma_{s_1} > \gamma_{s_2} \tag{6.4}$$

The next $|M|$ constraints ensure that every MD is assigned to some AP.

$$\sum_{s \in S} z_{sm} = 1 \quad \forall m \in M \tag{6.5}$$

The next set of constraints force $x_{s_1s_2f}$ to assume the value 1 if and only if both AP s_1 and s_2 are installed and assigned frequency f.

$$x_{s_1s_2f} \leq y_{s_1f} \quad \forall s_1, s_2 \in S, f \in F | s_1 < s_2 \tag{6.6}$$

$$x_{s_1s_2f} \leq y_{s_2f} \quad \forall s_1, s_2 \in S, f \in F | s_1 < s_2 \tag{6.7}$$

$$x_{s_1s_2f} \geq y_{s_1f} + y_{s_2f} - 1 \quad \forall s_1, s_2 \in S, f \in F | s_1 < s_2 \tag{6.8}$$

Suppose MD m_1 is associated with AP s_1. The throughput for m_1 is related to the signal-to-noise ratio at m_1, as well as the total number of MDs associated with s_1 (see [16]). The signal strength at m_1 is given by $P_{s_1}\gamma_{s_1m_1}$ and the noise at m_1 is given by $\sum_{s_2 \in S \setminus \{s_1\}} P_{s_1}\gamma_{s_1m_1} + N_{m_1}$. Let t_{sm} denote the downlink signal strength at m for the $s-m$ association. Then

$$t_{sm} = P_s \gamma_{sm} z_{sm} \quad \forall s \in S, m \in M \tag{6.9}$$

If m is not associated with s then both z_{sm} and t_{sm} will be zero. Let u_{sm} denote the interference at m experienced by the $s-m$ association. Then

$$\begin{aligned} u_{sm} = & \sum_{s_2 \in S \setminus \{s\}: s < s_2} \sum_{f \in F} P_{s_2}\gamma_{s_2m} x_{ss_2f} + \\ & \sum_{s_2 \in S \setminus \{s\}: s_2 < s} \sum_{f \in F} P_{s_2}\gamma_{s_2m} x_{s_2sf} + N_m \quad \forall s \in S, m \in M \end{aligned} \tag{6.10}$$

Note that only the other APs that are assigned the same frequency as s will contribute to the noise at m. Let v_m denote the total number of MDs assigned to the same AP as MD m. Then v_m can be determined by the following inequalities:

$$\sum_{m_1 \in M} z_{sm_1} - |M|(1 - z_{sm}) \leq v_m \quad \forall s \in S, m \in M \tag{6.11}$$

$$v_m \leq \sum_{m_1 \in M} z_{sm_1} + |M|(1 - z_{sm}) \quad \forall s \in S, m \in M \tag{6.12}$$

The boundary conditions are as follows:

$$w_s \text{ binary} \quad \forall s \in S \tag{6.13}$$
$$y_{sf} \text{ binary} \quad \forall s \in S, f \in F \tag{6.14}$$
$$z_{sm} \text{ binary} \quad \forall s \in S, m \in M \tag{6.15}$$
$$x_{s_1 s_2 f} \text{ binary} \quad \forall s_1, s_2 \in S, f \in F | s_1 < s_2 \tag{6.16}$$
$$NS \text{ integer} \tag{6.17}$$
$$P_s \in P \quad \forall s \in S \tag{6.18}$$

Since we seek a maximum capacity, the objective function is

$$\text{Maximize} \quad \sum_{m \in M} \log \frac{1 + \sum_{s \in S} \frac{t_{sm}}{u_{sm}}}{v_m} \tag{6.19}$$

The capacity maximization WLAN design model defined by (1)-(19) is an integer nonlinear mathematical program. The objective function is nonlinear and constraints of type (9) and (10) are nonlinear. If the transmit power (P_s), for each AP is fixed, then the constraints are linear. Design under a fixed budget constraint may be achieved by fixing NS, the number of APs. A channel assignment model is obtained by fixing the w_s variables for all APs. When the transmit power is fixed, the objective function is composed of the sum of terms of the form $g(a,b,c) = [\log(1+a/b)]/c$. It can be shown that the Hessian of $g(a,b,c)$ is neither positive nor negative semi-definite, and therefore the function is of undetermined concavity. Hence, (19) is neither concave nor convex. A plot of $g(.)$ along a straight line from $(2.0e-4, 1.0e-6, 9)$ to $(1.0e-6, 1.0e-6, 24)$ is illustrated in Figure 6.4. Of all the models in the literature, the model of Bahri and Chamberland [2] is most

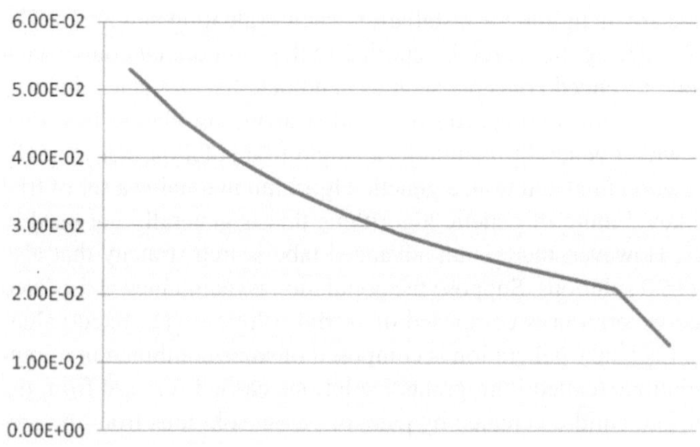

Fig. 6.4 The Function g(a,b,c) Plotted Along a Straight Line

closely related to (1)-(19). The objective of their model is simply minimize NS. In addition, they assume that the data rate between each AP and the associated MD is fixed. Both models are nonlinear integer programs with a large number of integer variables.

6.3.1 Solution Techniques for the Capacity-Oriented Model

There are no general algorithms for integer nonlinear non-convex mathematical programs. However, there are several powerful metaheuristics, e.g. tabu search, simulated annealing, and genetic algorithms, that can be applied to this difficult design problem. Both tabu search and simulated annealing use the concept of neighborhood search. It is very easy to obtain a feasible solution for (1)-(18). Simply choose a value for NS, and then select NS APs for installation. Next, assign a frequency, y_{sf}, and a transmit power, P_s, to each selected AP. Given these values, the other variables will be uniquely determined. One way to obtain a neighboring solution is to change a frequency or to replace one AP with another.

Using the algorithm described in Hillier and Lieberman p. 616 ([7]) as a guide, a simple tabu search algorithm was developed for this model. Given an initial solution represented by $w_s, \forall s \in S$ and $y_{sf}, \forall s \in S, f \in F$, a neighbor is determined by a pair of APs, say s_1 and s_2, where $w_{s_1} = 1$ and $w_{s_2} = 0$. That is, s_1 is selected and s_2 is not selected. The neighbor has s_2 selected and s_1 not selected. Generally, s_2 will be located near s_1. The frequency assigned to s_2 will be the frequency that results in the largest capacity for the design. With only three or four non-overlapping frequencies, this can be quickly evaluated. A finite length tabu queue is maintained to hold the APs recently removed from the solution. Once the best neighbor has been determined, then the departing AP is placed at the tail of the tabu queue. Once the tabu queue is full, entries are removed from the head prior to an insertion at the tail. APs in the tabu queue are ineligible for installation when evaluating neighbors. The best design discovered during the search is reported as the solution. Of course, tabu search can implement advanced concepts such as candidate list strategies, intensification approaches, diversification approaches, and strategic oscillation (see [5]). Unlike tabu search, which generally maintains a single trial solution that is modified in a search for a good final solution, a genetic algorithm maintains a set of trial solutions. This is a key feature of genetic algorithms that is generally not used in other metaheuristics. However, there is an advanced tabu search strategy that also uses a population of trial solutions. Suppose the population is maintained at the fixed size of p. Then each generation is composed of p trial solutions (i.e. fixed values for both w_s and y_{sf}). Each new generation is composed of some combination of randomly generated solutions (called immigrants), solutions carried forward from the previous generation, and children created by pairs of parent solutions from the previous generation. The procedure runs for a fixed number of iterations (generations) and the best trial solution found is provided. In our genetic algorithm, the bottom third of the population is dropped in each generation and the top two thirds become potential parents. A pair of parents is selected at random from the parent pool. Each parent has NS APs selected for installation along with assigned frequencies. Two children are created by randomly selecting approximately half of the APs from one parent and approximately half from the other. If *NS* is an even number, then each child receives 50% of the solution from each parent. The APs retain their frequency assignment from the parents. Of these four solutions, two parents and two children, only the best two with respect to the Shannon capacity estimate are placed into the next generation. The population is completed with immigrants and any single parents left over if $\lceil 2p/3 \rceil$ is odd. This strategy ensures that the best solution in each generation is at least as good as that of the previous generation.

6.3.2 Empirical Analysis

Four procedures have been implemented in software and empirically evaluated using a 200 ft. x 200 ft. site. The tabu search procedure and the genetic algorithms described above were coded along with a simple neighborhood search. We also coded a complete enumeration procedure so that an optimum for a given set of MD locations would be known. For each problem, the value of *NS* is fixed as well as the power for each AP. Hence, the resulting model has linear constraints and a nonlinear objective function. Thirty-six test cases were randomly generated with various values for $|S|$, *NS*, $|M|$, and $|F|$ as illustrated in Table 6.5. The first line in the table gives the results for a problem having 20 potential APs on a grid of the site. The number of APs to be installed is 5, the site contains 30 MDs also on a grid, and the designer is restricted to the use of 2 non-overlapping frequencies. The attenuation values were calculated as illustrated in Appendix A. The optimal capacity is scaled to 100. The neighborhood search was either first or tied for first 26 times. Tabu search came in second with 22 first place positions while the genetic algorithm was in first place only 10 times. For this test run, the neighborhood search seems to do slightly better than our tabu search procedure which is slightly better than the genetic algorithm. All procedures performed fairly well on this test set. In the 108 runs, the worst cases were still within 11% of an optimum.

There are actually two types of WLAN design problems: the static problem and the dynamic problem. In the static problem, MD locations are known and do not change. The static problem would be found in an office environment with cubicles in fixed locations. In the dynamic problem, the MD locations are not fixed, and mobile devices may be located at random coordinates. For example, guests in a hotel lobby may exhibit dynamic as opposed to static behavior. For the dynamic problem, we seek a design that maximizes the expected capacity over a large sample of potential MD locations. The model (1)-(19) is for the static case; however, we have also designed an experiment to illustrate its use for the dynamic case.

Unlike the previous problems reported on in Table 6.5 where all MDs were located on grid points, the MDs for the second experiment were randomly distributed throughout the 200 x 200 site using a Uniform distribution. Thirty test cases for each of the thirty-six problem parameters listed in Table 6.5 have been generated. The solutions provided for the initial problem by the four algorithms were evaluted

Table 6.5 Scaled Capacity Values for Fixed MD Locations

APs	Selected	MDs	Frequencies	Enumeration	GeneticAlg	NeighborSearch	TabuSearch
20	5	30	2	100	99	100	98
20	5	30	3	100	97	100	100
20	5	40	2	100	99	99	98
20	5	40	3	100	99	93	99
20	6	30	2	100	96	100	100
20	6	30	3	100	100	100	99
20	6	40	2	100	97	100	98
20	6	40	3	100	100	100	100
20	7	30	2	100	99	100	99
20	7	30	3	100	100	100	99
20	7	40	2	100	98	100	100
20	7	40	3	100	98	99	100
27	5	30	2	100	96	100	100
27	5	30	3	100	100	100	100
27	5	40	2	100	93	100	99
27	5	40	3	100	100	98	99
27	6	30	2	100	92	99	91
27	6	30	3	100	99	100	99
27	6	40	2	100	93	99	99
27	6	40	3	100	100	100	100
27	7	30	2	100	99	93	100
27	7	30	3	100	99	99	99
27	7	40	2	100	97	98	100
27	7	40	3	100	94	96	99
34	5	30	2	100	95	100	89
34	5	30	3	100	94	97	100
34	5	40	2	100	95	98	100
34	5	40	3	100	95	98	99
34	6	30	2	100	97	99	99
34	6	30	3	100	100	100	98
34	6	40	2	100	94	98	98
34	6	40	3	100	98	100	100
34	7	30	2	100	92	98	89
34	7	30	3	100	93	99	94
34	7	40	2	100	98	99	99
34	7	40	3	100	96	96	100

over the 36 x 30 = 1080 problems, and the average capacity was calculated. The solutions from the complete enumeration procedure were again scaled to 100 and the results are illustrated in Table 6.6. Since all MD locations are random, some solutions are better than 100. For this experiment, the genetic algorithm gave the best performance for about half of the problems. However, none of these algorithms are tailored for the dynamic problem, and this is a fruitful area for further research.

Table 6.6 Scaled Capacity Values for Random MD Locations

APs	Selected	MDs	Frequencies	Enumeration	GeneticAlg	NeighborSearch	TabuSearch
20	5	30	2	100	129	97	108
20	5	30	3	100	129	102	99
20	5	40	2	100	103	96	99
20	5	40	3	100	115	81	101
20	6	30	2	100	102	95	95
20	6	30	3	100	93	98	96
20	6	40	2	100	94	99	98
20	6	40	3	100	75	104	87
20	7	30	2	100	99	105	100
20	7	30	3	100	111	101	117
20	7	40	2	100	102	114	103
20	7	40	3	100	103	108	90
27	5	30	2	100	89	83	87
27	5	30	3	100	129	117	124
27	5	40	2	100	121	97	109
27	5	40	3	100	73	74	84
27	6	30	2	100	116	101	116
27	6	30	3	100	118	96	104
27	6	40	2	100	115	115	110
27	6	40	3	100	99	95	98
27	7	30	2	100	134	102	112
27	7	30	3	100	105	105	103
27	7	40	2	100	94	84	93
27	7	40	3	100	105	121	98
34	5	30	2	100	108	97	94
34	5	30	3	100	103	99	102
34	5	40	2	100	136	94	103
34	5	40	3	100	66	71	85
34	6	30	2	100	106	105	110
34	6	30	3	100	98	102	84
34	6	40	2	100	122	105	108
34	6	40	3	100	115	122	123
34	7	30	2	100	89	83	75
34	7	30	3	100	93	84	78
34	7	40	2	100	106	109	105
34	7	40	3	100	117	116	101

6.4 Summary and Conclusions

One of the most difficult issues involved in the design of WLANs is obtaining good attenuation data. In a limited range ISM spectrum environment, radio signals are affected by the type and location of office furniture, the location and type of construction materials used for walls and floors, and by the interference caused by other radio frequency devices located in their close proximity. This issue is addressed in Chapter 3, Wireless Channel Modeling. Given that this type of data can be obtained, estimated, or measured, there are several decisions that arise in the design of WLANs that may be modeled as integer programming problems. One model involves the assignment of frequencies to a fixed set of APs. Another is to find the minimum number and location of APs that will provide coverage to a given area. The most sophisticated model is our new formulation to determine the location, power, and frequency assignment of a fixed number of APs to maximize the capacity of a new WLAN. Our formulation is for the static problem with known MD locations. Research is needed to develop good solution algorithms for this problem. The dynamic version of this design problem arises when the precise location and number of MDs requesting service varies by day and time of day. In this case, the best design is one that maximizes the expected capacity over a large number of potential MD configurations. Hence, a robust design is preferable for the dynamic case and research is also needed for this version of the problem.

Appendix A: Attenuation Calculation

Several models are available to determine the attenuation between a pair of locations. The path loss is the signal attenuation given in dB, and is defined as the difference between the transmitted power at the source location and the received power at the destination location. Many factors affect path loss between two locations in a particular setting, including distance, the presence and nature of obstacles, interference with other signals or reflected copies of the signal, etc. The presentation that we follow in this Appendix is based on [16] where path loss (*PL*) is defined on p. 108, equation 4.5 by

$$PL = 10\log\frac{P_t}{P_r} \tag{A.1}$$

where P_t is the transmitted power, and P_r is the received power. This equation can be rewritten as

$$P_r = P_t 10^{-\frac{PL}{10}} \tag{A.2}$$

Path loss can also be calculated directly between a pair of locations, given in [16] on p. 161, equation 4.93 by

$$PL = PL_0 + 10n\log\frac{d}{d_0} + X_\sigma \tag{A.3}$$

where PL_0 is the known path loss between locations at a distance of d_0, d is the distance between the source and destination location, n is a constant determined by the medium of transfer, and X_σ is a random variable representing noise. For simplicity, the remainder of this description we assume $X_\sigma = 0$ and that path loss is simply a function of distance. In practice, the other factors mentioned earlier would contribute to path loss, however a detailed presentation is outside of the scope of this presentation.

Rappaport [16] gives $n = 3.27$ in Table 4.7 on p. 165, and we use that number. Rappaport also gives a value for PL_0 for a $d_0 = 1$ meter. This value can be calculated by working backwards Example 4.11 on p. 166, giving a value for PL_0 of 31.5 dB. As we are working in feet, we had to convert this value into the PL at 1 ft.

$$PL(1\text{ft}) = PL(1\text{m}) + 32.7\log\frac{1\text{ft}}{1\text{m}} \tag{A.4}$$

Therefore, the PL at d (measured in feet) from the source is given by

$$\begin{aligned} PL &= PL(1\text{ft}) + 32.7\log\frac{d}{1\text{ft}} \\ &= 14.6 + 32.7\log d \end{aligned} \tag{A.5}$$

Combining (A.2) and (A.5) gives

$$\begin{aligned} P_r &= P_t 10^{-\frac{14.6+32.7\log d}{10}} \\ &= P_t 10^{-1.46} d^{-3.27} \end{aligned} \tag{A.6}$$

References

1. Adickes, M., Billo, R., Norman, B., Banerjee, S., Nnaji, B., Rajgopal, J.: Optimization of indoor wireless communication network layouts. IIE Transactions **34**, 823–836 (2002)
2. Bahri, A., Chamberland, S.: On the wireless local area network design problem with performance guarantees. Computer Networks **48**, 856–866 (2005)
3. Fortune, S.F., Gay, D.M., Kerninghan, B.W., Landron, O., Valenzuela, R.A., Wright, M.H.: WISE design of indoor wireless systems: Practical computation and optimization. IEEE Computational Science and Engineering **2**(1), 58–68 (1995)
4. Geier, J.: 802.11a/b site survey: A testimonial. Retrieved on August 14, 2009 at "http://www.wi-fiplanet.com/columns/article.php/1479831" (2009)
5. Glover, F.: Interfaces in Computer Science and Operations Research: Advances in Metaheuristics, Optimization, and Stochastic Modeling Technologies, chap. 1. Kluwer Academic Publishers, Boston, MA (1997)
6. Hata, M.: Empirical formula for propagation loss in land mobile radio service. IEEE Transactions on Vehicular Technology **29**, 317–325 (1980)
7. Hillier, F., Lieberman, G.: Introduction to Operations Research. 9th edn. McGraw Hill, New York, New York (2010)
8. Hills, A.: Large-scale wireless LAN design. IEEE Communications Magazine **39**(11), 98–104 (2001)
9. Kauffmann, B., Baccelli, F., Chaintreau, A., Mhatre, V., Pagagiannaki, K., Diot, C.: Measurement-based self organization of interfering 802.11 wireless access networks. In: IEEE INFOCOM, pp. 1451–1459 (2007)
10. Leung, K., Kim, B.: Frequency assignment for IEEE 802.11 wireless networks. In: IEEE Vehicular Technology Conference, vol. 3, pp. 1422–1426 (2003)
11. Longo, F.: WLAN design solution. Retrieved on July 15, 2009 at "http://www.virtualacquisitionshowcase.com/docs/2007/OPNET-Brief.pdf" (2009)
12. Mateus, G., Loureiro, A., Rodrigues, R.: Optimal network design for wireless local area network. Annals of Operations Research **106**, 331–345 (2001)
13. Mishra, A., Banerjee, S., Arbaugh, W.: Weighted coloring based channel assignment for WLANs. Mobile Computing and Communications Reviews **9**(3), 19–31 (2005)

14. Panjawni, M.A., Abbott, A.L., Rappaport, T.S.: Interactive computation of coverage regions for wireless indoor environments. IEEE Journal on Selected Areas in Communications **14**(3), 420–430 (1996)
15. Proxim Wireless: ORiNOCO AP-700 access point. Retrieved on September 15, 2009 at "http://www.proxim.com/products/ap_700/" (2009)
16. Rappaport, T.: Wireless Communications: Principles and Practices. 2nd edn. Pearson Education (Singapore) Pte. Ltd., Indian Branch, 482 F.I.E. Patparganj, Dehli 110 092, India (2002)
17. Rodrigues, R., Mateus, G., Loureiro, A.: On the design and capacity planning of a wireless local area network. In: Network Operations and Management Symposium, pp. 335–348 (2000)
18. Sherali, H., Pendyala, C., Rappaport, T.: Optimal location of transmitters for micro-cellular radio communication system design. IEEE Journal on Selected Areas in Communications **14**(4), 662–673 (1996)
19. Skidmore, R., Verstak, A., Ramakrishnan, N., Rappaport, T., Watson, L., He, J., Varadarajan, S., Shaffer, C., Chen, J., Bae, K., Jiang, J., Tranter, W.: Towards integrated PSEs for wireless communications: Experiences with the S^4W and SitePlanner projects. Mobile Computing and Communications Review **8**(2), 20–34 (2004)
20. Vanhatupa, T., Hännikäinen, M., Hämäläinen, T.: Evaluation of throughput estimation models and algorithms for WLAN frequency planning. In: Computer Networks, vol. 51, pp. 3110–3124 (2007)

Chapter 7
Spectrum Auctions

Karla Hoffman

Abstract Most developed countries allocate radio spectrum by auction. The Simultaneous Ascending Auction (SAA) has proven to work well for this application. Recently, new designs that allow package bidding have been proposed. These designs have only been tried in the past few years. We first provide some historical background regarding the allocation of spectrum, describe the use of the SAA design and its modifications over the past 15 years, and then highlight the new advances in combinatorial auction designs and their use for the allocation of spectrum.

7.1 Introduction

Many countries around the world use auctions for the allocation of radio spectrum. In order to understand the evolution of auctioning radio spectrum, one must first begin by explaining what spectrum is and how spectrum had been allocated prior to its being auctioned. Radio spectrum is the part of the electromagnetic spectrum used to transmit voice, video and data. It uses frequencies from 3 kHz to 300 GHz. Thus, it includes spectrum use for radio and TV broadcasts, paging, wireless and satellite communications, and many other important telecommunication applications. In general, most countries allocate the rights to use these airwaves as a license to provide a specific service (e.g. radio broadcast), using a specific portion of the spectrum band (e.g. the use of the 15MHz of spectrum between 950MHz and 965MHz in a specified region (e.g. a state, province or nationwide). There can be a number of restrictions on its use including the strength of the broadcast signal, the assurance that it does not create interference with adjoining spectrum regions, and other public service goals such as requiring that the provider be able to reach a certain percent-

Karla Hoffman
George Mason University, Systems Engineering and Operations Research Department, Fairfax, Virginia 22030, USA, e-mail: khoffman@gmu.edu

J. Kennington et al. (eds.), *Wireless Network Design: Optimization Models and Solution Procedures*, International Series in Operations Research and Management Science 158,
DOI 10.1007/978-1-4419-6111-2_7,

age of the population or a certain physical area. Spectrum is considered a national resource and is therefore allocated with concepts of social welfare in mind.

This chapter begins by first outlining some auction terms that will be used throughout the chapter. It then describes some alternative methods for spectrum allocation other than auctions and outlines the reasons for the conversion from administrative procedures to auctions. We describe in detail the most commonly used auction design - the simultaneous ascending auction (SAA) design and show how specific rules have been altered to improve its performance. Next we detail some of the newer designs that allow bidders to package licenses together and indicate a value for the "all or nothing" package. These designs remove the risk that bidders might win only part of what they need and pay too much for the resulting sub-package. We detail where these designs have been used and highlight their successes and shortcomings.

7.2 Some Auction Terms and Mechanisms

We begin with a very short discussion of auction design alternatives. Here we will present the terms for one-sided auctions since there have been no standardized spectrum exchanges (i.e. formal mechanisms for the *trading* of spectrum among license holders). We will also restrict our attention to the case where there is a single seller and multiple buyers, since that is the case where regulators are seeking to allocate licenses for spectrum for use by telecommunications companies. We will also assume that there are multiple items being sold and that, for at least some of the buyers, a collection of items are needed in order to have a viable business plan — they need to either aggregate licenses in order to either acquire sufficient bandwidth to offer specific services or acquire a sufficient geographic footprint. Acquiring spectrum licenses in multiple regions avoids roaming charges and allows economies of scale in terms of the build-out and marketing costs. Increases in bandwidth can increase the types of offerings carriers can provide. Thus, this chapter will highlight both multiple-item and *package-bidding auction* mechanisms that allow buyers to procure *collections* of the items that are being sold. These designs should be sufficiently general to allow bidders to express a value on a package where the collection of items may have a value greater than the individual items (i.e. the goods are *complements*), as well as on a package where a buyer can express a quantity discount for buying more of the good (i.e. the goods are substitutes).

Why does a government sell or lease valuable resources via an auction mechanism? Often the government moves to a market-clearing framework after many failed attempts at administrative negotiation with the entities that desire the use of the resources. The results of these negotiations are fraught with administrative costs and the likelihood that the items will be awarded to those that have the greatest political access rather than those that value the items the most. In many such instances, the government has no mechanism to assess the true value of the assets being leased and the public or its representatives perceive that the government is not receiving full value. Auctions allow greater participation, allow the market to set the price, and are generally perceived as fair. Often, spectrum auctions were only considered when other attempts at allocation (administrative evaluation and lotteries) failed. We will say more about this in the section on the history of spectrum auctions.

In an auction with a single seller and multiple buyers, the potential buyers wish to determine the minimum price that they must pay given that they must compete with others for the ownership of this good. From the seller's perspective, submitting goods to an auction may increase the number of buyers, thereby increasing the potential for competitive bidding and higher selling prices. Thus, an auction is a mechanism that allows the market to set the price. Since all buyers must adhere to the same set of rules, auctions are perceived as both less haphazard and fairer than if the price were set by bilateral negotiations. Most importantly, if the rules are well designed, the result will have the goods given to the entity that values them the most. When the design also encourages significant participation, the revenue provided to the seller is likely to be reasonably high.

We begin by first classifying auctions into a number of major types. One of the simplest mechanisms is the *first-price sealed-bid auction*. In this design, all bidders submit their bids by a specified date. All bids are examined and compared and the highest bid is awarded the object (or collection of objects) and pays the amount that was bid. A problem with the first-price sealed-bid auction is that the winning bidder may suffer a "*winner's curse*", i.e. the bidder may pay more than was necessary to win since the second highest bid price was far less than the winning bid amount. For this reason, first-price sealed-bid auctions encourage bidders to shave some amount from the highest amount that they are willing to pay in order not to pay more than is necessary.

An alternative is the *second-price sealed bid auction* whereby the bidder that has submitted the highest bid is awarded the object, but he pays only slightly more (or the same amount) as that bid by the second-highest bidder. In second price auctions with statistically independent private valuations[1], each bidder has a dominant strategy to bid exactly his valuation. The second price auction is often called a *Vickrey auction* [49].

Often the value of the good is not completely known. Instead, there is a *common component* to the bid value — that is, the value of the item is not independent of the others bidders. Instead, the value that others place on the asset is relevant. In such situations, each agent has partial information about the value. Many high-stakes auctions, such as antique, art and horse auctions, as well as government auctions for oil exploration, land use, and spectrum fall into this class.

When the bidders need to obtain information about the common component of the item's value, then ascending bid auctions, often known as *English* (or *open outcry*) *auctions* are used. Here, the auctioneer announces a relatively low starting price and bidders respond. A standing bid wins the item unless another higher bid is submitted. All bidders observe the previous bids and decide whether to submit a higher new bid. When someone responds with a higher bid, the auctioneer increases the

[1] In a private value auction, each bidder knows his value of the good and this information is both private and independent of the value of other bidders.

price. The auction ends when no one is willing to top the standing high bid. An alternative auction design is the *Dutch auction* whereby the price is set very high and the price is gradually lowered by a clock until it is stopped by some bidder that is willing to buy some or all of the (identical) items up for sale at the current price. The quantity bought by this bidder is deducted from the total amount available and the auction resumes with the auctioneer continuing to reduce the price until another bidder indicates a wiliness to buy some quantity of the item at the announced price. The name derives from the fact that tulips were sold by the Dutch via this mechanism. This auction mechanism is still employed for many agricultural products where many identical (fungible) items are being sold and the bidders determine both the price and the lot size.

Finally, for a large group of identical objects, a variation of the ascending auction is one in which the auction price ascends until there are no new bids. At the end of the auction, all buyers pay the lowest amount bid by any of the winning bidders. Such auctions are called *all-pay one-price auctions*, or *uniform price* auctions.

Another important auction format issue is whether the auctioneer sells each of the items sequentially or auctions all items simultaneously. With sequential auctions, bidders must guess the resulting prices of future auction items when determining how to respond to prices in the current auction. If a bidder needs multiple items and the items are auctioned sequentially, then the bidder is faced with the issue of risking winning only part of what is needed. One mechanism that has been used to overcome this problem is to first auction related items individually (e.g. auction each chair and the table of a dining room set separately) and then ask if there is someone who is willing to pay more for the set than the sum of the individual items. But, this approach assumes that there is a unique package that may have a greater value than the individual components. Alternatively, all items can be auctioned *simultaneously*. We present one such auction design: *the simultaneous ascending auction* (SAA) in detail because that design has been the most popular in auctioning spectrum.

We now specify a few additional rules of an auction mechanism that can have significant impact on its success. Firstly, an ascending auction can have discrete rounds or bidders may be allowed to provide bids continuously. With the continuous approach, there must be a stopping rule that determines when the auction ends. Auctions, such as eBay are continuous auctions and specify a fixed time when the auction will end. Fixed stopping times encourage bidders to stay silent until very late in the auction and then present last minute bids in the hopes that other bidders do not have time to respond. Such bidding activity is called *sniping*. Conversely, if the auction has discrete rounds, then one may need *activity rules* that force bidders to participate throughout the auction. These activity rules can help price discovery. In an ascending multi-round auction design, bidders see the price of all items during the auction and adjust their strategies and business plans as the auction progresses.

Other questions in setting up an auction design are how to set the opening bid price (in an ascending auction) and, in either sealed-bid or open-outcry auctions, whether or not to set a reserve price. If the auction does have a reserve price - for fear that the item might otherwise be sold at too low a price - the auctioneer must also decide whether to have that reserve price revealed before the auction begins or have it disclosed only when it need be enforced.

In many high-stakes auctions, prior to the beginning of the auction, the auctioneer has a qualification stage. In this stage, bidders must prove that they are qualified both monetarily and have the skills to fulfill whatever obligations go along with the leasing/owning of the property.

As stated earlier, throughout this chapter, we will assume that we are seeking an auction format that is appropriate for problems where bidders need to procure a collection of items in order to have a viable business plan. When describing the history of spectrum auctions, we will highlight the success or failure of specific designs at satisfying some or all of the possible goals of a telecommunications regulator. Such goals may include:

1. The bidders understand what they are buying, i.e. the property rights are well defined. A fully-described property right will include the ways in which the item can be used, the time-period of use, the termination process for misuse or lack of payment, the right of the owner/lessee to sub-let/sell the right, the obligations of the lease holder, etc.
2. Bidders are able to, through their bids, announce the entire collection of objects that they need.
3. The auction results in reasonable revenue to the seller.
4. The auction results in an efficient outcome i.e. all items are collectively allocated to the bidders that value these items the most.
5. The auction is perceived as fair.
6. The auction ends in a reasonable amount of time.
7. The auction has limited transaction costs, i.e. the rules are not so difficult or the bidding so complicated that a straightforward bidder finds it difficult to participate.
8. The auction design discourages gaming and signaling.
9. The auction allows price discovery, since such auctions often bring more participation.
10. The auction is computationally feasible and scalable.
11. The auction satisfies, through its rules, any social welfare outcomes that are desired by the public entity selling or leasing the item.

It is not possible to have all such attributes simultaneously. For certain applications, some of these goals will be more important than others. For more on what really matters in auction design, we recommend the paper by Klemperer [32] who maintains:

> What really matters in auction design are the same issues that any industry regulator would recognize as key concerns: discouraging collusive, entry-deterring and predatory behavior. In short, good auction design is mostly good elementary economics.

In addition, the auction mechanism should consider any application-specific issues that might arise. For example, an auction design might consider how market power impacts the outcome, whether there will be sufficient participation, and whether the outcome will satisfy specific social welfare goals. In certain situations, there may need to be a transition period that allows the market to adjust to a change in the way rights are allocated. One may need to consider the associated rights that a bidder would need to be able to use the right being sold or leased in the auction. For expensive goods, buyers may need to pay for the lease over a period of time. The money obtained from the auction may need to be designated for a specific use in order for the government to obtain the approval of all constituents. The auction design may also need to satisfy other social goals specific to the application (e.g. increasing competition, incentivizing innovation, assuring access to most or all of the citizenry, precluding interference with other licenses and technologies, assuring that international treaties are satisfied).

The allocation of spectrum has many of the social-welfare issues described above and the design of the auction must keep these goals in mind. In the next two sections, we will first detail the New Zealand auction — the very first auction to allocate spectrum. Then, we describe the evolution of spectrum allocation in the U.S. from administrative procedures to auctions and then describe the simultaneous ascending auction in detail. This auction design is now used throughout the world and is based on concepts from mathematical game theory. In this chapter we present a summary of this evolution, for more reading on this subject, we recommend: McMillan [39] on the history of markets, Krishna [35] on auction theory, Steiglitz [48] on the success and pitfalls of eBay auctions, Klemperer [33] on auction theory and practice, and Milgrom [41] for a more detailed history and theoretical underpinnings for the simultaneous ascending auction design chosen by the US Federal Communications Commission (FCC).

7.3 The First Spectrum Auction: The New Zealand Experiment

The first auction of radio spectrum took place in 1990 in New Zealand. An economic consulting firm, NERA, recommended using a second-price sealed-bid auction design. This auction was for multiple nationwide television licenses, each of 8 MHz. Each license was considered essentially identical and a separate sealed-bid auction was held for each license where all bids for all licenses were submitted at the same time. Thus, participants in the auction needed to submit a separate bid simultaneously for each license. Bidders who only wanted one license were uncertain as to how to bid. Do they bid in only one auction, or in all auctions? If only one auction, which one? If many entities bid in some auctions but not all, then in the auction where there is significant competition the license is likely to cost more than some other (identical) license that had less competition. The results of the New Zealand auction illustrate precisely the problem that bidders encountered: one bidder bid on all six licenses but won only one, while another bid for just one license, won that license, and paid less than the bidder who bid on all six licenses. The license prices ranged from NZ$410,000 to NZ$100,000. Since the New Zealand government published the bid amounts of each bid, there was a public outcry when it was disclosed that a bidder was willing to pay NZ$2,371,000 for a license but only paid NZ$401,000 because of the second-price design.

McMillan [38] chronicles two other New Zealand auctions where bidders paid far less than their bid price. In one auction, a bidder placed a bid for NZ$100,000 but paid only NZ$6 (the second-highest bid) and in another, a bidder placed a bid of NZ$7Million and paid only NZ$5,000. In 1991 and continuing through 1994, the New Zealand government reverted to first-price sealed bid auctions preferring winners to be exposed to the winner's curse rather than to have the government receive virtually nothing for licenses.

As Milgrom [41] points out, there were a number of auction design alternatives that the New Zealand government could have chosen instead of the design chosen. For example, within the sealed-bid approach, each bidder could have been required to submit a demand schedule indicating, for each alternative amount of bandwidth, the amount it was willing to pay. In this design, all licenses would have been considered simultaneously within a single auction. Klemperer argues that an alternative design, namely the Dutch auction could have been used, whereby the auction starts at a high price and the price is reduced until a bidder indicates that it is willing to buy licenses at that price. That bidder then gets to announce the quantity it wants and the auction continues with the remaining licenses. These suggested alternatives highlight that there are many auction choices and one must consider the pros and cons of each. In 1994, the U.S. Federal Communication Commission (FCC) chose a truly new design for its first spectrum auction.

7.4 Evolution of the US Simultaneous Ascending Auction (SAA) Design

Prior to the allocation of spectrum by auction, the FCC allocated spectrum through comparative hearings whereby regulators would review and compare applications provided by a variety of private enterprises in order to determine which of the applicants would put the spectrum to its best public use. This administrative process was extraordinarily lengthy. The process required that the petitioner hire a large staff of lawyers, lobbyists, engineers and accountants to justify its claims that it is the most qualified to provide the services. The regulators, as well, needed large staffs to be able to first create the comparative criteria for applicants to use when developing their applications and then to evaluate the claims made. There were often lengthy appeals processes and court battles over these licenses. This administrative process is often referred to, in the economic literature, as a "beauty contest" because of its lack of transparency and because the decisions were based on the often not fully understood preferences of a small group of people.

By 1982, the FCC and Congress wanted to encourage the use of a new technology: cellular phones. Congress worried that the administrative process would delay the introduction of these technologies and therefore passed legislation that required that cellular licenses be allocated by lottery. Specifically, all applicants who could provide documentation that they had the funding to build out a cellular business would be placed into a pool. The government would choose among this applicant pool randomly.

This lottery process allocated the licenses very quickly - the essential goal of the legislation. However, the potential for large gains from winning a license resulted in over 400,000 applications for cellular licenses and many of the winners were speculators with no experience in telecommunications. A secondary market ensued for the active re-selling of licenses. But, for those who wanted to create a nation-wide cellular license, it was difficult to accumulate the many small regional licenses through bilateral negotiations with multiple sellers. In addition, a flurry of news reports highlighted specific people who made substantial profits by applying for a license but with no experience or interest in being part of the telecommunications industry[2].

To correct this revenue loss to the US government and to assure that the licenses were awarded quickly to those that would build-out the new technologies, Congress, in 1993, authorized the licenses of spectrum by auction. It specified that the first auction must take place by July 1994. The FCC was instructed to create procedures that:

A. Encourage the development and rapid deployment of new technologies, products and services for the benefit of the public, including those residing in rural areas, without administrative or judicial delays;
B. Promote economic opportunity and competition and ensures that new and innovative technologies are readily accessible to the American people by avoiding excessive concentration of licenses and by disseminating licenses among a wide and variety of applicants, including small businesses, rural telephone companies and businesses owned by members of minority groups and women;
C. Recover for the public a portion of the value of the public spectrum resource made available for commercial use and avoidance of unjust enrichment;
D. Promote efficient and intensive use of the electromagnetic spectrum; and
E. Ensure that, in the scheduling of any competitive bidding process, an adequate period of time is allowed, before issuance of bidding rules, to permit notice and comment on proposed auction procedures, and, after issuance of bidding rules, to ensure that interested parties have a sufficient time to develop business plans, assess market conditions and evaluate the availability of equipment for the relevant services[3].

With these goals in mind, the FCC began obtaining advice from game-theoretic economists about auction design. John McMillan was hired as a consultant to the FCC while a number of economists provided written filings to the FCC. Evan Kwerel, an economist within the FCC, working with the FCC Chairman Reed Hundt,

[2] One such group consisted of dentists in Massachusetts who won a license and immediately sold it for $41 million. For more on the history of the FCC auctions and their predecessor administrative and lottery approaches, see Milgrom [41] and McMillan [39].

[3] Section 309(j) of the Communications Act, 47 U.S.C.

focused on getting it right the first time, since as he stated in the preface to Milgrom [41]:

> It was my view that whatever method was used in the first FCC auction, if it appeared successful, would become the default method for all future auctions.

The FCC hoped to encourage participation by new entrants while simultaneously allowing existing cellular licensees to acquire holdings that were complementary to their existing holdings. Since different business plans required very different aggregations, it was not clear how to optimally subdivide the country and the sequential auctioning of different lots would hamper a bidder's ability to aggregate licenses of complementary regions.

The FCC chose a novel auction design by choosing to auction all licenses simultaneously. Bidders could bid on any item until the bidding stopped on all items. Within each round, the auctioneer examines all bids on a given item, and announces that a bidder is "provisionally winning" if that bidder has the highest bid on that item in that round. A "provisionally winning bidder" is responsible for paying his bid price unless outbid by another bidder in a later bidding round. When there are two bidders with the same high bid, the provisionally winning bidder is chosen randomly among those with the high bid.

Thus, bidders examined the current round prices of all of the licenses and choose a set of licenses that fit their business plan. Prices of substitutable licenses would remain relatively similar, and a bidder had the flexibility to alter their business plans as licenses rose in value. Although this auction design did not allow bidders to package licenses into "all or nothing" bids, bidders had the flexibility to move among different band plans and collect a sufficient group of licenses to develop usable business plans.

Another novel feature of this design was the concept of an *activity rule* whereby each bidder is required to participate throughout the auction. The activity rule was based on a "bidding unit" concept whereby a bidding unit relates to both the population and the spectral size (as measured in megahertz) of the band (often referred to as "MHz pops"). Thus, licenses were ranked by size, a license covering New York, for example, had more bidding units associated with it than a license covering a rural area in Montana. A bidder had to secure bidding units prior to the start of the auction by way of a refundable upfront payment. Based on this upfront payment, bidders were "eligible" to bid on no more than a given number of bidding units. The bidder was then forced to consider all of its bids and assure that the number of bidding units associated with the bids placed in a round met an activity requirement (normally stated as a required percentage of a bidder's eligibility). If the bid did not meet this requirement, then the bidder was threatened with a reduction in its eligibility, thereby limiting the number of bidding units that it could bid on in the future. By forcing bidders to bid each round or lose eligibility, the relative value of items (i.e. price information about the items for lease) was provided to bidders each round. There were also penalties imposed for bidders that defaulted on a winning bid. Thus, as the auction progresses and prices rise, the activity rule forces bidders to place binding pledges at the current prices or leave the auction. This rule precludes bid sniping (staying silent until very late in the auction and then presenting last minute bids in the hopes that other bidders do not have time to respond).

Until recently, the FCC auctions were completely transparent. That is, bidders and other interested parties could see the identities and bids of all bidders in each round. Thus, all participants could verify that the rules were followed. If problems existed, they could be found and resolved. It was argued that transparency was essential, both because of the newness of the auction design and because bidders needed to establish values based on a number of technological issues: they needed to be sure that their customers could roam onto other networks that used similar technologies and that they needed to understand the interference issues that might arise based on who owned the licenses adjacent to the considered license. However, such transparency gave bidders valuable price information at no cost and provided mechanisms for colluding and bid signaling. Recently, the auction rules were changed so that only the provisionally winning bid amounts are disclosed. Bidder identities are no longer revealed until the end of the auction. For more on the gaming that is possible when bidder identities are provided see Cramton and Schwartz [16].

The legislation required that the FCC encourage participation by small businesses, minorities and new entrants. Instead of creating set-asides for such groups, the auction design chose to provide such groups with bidding credits. The idea was that small bidders might find it difficult to acquire the capital to compete with major telecommunications companies. To overcome this handicap, small companies and minority and women-owned businesses (referred to as "designated entities") would be given a "credit". If a designated bidder won a license at a given price they would only pay some fraction (e.g. 90%) of the price bid. In this way, both small and large bidders would be bidding within the same auction, thereby encouraging participation and competition. In the FCC auctions, the bidding credits have ranged as high as 40% and as low as 10%.

The FCC was also concerned with limiting the amount that a single participant could acquire in the auction. There was fear that "deep pocket" bidders could become a monopoly (or duopoly) player in important markets. To assure that this was not possible, a spectrum cap was placed in each geographic area. Spectrum caps assured that incumbents were significantly inhibited from bidding in regions where they already had large holdings and insured that, in some major markets, the auctions would end with new entrants.

The Simultaneous Ascending Auction (SAA) design has proven extraordinarily successful and been adopted worldwide, resulting in revenues in excess of $200 billion dollars (Cramton [11]). The rules were originally proposed by Paul Milgrom, Robert Wilson and Preston McAfee and have been refined over time when problems arose with the original design. We highlight some of the problems and the refinements to the rules.

At first, the FCC allowed bidders to bid any whole dollar amount above a minimum acceptable bid amount on any license during a given round. This bid flexibility was intended to allow bidders to increase the speed of the auction by pushing prices up quickly when it was known that the opening price of a given license was considerably lower than its eventual selling price. However, this flexibility also allowed bidders to signal to other bidders and to punish bidders if they did not agree to collude. Specifically, bidders were able to engage in signaling by attaching identifiers in the trailing digits of their bids (called "code bids") to signal another bidder that it would be punished if it continued to bid on a certain license. A bidder could place a very large bid price on a license that it did not want, but knew another bidder wanted, and then subsequently withdraw the bid, thereby signaling that it was willing to stay off the rival's properties if the other bidder reciprocated by staying off the signaler's most desired licenses. In order to stop this behavior, the FCC implemented "click-box bidding" whereby the bidder could only click on the price set by the FCC. In order to move the auction along, the FCC allowed bidders to place a limited number of "jump bids", i.e. the bidders could choose one of nine possible increment bids — the minimum increment or any of eight other increments, each 10% above the minimum increment. In more recent auctions, the FCC has limited jump bidding to fewer choices, still concerned that the jump bids could be used as signals.

The original design allowed withdrawal of bids in order to allow a bidder who only won some of what they wanted to withdraw from this partial aggregation. However, as already discussed, bidders could and did bid on a rival's license in one round and withdraw it in the next as if to say, "I will stop bidding on your license if you stop bidding on mine." To preclude this signaling, the FCC has severely limited the number of withdrawals that are allowed throughout the auction. The changes in these rules (click-box bidding, and limited withdrawals) have limited the amount of signaling, but they did not completely eliminate the problem. In a further attempt to deter such signaling, the FCC has recently removed the reporting of bidder identities each round making it nearly impossible to single out and punish any given bidder.

Other enhancements that have taken place since the inception of the SAA auction design include: (a) eliminating installment payments which encouraged speculative bidding, (b) using license-specific bid increments whereby items with significant bidding activity have larger bid increments thereby speeding the auction along, (c) increasing the number of rounds per day, and (d) setting the minimum opening bid amounts higher. For more on the history and issues with the SAA, see Milgrom [41], Cramton [12], Porter and Smith [46], and Wilkie [50].

One major shortcoming of the above design is that while it does allow bidders to bid on multiple items at the same time, it does not allow a bidder to describe the synergy value of a package of licenses. When the winner of each item is determined independently, bidding for a synergistic combination is risky. Without the ability for a bidder to state that his value for a collection of items is greater than the value of the sum of the individual items, an *exposure problem* can exist. The bidder may fail to acquire key pieces of the desired combination, but pay prices based on the synergistic value. Alternatively, the bidder may be forced to bid beyond his valuation on some pieces in order to secure the synergies and reduce its loss from being stuck with only the less valuable pieces. Bidding on individual items exposes bidders seeking synergistic combinations to aggregation risk. To mitigate this exposure problem, the FCC allowed bidders the option of withdrawing bids a limited number of times, so they could back out of failed aggregations. Another feature that mitigated the exposure problem was that there were multiple bands of spectrum that could be substituted (bands of spectrum close to one another in the same region). An active secondary market allowed trading to correct for incomplete package formations.

Recently, the FCC has considered altering the overall auction design. In 2000, it proposed a simultaneous multi-round ascending *package-bidding* design for the 700 MHz band of spectrum (the spectrum freed due to the conversion from analog to digital television). The package-bidding design proposed for that auction was not scalable but was easily computable for an auction of 12 licenses. The spectrum for this auction was based on the conversion of analog TV channels to digital, and the auction was postponed because that conversion took longer than originally anticipated. By 2008, when the conversion was imminent, considerably more spectrum had become available and the FCC decided to include other spectrum bands within the auction. In deciding the bandplan, the FCC chose configurations for the different bands that made the movement from one band to another difficult.

Specifically, the FCC decided to divide these spectrum bands into various regional configurations: one band was a nationwide license, one band had twelve regional licenses, two bands had 176 licenses and one had 734 licenses, yielding a total of 1099 licenses[4]. For this large number of disparate license configurations, the package-bidding auction postulated in 2000 was not feasible. The number of alternative packages that a bidder could create was exponentially huge and unworkable. Instead, the FCC chose a modified package-bidding design where only the licenses on one band with twelve regions (the "C" block) could be packaged and, even on this band, the FCC pre-configured the allowable packages. In each of the other bands, no packaging was allowed.

Other rules changes for this auction included: (a) having only two possible jump increments, (b) making all bidding anonymous (except for the announcement of the provisionally winning bids), (c) forcing "open access" rules for the C band, and (d) adding multiple rules for the nationwide license that related to public-safety use.

The results of this auction were mixed. The revenue from this auction exceeded $19 billion. However, the nationwide license did not reach the reserve price and remained unsold, probably because of the onerous public-safety rules surrounding that license. AT&T and Verizon won most of the licenses being auctioned, thereby strengthening their holdings in the cellular business — the cap on spectrum holdings had been eliminated prior to this auction. The varying sizes of the blocks auctioned made it particularly difficult to move across bands and thereby hindered the substitutability property that had been prevalent in prior auctions. Because the blocks were partitioned differently with different technology rules and specifications, it was difficult for bidders to substitute seemingly identical spectrum as prices rose. The result was that bands with similar characteristics, in terms of MHz and regions, sold for significantly different amounts. Cramton [15] argues:

> In most instances, spectrum lots covering the same region in adjacent frequencies are nearly perfect substitutes. The bidder primarily cares about the quantity of spectrum in MHz it has

[4] This auction was relabeled Auction 73 whereas the original 12-license combinatorial auction was labeled Auction 31.

in the region, rather than the exact frequency location. Moreover to minimize interference problems, bidders prefer contiguous spectrum within a region. In this setting, it makes sense in the initial stage to auction generic spectrum.

Pressure from varying groups argued against having all bands have the same regional configuration and the end result was that bidders found it difficult to move from band to band during the auction.

In addition, unlike many past FCC auctions, this auction saw few new entrants winning large holdings of spectrum. Both the non-substitutability of the bands and the FCC removal of the spectrum caps contributed to the result that the two largest cellular carriers acquired significant new spectrum holdings. By comparison, the Canadian government chose, in its 2008 SAA-design 3rd Generation spectrum auction, to set aside specific bands for designated entities and to also allow these designated entities to bid on blocks of spectrum that were not set aside. The Canadian auction ended with prices across all blocks being valued similarly.

We point out the differences in results between the Canadian and FCC auctions to suggest that it is not only the auction mechanism itself, but also the band plan and the rules associated with participation that can impact the success of the auction. In making such decisions, the regulator first determines how much spectrum will be available for auction, determines how substitutable the spectrum is (e.g. are there any encumbrances on some of the spectrum? Are there any problems with interference for the spectrum closest to already allocated spectrum?). Next, the regulator must determine what the allowable uses will be and then, one must decide the appropriate regional specification and bandwidth size for each license. Finally, the regulator must determine the auction mechanism and the specific rules within that auction design.

As described earlier in this chapter, the New Zealand government first tried a second-price sealed-bid auction (without reserve prices), then a first-price sealed-bid auction. Not being happy with the results of either of these two approaches, the government went to an SAA auction in 1994. Similarly in 1993, the Australian government had some adverse experience in the auctioning of licenses for satellite-television services. A first-price sealed bid auction was used to auction two satellite-television service licenses. The winners of the two licenses had bid very high prices, won the licenses and then defaulted. When the licenses were re-allocated to the next highest bidders, they too defaulted and a cascade of defaults ensued until a bidder was found that could pay for his/her bid. The final prices actually paid were on the order of 100 million dollars less than the original bids. The Australian government decided to use a simultaneous ascending-price multiple-round auction mechanism (with bid default penalty mechanisms) in 1997, and continues to use this auction format for spectrum license assignment. Due to these and other similar experiences with sealed-bid auctions, most countries today use some form of the simultaneous ascending package bidding design.

But SAA designs are not foolproof. In 2000, the German government held an auction for third-generation (3G) spectrum use using a SAA auction design. Although the auction brought $45 billion to the German government, these prices (set during the dot-com bubble) were perceived to be too high and the entities that were awarded the licenses found it difficult to obtain financing for the build-out of the networks. A similar result occurred in the UK auction for 3G use held during the same period. Thus, what looked like successes at the end of the auction may not have satisfied the government's long-term goal of quick build-out of new cellular technologies. One problem with these auctions was that they were the first auctions for 3G spectrum in Europe. If major telecommunications corporations did not get licenses in these countries, then their ability to have a large footprint throughout Europe would be impaired. Thus, the timing of auctions can have a significant effect on outcomes.

There are those that argue that one problem with ascending auctions is that bidders set their valuations within the auction based on the price signals of others. Public outcry auctions encourage such unfettered exuberance, since the auctioneer represents the seller who wishes to maximize revenue. However, when the government is selling assets for social welfare goals, overbidding can be detrimental to these goals. For more on the results of the European Auctions and their differing results, see Cramton [11] and Klemperer [33].

To conclude: a regulator must consider many issues when designing a spectrum auction. First, one must be careful to consider the consequences of the band plan. Slight changes in auction rules can lead to unintended consequences. Over the 15 years that the FCC has been involved in auctions, the rules have changed to accommodate new circumstances or to reduce any observed bidder actions that were believed to be harmful to the outcomes.

Today, spectrum auctions take place throughout the world. Sometimes the decision to move from administrative processes to auctions occurs when a scandal associated with the administrative process forces the country to create a more transparent approach to the allocation; other reasons may include that the regulator wants to assure that the public receives a portion of the value of the resource made available for commercial use; that the new technologies reach the public quickly, or that there is sufficient competition within the market.

Recently, newer auction designs are being tested both in laboratories and by regulators. We detail the two designs that have been proposed for spectrum use.

7.5 Combinatorial Auction Designs and their Use for Spectrum Allocation

One of the strongest arguments against the simultaneous multi-round auction is that the design has an exposure problem whereby a bidder only wins part of what he needs. In a simultaneous ascending-round auction, the bidder may be stuck with items that are valued at less than the price won because the other complementary goods were not won. To overcome the exposure problem, many have proposed auction designs that allow bidders to construct packages of subsets of the items. Such auctions are called combinatorial auctions. In this chapter, we will only discuss auction designs that have been used to allocate spectrum. For the interested reader, we suggest the following papers that describe the use of combinatorial auctions in other applications: bus services for the London Transportation System (Cantillon and Pesendorpher [9], supplying meals to school children in Chile (Epstein et al. [23]), supply-chain procurement auctions (for a Mars Candy Company Auction see Hohner et al. [30]; for Motorola's Procurement Process see Metty et al. [40]; for Sears' Combinatorial Auction, see Ledyard et al. [37]). For a survey of combinatorial auctions used in the power industry, see Kaglannas et al. [31]. For a detailed discussion of alternative package-bidding auction designs and their applications see the text *Combinatorial Auctions* edited by Cramton et al. [17].

7.5.1 Sealed-bid Combinatorial Designs

A relatively simple idea is to allow package bids to be submitted within a sealed-bid framework. In this setting, one must determine the set of bidders that maximizes the revenue to the seller. Since package bids often overlap, determining the collection of package bids that do not overlap and maximize revenue is a combinatorial optimization problem known as the *winner determination problem.* Thus, each bidder provides a collection of bids to the auctioneer and the auctioneer selects among these bids the collection of bids that maximizes revenue. The simplest of the formulations for the winner determination problem can be specified mathematically as:

$$[\text{WDP}] \quad \max \sum_{b=1}^{\#Bids} BidAmount_b x_b$$

subject to

$$Ax = 1 \tag{7.1}$$

$$x \in \{0,1\} \tag{7.2}$$

where x_b is a zero-one variable indicating if bid b is winning or not; and A is an m x n matrix with m rows, one for each item being auctioned. Column A_j describes package bid j, where $A_{ij} = 1$ if package j contains item i, and zero otherwise. Constraint set (7.1) specifies that each item can be assigned exactly once. Each of the n columns represents a bid where there is a one in a given row if the item is included in the bid and zero otherwise. Constraint set (7.1) are equations rather than $\leq$ constraints since the regulator provides m bids (one for each item) at a price slightly below the minimum opening bid price. In this way, the regulator will keep the item rather than allow it to be won by a bidder at less than the opening bid price.

One issue in a combinatorial auction design is the bidding language used. A simple idea is that the bidder submits a collection of package bids and the bidder can win any combination of these bids as long as each item is awarded only once, referred to as the OR language[5]. The problem with this language is that it creates a new type of exposure problem - that of winning more than the bidder can afford. When multiple bids of a single bidder can be winning, it is incumbent on the software to highlight the maximum exposure to the bidder. This calculation requires that a combinatorial optimization problem be solved for each bidder that calculates the dollar exposure, creating new computational issues for the auctioneer.

Fujishima et al. [25] propose a generalization of this language called OR* that overcomes this problem. Namely, each bidder is supplied "dummy" items (these items have no intrinsic value to any of the participants). When a bidder places the same dummy item into multiple packages, it is telling the auctioneer that he wishes to win at most one of these collections of packages. This language is fully expressive as long as bidders are supplied sufficient dummy items. We note that this language is relatively simple for bidders to understand and use, as was shown in a Sears Corporation supply-chain transportation auction. In that auction, all bids were treated as "OR" bids by the system. Some bidders cleverly chose a relatively cheap item to place in multiple bids thereby making these bids mutually exclusive, see Ledyard et al. [37]). There have been a number of alternative bidding languages that have been proposed. We refer the reader to the research of Boutilier et al. [6], Fujishima et al. [25] and Nissan [42] for complete descriptions of the alternative languages.

A simple bidding language to evaluate from a theoretical game-theory perspective, and the one that all current spectrum bidding software uses specifies that all bids of a given bidder are mutually exclusive. This language also removes the dollar exposure problem, since the maximum that a bidder can possibly pay is the highest bid amount of any of its bids. The problem with this language is that it places a new burden on the bidder: the bidder is forced to enumerate all possible combinations of packages and their associated value. Clearly, as the number of items in an auction increase, the number of possible bids goes up exponentially. When the XOR bidding language is used the winner determination problem becomes:

$$[\mathrm{WDP_{XOR}}] \quad \max \sum_{b=1}^{\#Bids} BidAmount_b x_b$$

subject to

$$Ax = 1 \tag{7.3}$$

$$\sum_{b \in S_B} x_b \leq 1 \qquad \text{for each Bidder } B, \tag{7.4}$$

$$x \in \{0,1\} \tag{7.5}$$

where S_B is the set of bids of bidder B, and constraint set (7.4) specifies that at most one of these bids can be in the winning set.

For a first-price sealed-bid combinatorial auction, one need only solve the winner determination problem to determine the winning bidders and the amounts that each

[5] The formulation WDP has the assumption of the OR language. Below we present the formulation when bids of a given bidder are mutually exclusive, labeled $\mathrm{WDP_{XOR}}$.

pays, since they pay that which they bid. But, as with the case of a single item auction, the bidder can experience the winner's curse of having paid too much for the items won in a first-price sealed-bid combinatorial auction.

We now consider second-price sealed bid auctions when bidders can bid on all or nothing packages In this case, we again look toward the Vickrey solution and generalize the idea when there are multiple items and the bidder wants to win "all or nothing" packages. To determine the winning bidders, one solves the same winner determination problem as one does for the first-price sealed-bid case but the winners do not necessarily pay what they bid. Instead, one calculates the marginal value to the seller of having this bidder participate in the auction. To do this, for each winning bidder, one calculates the revenue that the seller would receive when that bidder participates in the auction and when that bidder does not, i.e. when none of the bids of this bidder are in the winner determination problem. The difference in the two objective function values is known as the *Vickrey-Clarke-Groves discount*[6], and the bidder pays his bid price minus the discount. When winners pay this amount, the auction is known as the Vickrey-Clarke-Groves (VCG) Mechanism.

Although it can be shown that the VCG mechanism encourages truthful bidding, it is almost never used in practice. For a complete list of reasons for it being impractical, see Rothkopf [47] and Ausubel and Milgrom [3]. In essence, the prices provided by this mechanism may be very low. Worse yet, when items have complementary values — i.e. the package is worth more to the bidder than the sum of the values of the individual items — the outcome may price the items so low that there is a coalition of bidders that would prefer to renege on the auction and negotiate privately with the seller, and the seller may respond by reneging on the sale since both the seller and the coalition of buyers will be better off. Ausubel and Milgrom [2] argue that prices should be set high enough so that no such coalitions exist. In game theoretic terms, the prices are set such that the outcome is in the core of a coalitional game. These authors introduced an auction design known as the "ascending proxy auction" in which the bidders provide all bids as if in a sealed-bid auction. Each bidder is provided with a proxy that bids for the bidder in a straightforward manner during an ascending auction. The proxy only announces bids that maximize the bidder's profit (i.e. bid price - announced price) in any given round. The auction continues as an ascending package-bidding auction until, in some round, there are no new bids. Thus, the auction simulates through proxy bidders an ascending auction where the increment in each round is infinitesimally small and each bidder, through the use of its proxy, bids in a straight-forward manner. This auction design is very similar to the iBundle design of Parkes [43].

Hoffman et al. [29] provide a computational approach toward speeding up the calculations associated with this auction design and Day and Raghavan [20] provide an elegant mechanism to obtain minimal core prices directly. The direct mechanism of Day and Raghavan sequentially solves winner determination problems to determine losing coalitions that could supply more revenue to the seller at the current

[6] The three authors, Vickrey [49], Clarke [10] and Groves [27] wrote separate papers each producing certain attributes that this auction design has as it relates to incentivizing bidders to reveal their truthful value of the goods demanded.

prices. To begin, the winner determination problem is solved with the bids of losing bidders at their announced amounts bid (i.e. treated as first prices), and the bid prices of any winning bidder's bids are set at the amount bid minus the VCG discount for that bidder. We refer to this winner determination problem as $\text{WDP}_{\text{Core}(\tau)}$ and state it as follows:

$$[\text{WDP}_{\text{Core}(\tau)}]$$

$$\max z^{\tau} = \sum_{B \in NW} \sum_{b \in S_B} BidAmount_b x_b + \sum_{B \in W} \sum_{b \in S_B} (BidAmount_b - Discount_B^{\tau}) x_b$$

subject to

$$Ax = 1 \tag{7.6}$$

$$\sum_{b \in S_B} x_b \leq 1 \qquad \text{for all Bidders } B, \tag{7.7}$$

$$x_b \in \{0,1\} \qquad \text{for all bids of all Bidders.} \tag{7.8}$$

This is the first iteration of an iterative algorithm, so $\tau = 0$. S_b is the set of bids of a given bidder B, W is the set of winning bidders and NW is the set of non-winning bidders, and $Discount_B^{\tau}$ is the VCG discount for winning bidder B. If the solution to this optimization problem yields revenue, z^0, greater than the VCG mechanism would provide, then the prices of the winning bid set are raised so that the total price paid by winning bidders is equal to this new revenue. We denote C^t as the set of winning bidders in $\text{WDP}_{\text{Core}(\tau)}$. To determine the new prices to be considered in iteration $\tau + 1$, one must be sure that any winning bidder that forms part of this blocking coalition does not have its price raised from its prior price since it would not be willing to join a coalition if it were to lose revenue relative to its prior offer by the seller. We therefore have the following linear optimization problem to solve at each step t of the algorithm:

Cutting Plane Approach for Finding Prices (Stage t):

$$\min \Theta(\pi)^t = \sum_{j \in W} \pi_j^t$$

subject to

$$\sum_{j \in W \setminus C^{\tau}} \pi_j^{\tau} \geq z(\pi^{\tau}) - \sum_{j \in W \cap C^{\tau}} \pi^{\tau} \qquad \tau \leq t, \tag{7.9}$$

$$\pi_j^{VCG} \leq \pi_j^{\tau} \leq \pi_j^{bid} \qquad j \in W, \tag{7.10}$$

where π_j^t will be the prices that will eliminate the coalition found in $\text{WDP}_{\text{Core}(\tau)}$ for $\tau = t$. Each of the constraints in (7.9) is a "cut" that forces the prices higher at each iteration t. The constraint assures that the total price paid is greater than or equal to the revenue $z(\pi^t)$ for each $t \in T$ and that the price paid by any winning bidder that is part of the coalition found at iteration τ is not more than the offer price of the prior round. The set of constraints (7.10) assures that the price paid by each bidder is bounded below by its VCG price and above by its maximum bid price (value).

Based on the prices generated in the above linear optimization problem, one calculates the *discount* for bidder j as the difference between the value of the winning bid and the price at this step of the algorithm. All bids of winning bidders are discounted by this amount and the winner determination problem ($WDP_{Core(\tau)}$) is again re-solved. The process continues until one can find no coalition that can increase revenue to the seller. This algorithm therefore finds prices for each winning bidder that are in the core. Since there may be many such minimum core prices, Day and Milgrom [18] suggest that one choose - in order to encourage sincere bidding - the minimum core prices that are closest in Euclidean distance from the VCG prices. Alternatively, Erdil and Klemperer [24] argue for a different set of minimum core prices that are based "on a class of 'reference rules' in which bidders' payments are, roughly speaking, determined independently of their own bids as far as possible".

These core-selecting second-price sealed-bid mechanisms are relatively new. They have the following nice properties: They are in the core; they eliminate the exposure problem; and, they encourage bidders to bid sincerely. As with all sealed-bid auctions, they make collusion and punishment for not adhering to tacit agreements extremely difficult. In 2010, Portugal is scheduled to hold a core-selecting second-price auction (see announcement on Portuguese website anacom.pt).

There are, however, negatives associated with this auction design in that it puts a significant burden on the bidders. Each bidder needs to assess for each possible combination of items whether it is a package that it would like to acquire and then, for those that make sense for the bidder, provide a value for the package. In addition, such mechanisms do not provide any information about how the packages submitted might "fit" with packages submitted by other bidders. To overcome these problems, a number of authors have suggested new package-bidding designs that replicate many of the features of the simultaneous ascending auction.

7.5.2 Two Combinatorial Ascending Auction Designs

7.5.2.1 General Ascending Package-bidding Design:

In 2000, the FCC considered a package-bidding auction that resembled the SAA auction in many respects. It had most of the same rules as the SAA except that it allowed bidders to submit package bids. Each round a winner determination problem was solved to determine the winners with the constraint that at most one bid of any bidder could be in an optimal solution.

In an ascending auction, the auctioneer must announce prices for the next round. Since packages, rather than individual items may be winning, one needs to be able to allocate the package price among the individual elements that make up that package. If one has such individual item prices, then one can determine the price of any new package as the sum of the items that make up the package. However, since there is often a synergy value associated with the package, any such pricing approach cannot do this disaggregation so that there are minimal distortions from the true prices and so that prices do not fluctuate each round.

The reason for wanting such prices relates to a problem unique to package bidding auctions. The problem is labeled the *threshold problem*, whereby smaller bidders may have difficulty overcoming the high bid of a bid on a large package. An extreme example is when a bidder creates a package of *all* items and places a high price on that package bid. Each individual bidder need not make up the total shortfall between the collection of small bids and the large bidder, but rather they must *collectively* overcome the bid price and they need to know how much of the shortfall each should be paying. Essential to overcoming this threshold problem is providing good price information and having activity rules that force participation.

A number of pricing algorithms have been suggested that are based on the dual-price information that arises from solving the linear-programming relaxation to the integer programming problem. However, these prices are only approximations to the "true" dual prices of a combinatorial optimization problem, and are often call *pseudo-dual prices*. See Bikhchandani and Ostroy [5], Kwasnica et al. [36], Dunford et al. [21], and Bichler et al. [4] for more on pricing in combinatorial auctions settings.

The FCC combinatorial software uses the following pseudo-price estimates, although there are many others suggested in the literature. The pseudo-price estimates are obtained by solving the following optimization problem, Pseudo-Dual Price Calculation (PDPC):

$$[\text{PDPC}] \quad \min z_\delta^* = \sum_{j \in B \setminus W} \delta_i$$

subject to

$$\sum_{i \in I^j} \pi_i + \delta_j \geq b_j, \qquad \forall j \in B \setminus W, \tag{7.11}$$

$$\sum_{i \in I^j} \pi_i = b_j, \qquad \forall j \in W, \tag{7.12}$$

$$\delta_j \geq 0, \qquad \forall j \in B \setminus W, \tag{7.13}$$

$$\pi_i \geq r_i, \qquad \forall i \in I, \tag{7.14}$$

where j is the set of bids, I is the set of items, I^j is the set of all bids of bidder j, and r_i is the reserve price on item i. Constraint set (7.12) assures that the sum of the bid prices π_i of the items i in the winning bid is equal to the winning bid amount. Constraint set (7.11) together with the objective function tries to enforce dual feasibility as much as possible, where the δ_j's represent the deviation from dual feasibility[7]. Constraint set (7.13) maintains the non-negativity of these violations,

---

[7] In *integer linear optimization*, there can be a duality gap whereby the primal and the dual problems do not provide the same values. In this case, if one uses the prices directly from the linear programming relaxation, one would obtain prices that would overestimate the prices needed. In

and constraint set (7.14) forces the prices on all items to be at least the minimum price set by the auctioneer.

Since the solution to this problem is not necessarily unique, the algorithm tries — by solving a second optimization problem — to reduce the fluctuations in prices from round to round. To accomplish this, an optimization problem with a quadratic objective function and linear constraints is solved. The objective function that applies exponential smoothing to choose among alternative pseudo-dual prices with the additional constraint on the problem that the sum of the slack variables equals z^*_δ (the optimal value of the Pseudo-Dual Price Calculation). The objective function minimizes the sum of the squared distances of the resulting pseudo-dual prices in round t from their respective smoothed prices in round $t-1$. Thus, it is the pseudo-dual price of item i in round t. The smoothed price for item i in round t, is calculated using the following exponential smoothing formula: $p^t_i = \alpha^t_i + (1-\alpha)p^{t-1}_i$; where, p^{t-1}_i is the smoothed price in round $t-1$, $0 \le \alpha \le 1$, and p^0_i is the minimum opening bid amount for item i. The following quadratic program (QP) will find the pseudo-dual price, for each item i in round t that minimizes the sum of the squared distances from the respective smoothed prices in round $t-1$ while assuring that the pseudo-dual prices sum up to the provisionally winning bid amounts:

$$[\text{QP}] \quad \min \sum_{i \in L} (\pi^t_i - p^{t-1}_i)^2$$

subject to

$$\sum_{i \in I^j} \pi^t_i + \delta_j \ge b_j, \qquad \forall j \in B^i \backslash W^t, \tag{7.15}$$

$$\sum_{i \in I^j} \pi^t_i = b_j, \qquad \forall j \in W^t, \tag{7.16}$$

$$\sum_{j \in B^i \backslash W^t} \delta_j = z^*_\delta, \tag{7.17}$$

$$\delta_j \ge 0, \qquad \forall j \in B^i \backslash W^t, \tag{7.18}$$

$$\pi^t_i \ge r_i, \qquad \forall i \in I. \tag{7.19}$$

Note that problem (QP) has the same constraints as [PDPC], but has added the additional restriction (7.17) that the sum of the δ_j's is fixed to the value z^*_δ, the optimal value from [PDPC].

Experimental testing by Kwasnica et al. [36] has shown — when bidders have complementary values for packages — that package-bidding auctions that use pseudo-dual prices result in more efficient outcomes. However, laboratory experiments also found negatives to such designs. In testing by Goeree et al. [26] and Brunner et al. [8], results showed that such auctions took longer to complete and bidders found it difficult to construct packages that fit well with the packages of

---

the context of this application, overestimating prices could lead to inefficient outcomes. Thus, we make approximations that assure that the prices are consistent with the revenue obtained from the integer optimization.

other bidders. In addition, bidders were uncomfortable with the fact that prices fluctuated throughout the auction, see Dunford et al. [21] for more about the price fluctuations using pseudo-dual prices. The complexity of constructing bids increased with the number of items. Similarly, the computations required by the auctioneer also increase. Scalability of an auction design is an issue for a regulator that wants to use the same auction design for multiple auctions. For this reason, Goeree et al. [26] suggested that the FCC use the less complex hierarchical design for a single block of the 700 MHz auction. Here there were only two choices, the bidder either bids on all of the states as a single (nationwide) license, or one bids on any of the eight regional licenses. If the sum of the eight regional auction bids eclipses the top bid for the national license, the spectrum will be sold in those pieces. Otherwise, it will be sold as one nationwide license. This hierarchical approach predetermines the composition of the packages and through that packaging makes computation easy but does not provide the general flexibility of the more complex package-bidding designs.

7.5.2.2 Clock Designs:

In 2003, Porter et al. [45] studied a *combinatorial clock design* in laboratory settings and found that it yielded high efficiency and was better liked by the participants then their package-bidding ascending auction design. One can think of a clock design as similar to the Dutch Auction in which there is a clock for identical items and at every clock "tick" a price for each item is announced and each bidder has to agree to the new (higher) price for that item by continuing to bid on it. Once the price gets too high, the bidder acknowledges that he is no longer interested in that item. In a combinatorial setting, Porter et al. suggested a clock for each item in the auction. This auction is similar to the ascending non-package bidding design but allows bidders to express their bids as "all or nothing package bids". There is no concept of a "provisionally winning bidder" and a winner determination problem is not solved to determine if the packages fit together. Instead, prices increase on any item for which there is more demand than supply, i.e. for which there is more than one bidder bidding on that item. Since there are no provisional winners, in every round all bidders must rebid on any of the items that they wish to procure. The only information provided to bidders at the end of each round is the quantity demanded for each item and the price for the next round. At the end of a round, the auctioneer examines if there is excess demand on any item. If there is no excess demand, then this stage of the auction ends and those that bid in the last round win exactly what they bid.

It can happen that — because there are no provisionally winning bids — an item that had excess demand in the prior round now has excess supply since more than one bidder chose to no longer bid on that item. If at the end of the auction, there are items for which there is excess supply, Porter et al. [45] collect all bids submitted during the auction, treat all bids of a given bidder as mutually exclusive, and solve the winner determination problem. If the solution to the winner determination problem has supply exactly equaling demand, then the auction ends and the winners pay the prices bid. If not, the authors solve for pseudo-dual prices based on the winner determination problem outcome and restart the clock at these new (lower) prices. The process continues until supply exactly equals demand. This novel clock design was tested in the laboratory and found to have improved efficiency over the more general package-bidding designs and over the non-package SAA design. Bidders liked the design, because it provided price discovery, bidders only had to submit a single bid each round and prices were monotonically increasing throughout a clock stage. To our knowledge, this auction design has not been used for a spectrum auction. We present it here because it is similar to the combinatorial auction design described below which has been used for spectrum allocation.

Ausubel, Cramton and Milgrom [1] suggest a similar clock design, which they called the *clock-proxy design*. During the clock phase of the auction, the process is very similar to the Porter et al. [45] design except that they generalize the design to fungible items. That is, items that are nearly identical are treated as identical during this phase of the auction and grouped together. Items that are fungible have a single clock, and there may be a number of different clocks for different items (e.g. bandwidth may be fungible but regions are not). For each clock, bidders specify the quantity (i.e. the amount of bandwidth) they desire at a given price. Prices increase whenever demand is greater than supply. All bidders must rebid on any item that they wish to procure in each round. The only information provided to bidders at the end of each round is the quantity demanded for each item and the price for the next round. They suggest an activity rule that is based on a "price times quantity" basis, referred to as the *revealed-preference activity rule*. This rule removes some parking strategies[8] that were observed in the strictly quantity-based (MHz-Pop) rule.

Another novel aspect of this design is that it allows bidders who find the increment set by the auctioneer too steep to be able to specify a price somewhere between the old price and the new price that indicates the maximum amount the bidder is willing to pay for that combination of items. In this way, the efficiency loss due to increment size is lessened. The clock stage ends when demand is less than or equal to supply on all items. In some cases demand could have dropped to below supply because more than one bidder dropped quantity simultaneously.

To rectify this problem, and to allow the specification of bids on packages that had similar values to the ones specified in a given round of the clock auction, bidders are allowed to submit additional bids so long as each of the bids satisfy the activity rules. These new bids as well as all of the bids provided in the clock stage are then the bids considered within a second-price sealed-bid second auction where all bids of a given bidder are treated as mutually exclusive. Winners are determined by solving the winner determination problem and the amount paid is the second price minimal core outcome as described in Day and Milgrom [18]. Klemperer [34] suggested a similar (non-package) approach to spectrum allocation with an ascending auction followed by a final first-price sealed-bid auction.

[8] Parking is bidding on an item, usually early in the auction, that you do not want but that has sufficient demand so that one can satisfy one's activity requirements without bidding up the licenses that one wants.

In 2007, Ofcom[9] in the United Kingdom ran its first combinatorial clock auction. This auction had virtually all of the features of the Ausubel, Cramton, Milgrom [1] clock-proxy auction design except that the activity rule was a quantity-based rule. Ofcom had its second combinatorial clock auction in 2008. Early in 2010, Ofcom will hold its third auction using this design. The upcoming auction is for valuable spectrum in the 2010 to 2690 MHz range. A novel feature of this upcoming auction is that the regulators did not have to decide on a band plan prior to the auction nor did they decide how much of the spectrum would be paired or unpaired. Instead of dividing up the bands into spectrum widths prior to the auction, bidders in the combinatorial clock auction specify the quantity of spectrum they desire (in terms of number of 5MHz blocks) and whether they want paired or unpaired spectrum[10]. At the end of the auction, based on the preferences of the winning bidders, the spectrum is allocated to assure that the bidders receive the bandwidth needed, is consistent with their paired and unpaired specifications and assures that their spectrum is contiguous. A second auction, known as the *allocation auction*, allows winning bidders to specify *precisely* the location of their spectrum holdings. This second auction allows these winning bidders to bid the amount they are willing to pay (in addition to the bid price paid in the prior auction) for a given location on the band plan. In this way, the auctioneer does not need to limit possible configurations and the bidders are able to express their needs through their bids.

Cramton examines the results of the first two Ofcom combinatorial clock auctions: the $10-40$ GHz auction that took place in February 2009 [13] and the L-band auction that took place in May 2008 [14]. He argues that both ended in efficient outcomes. However, the data presented shows that the final prices obtained in the sealed-bid component of each auction were far less than that obtained in the clock phase. Specifically, in the $10-40$ GHz auction, there were 10 bidders and each won some package in that auction. However, the revenue from the final clock prices was roughly *five times higher* than the prices actually paid at the end of the second-price sealed-bid phase. Specifically, the clock phase ended with total revenue of 6,836,000 British pounds, whereas the final amount paid by all bidders was $1,417,500$ pounds. One likely reason for the disparity in prices between the two phases of the auction is that only two of the eight bidders provided a wide range of supplementary bids in the sealed-bid phase (BT submitted 545 bids and T-mobile submitted 107 bids). All other bidders submitted less than 23 bids each, with five submitting less than five bids. Since the bidders were not provided with information about the package forms of the bids submitted during the clock phase, it was difficult for bidders to deduce what additional bids would fit with the existing bids from the clock phase.

In the L-band Auction, there were 17 unique items up for lease. Eight bidders participated in the auction. t the end of the clock phase, Qualcomm was the high

---

[9] Ofcom is the independent regulator and competition authority for the UK communications industries, with responsibilities across television, radio, telecommunications and wireless communications services (Ofcom: http://www.ofcom.org.uk/)

[10] Wireless technologies differ and some technologies require that there be an uplink and a downlink separated by a certain width (e.g. LTE) while other technology (e.g. WIMAX) that does not require this pairing.

bidder on all 17 lots and remained the high bidder after the sealed-bid phase. In this case, Qualcomm bid 20 million pounds in the final bid phase (slightly higher than its final clock bid of 18.6 million pounds) but paid only 8.33 million pounds. Again, the reason for Qualcomm receiving these licenses for *less than half* of what was needed to overcome the bidders in the clock phase was that the bids submitted in the sealed-bid phase did not fit well together. Indeed, the second-highest coalition (i.e. the coalition that determined what Qualcomm paid) consisted of a set of bids that did not cover six of the 17 licenses. Thus, Qualcomm paid nothing for these six licenses.

As Ausubel, Cramton and Milgrom [1] state, the purpose of the clock stage is for price discovery. The results of the only two auctions that have used this design indicate that the price signals provided in this stage significantly overestimate the prices actually paid at the end of the auction. Since the design does not allow bidders to understand how synergistic valuations are impacting the prices presented, it is hard for bidders to understand the common-value component of the bids expressed in the clock phase. Also, as Cramton [14] points out, the activity rule between the two phases might have hindered their ability to submit supplementary package bids. More likely, the bidders did not understand how submitting more bids could impact final prices. In the L-band auction, all bidders understood that Qualcomm had already won all licenses. They knew that if Qualcomm bid only one increment above its winning price in the clock phase that it was guaranteed to win all licenses in the sealed-bid phase. It is not clear that they understood that they could have significantly impacted how much Qualcomm paid by submitting supplemental bids.

More testing needs to be done to better understand what should be done when there is excess demand. Should one have a final-round sealed-bid phase? Would it be better to use the intra-round bidding to eliminate the excess demand, or — as Porter et al. [45] suggest — should one restart the clock after solving the winner determination problem with the packages submitted in the clock phase? If a final-round sealed-bid phase takes place, how much information do bidders need to be able to overcome the fitting and threshold problems?

Most countries still use the (non-package) SAA auction design although many are considering changing to some form of the combinatorial clock design. The clock design is especially useful when the licenses being offered are for different amounts of fungible bandwidth, and each license is a nationwide license. For such applications, very few clocks are needed and bidders treat bands of spectrum as fungible. In addition to the United Kingdom, this auction design has been used in Trinidad and Tobago. It is also being considered for use in upcoming auctions in The Netherlands, Denmark and Austria.

In the US and Canada, it is not clear how to design a clock-proxy auction when the licenses are for multiple small regions. In this case, the proxy stage of allowing multiple package bids may be difficult for bidders as there may be an enormous number of new bids that they may wish to submit. And, in order to keep the optimization problem tractable, the auctioneer may have to limit the number of supplemental bids any bidder can submit. The impacts on efficiency of such rules are unclear. Thus, there is significant uncertainty as to where this auction design might work best and where some other package-bidding design might be better.

Clearly, combinatorial auctions are more complex that the SAA or simpler clock designs. When Cramton [15] tested the combinatorial clock-proxy design, he used Ph.D. students who had taken advanced courses in game theory and auction theory. He discovered that even these bidders found the revealed-preference rule too complex especially as it related to composing new supplementary bids for the sealed-bid phase of the auction. Cramton suggests that bidders need help and suggested that the bidding system could provide a bidder with a set of bids that satisfy all constraints (both revealed preference and budget constraints). In addition, the system could provide additional bids that are closest to the bids a bidder wanted to submit but that violated the revealed-preference constraints. His experiments also showed that the revealed preference rule may, at times, be too restrictive either because bidders are faced with budget constraints or because bidders change their values over the course of the auction as a result of learning the common value component of a given item. We discuss further the need for bidder-aide tools in the next section.

7.6 The Need for Bidder-aide Tools

Since multi-item auctions are complex and require bidders to consider multiple alternative bid options, we believe that it important that the computer software associated with such auctions be easy to use and understand. Good graphical user interfaces help bidders to feel comfortable that they understand the current state of the auction (they have been able to find the current price information, the amount of bidding necessary to remain eligible, their dollar exposure based on what they have bid, the degree of demand for different items, etc.). The system must also provide easy ways for them to input their next bids and confirm that they have provided the system with the correct information. Computer interfaces for such processes continue to improve and to provide better ways of displaying information to the users through charts, graphs and pictures. We foresee continued improvement in this area.

These tools do not, however, help the bidder determine the optimal combination of items to bundle as a package and the optimal number of packages to supply to the system. Although all of us face similar decisions regularly — we are bombarded with a near limitless variety of purchasing opportunities for even the simplest of tasks, e.g., what to eat for dinner — when the stakes are high and the number of substitutes is limited, the problem of providing decision support tools becomes more pressing. Thus, there is a need for the auction community to create tools that will help narrow the scope of the search and suggest good packages based on the bidder-specific business plans and preferences.

As the number of items in an auction grows, there is an exponential growth in the effort related to bid preparation and communication. Optimization tools that translate business plans into profitable packages will lessen the bidder's burden. We refer the reader to a few papers that have begun work in this very important area. For some discussion on this topic, see Hoffman [28] for tools specific to the spectrum auction setting, and Boutelier [7] and Elmaghraby and Keskinocak [22] for tools for other applications. Also, Day and Raghavan [19] and Parkes [44] provide alternative ways of bidders expressing preferences that preclude the bidder specifying specific packages to the auctioneer.

In spectrum auctions, bidders usually think of valuing spectrum based on "MHz Pop" calculations (i.e. the population within the region times the amount of spectrum). A bidder-aide tool for this application might ask bidders to supply information in terms of their business plans: What geographic areas are their primary, secondary and tertiary markets? What is their minimum and maximum bandwidth needs? Do they prefer certain bands? How much population do they need to have a viable business? How much are they willing to spend per MHz population unit? Are their exceptions for certain important regions? With this information, the tool translates their needs into constraints for an optimization model that maximizes the bidder's profit, given the current purchase prices for the licenses of interest. Packages are then provided back to the user for evaluation. The process continues with the bidder having the ability to change his business plan, budget or detailed specifications within a plan. The tool re-optimizes and creates new packages. Once the bidder is satisfied with the outputs, he indicates the package or packages (depending on the auction design) that are best, and these are supplied to the auctioneer. Tools similar in nature are being used by the major supply-chain auction providers. They must always be designed for the specific application and include all of the constraints relevant to that business concern. However, with such tools, more entities are willing to participate in the auction thereby improving the overall competitiveness of the auction.

7.7 Conclusions

Auctions have proven quite successful in allocating spectrum. Auctions have brought greater transparency to the overall process and have allowed the spectrum to be allocated in a timely manner. Recently, auction theorists have been arguing for more complicated auctions that allow the packaging of licenses. One reason for this new interest is the ability of current software — based on sophisticated mathematical algorithms developed over the past 20 years — to be able to solve very large combinatorial optimization problems routinely. In addition, the game theorists and the experimental economists have highlighted issues with prior designs and provided new mechanisms that have nice theoretical properties. Because of the need for sophisticated allocation mechanisms, computer scientists, operations research analysts, experimental economists and game theorists are working together to determine new approaches for the allocation of these resources.

As broadband applications multiply, the wireless telecommunications industry clamors for more spectrum to satisfy the seemingly insatiable appetite for applications that require more and more bandwidth. Research into new auction designs, new bidding tools, new bidding languages and the testing of such approaches in laboratory settings makes this a growing area to study. Since much of the radio spectrum is already allocated, it is likely that spectrum exchanges that encourage the reuse of existing spectrum more efficiently and more flexibly will also be a growing area of research.

References

1. Ausubel, L., Cramton, P., Milgrom, P.: The clock-proxy auction: A practical combinatorial auction design. In: P. Cramton, Y. Shoham, R. Steinberg (eds.) Combinatorial Auctions, pp. 115–118. MIT Press, Cambridge, MA (2006)
2. Ausubel, L., Milgrom, P.: Ascending auctions with package bidding. Frontiers of Theoretical Economics **1**, 1–45 (2002)
3. Ausubel, L., Milgrom, P.: The lovely but lonely vickrey auction. In: P. Cramton, Y. Shoham, R. Steinberg (eds.) Combinatorial Auctions, pp. 1–40. MIT Press, Cambridge, MA (2006)
4. Bichler, M., Shabalin, P., Pikovsky, A.: A computational analysis of linear price iterative combinatorial auctions. Information Systems Research **20**, 33–59 (2009)
5. Bikhchandani, S., Ostroy, J.M.: From the assignment model to combinatorial auctions. In: Combinatorial Auctions. MIT Press, Cambridge, MA (2006)
6. Boutilier, C., Hoos, H.H.: Bidding languages for combinatorial auctions. In: Seventh International Joint Conference on Artificial Intelligence, vol. 17, pp. 1211–1217 (2001)
7. Boutilier, C., Sandholm, T., Shields, R.: Eliciting bid taker non-price preferences in combinatorial auctions. Proceedings of the National Conference on Artificial Intelligence, pp. 204–211. San Jose, CA (2004)
8. Brunner, C., Goeree, J., Holt, C., Ledyard, J.: Combinatorial auctioneering. Available at http://econ.arizona.edu/downloads/Seminars/HoltSp07.pdf (2006)
9. Cantillon, E., Pesendorfer, M.: Combination bidding in multiple unit auctions. Tech. Rep. 76, Harvard Business School (2001). Available at http://www.ecares.org/ecare/personal/cantillon/web/combination-bidding.pdf
10. Clark, E.: Multipart pricing of public goods. Public Choice **XI**, 17–33 (1971)
11. Cramton, P.: Spectrum auctions. In: M. Cave, S. Majumdar, I. Vogelsang (eds.) Handbook of Telecommunications Economics, pp. 605–639. Elsevier Science, Amsterdam, NL (2002)
12. Cramton, P.: Money out of thin air: The nationwide narrowband PCS auction. Journal of Economics and Management Strategy **4**, 267–343 (2006)
13. Cramton, P.: A review of the 10-40 GHz auction. Available at http://www.cramton.umd.edu (2008)
14. Cramton, P.: A review of the L-Band auction. Available at http://www.cramton.umd.edu (2008)
15. Cramton, P.: Spectrum auction design. Available at http://www.cramton.umd.edu (2009). Working Paper
16. Cramton, P., Schwartz, J.: Collusive bidding: Lessons from the FCC spectrum auctions. Journal of Regulatory Economics **17**, 229–252 (2000)
17. Cramton, P., Shoham, Y., Steinberg, R.: Combinatorial Auctions. MIT Press, Cambridge, MA (2006)
18. Day, R., Milgrom, P.: Core-selecting package auctions. International Journal of Game Theory **36**, 393–407 (2008)
19. Day, R., Raghavan, S.: Assignment preferences and combinatorial auctions. Available at http://users.business.uconn.edu/bday/index.html (2005). Working Paper
20. Day, R., Raghavan, S.: Fair payments for efficient allocations in public sector combinatorial auctions. Management Science **53**, 1389–1406 (2007)

21. Dunford, M., Hoffman, K., Menon, D., Sultana, R., Wilson, T.: Ascending combinatorial auctions. Tech. rep., Systems Engineering and Operations Research Department, George Mason University (2008)
22. Elmaghraby, W., Keskinocak, P.: Combinatorial auctions in procurement. In: T.P. Harrison, H.L. Lee, J.J. Neale (eds.) The Practice of Supply Chain Management: Where Theory and Application Converge, *International Series In Operations Research and Management*, vol. 62, pp. 245–258. Springer US, New York (2004)
23. Epstein, R., Henriquez, L., Catalan, J., Weintraub, G.Y., Martinez, C.: A combinatorial auction improves meals in chile. Interfaces **32**, 1–14 (2002)
24. Erdel, A., Klemperer, P.: A new payment rule for core-selecting package auctions. Journal of the European Economic Association (2009)
25. Fujishima, Y., Leyton-Brown, K., Shoham, Y.: Taming the complexity of combinatorial auctions: Optimal and approximate approaches. Proceedings of IJCAI, pp. 548–553 (1999)
26. Goeree, J.K., Holt, C.A., Ledyard, J.O.: An experimental comparison of the FCC's combinatorial and non-combinatorial simultaneous multiple round auctions. Tech. rep., Economics Department, The University of Virginia (2006). Prepared for the Wireless Telecommunications Bureau of the FCC
27. Groves, T.: Incentives in teams. Econometrica **61**, 617–631 (1973)
28. Hoffman, K., Menon, D., van den Heever, S.A.: A package bidding tool for the FCC's spectrum auctions, and its effect on auction outcomes. Telecommunications Modeling, Policy, and Technology **44**, 153–189 (2008)
29. Hoffman, K., Menon, D., van den Heever, S.A., Wilson, T.: Observations and near-direct implementations of the ascending proxy auction. In: P. Cramton, Y. Shoham, R. Steinberg (eds.) Combinatorial Auctions, pp. 415–450. MIT Press, Cambridge, MA (2005)
30. Hohner, G., Rich, J., Ng, E., Reed, G., Davenport, A., Kolagnanam, J., Lee, H.S., An, C.: Combinatorial and quantity discount auctions with mutual benefits at mars, incorporated. Interfaces **31**, 23–35 (2002)
31. Kaglannas, A., Askounis, D., Psarras, J.: Power generation planning: A survey from monopoly to competition. International Journal of Electrical Power and Energy Systems **26**, 413–421 (2004)
32. Klemperer, P.: What really matters in auction design. Journal of Economics Perspectives **16**, 169–189 (2002)
33. Klemperer, P.: Auctions: Theory and Practice. Princeton University Press, Princeton, New Jersey (2004)
34. Klemperer, P.D.: Auctions with almost common values. European Economic Review **42**, 757–69 (1998)
35. Krishna, V.J.: Auction Theory. Academic Press, London (2002)
36. Kwasnica, A.M., Ledyard, J.O., Porter, D., DeMartini, C.: A new and improved design for multi-object iterative auctions. Management Science **51**, 419–434 (2005)
37. Ledyard John, O., Olson, M., Porter, D., Swanson, J.A., Torma, D.P.: The first use of a combined-value auction for transportation services. Interfaces **32**, 4–12 (2002)
38. McMillan, J.: Selling auction rights. Journal of Economic Perspectives **8**, 145–182 (1994)
39. McMillan, J.: Reinventing the Bazaar: A Natural History of Markets. W.W. Norton and Company, New York (2002)
40. Metty, T., Harlan, R., Samelson, Q., Moore, T., Morris, T., Sorensen, R., Schneur, A., Raskina, O., Schneur, R., Kanner, J., Potts, K., Robbins, J.: Reinventing the supplier negotiation process at motorola. Interfaces **35**, 7–23 (2005)
41. Milgrom, P.: Putting Auction Theory to Work. Cambridge University Press, Cambridge, UK (2004)
42. Nisan, N.: Bidding and allocation in combinatorial auctions. In: EC '00: Proceedings of the 2nd ACM conference on Electronic commerce, pp. 1–12. ACM, New York, NY, USA (2000). DOI http://doi.acm.org/10.1145/352871.352872
43. Parkes, D.C.: iBundle: An efficient ascending price bundle auction. Proc. ACM Conference on Electronic Commerce, pp. 148–157 (1999)

44. Parkes, D.C.: Auction design with costly preference elicitation. Annals of Mathematics and AI **44**, 269–302 (2005)
45. Porter, D., Rassenti, S., Roopnarine, A., Smith, V.: Combinatorial auction design. In: Proceedings of the National Academy of Sciences, vol. 100, pp. 11,153–11,157 (2003)
46. Porter, D., Smith, V.: FCC auction design: A 12-year experiment. Journal of Law, Economics and Policy **3**, 63–80 (2007)
47. Rothkopf, M.: Thirteen reasons why the vickrey-clarke-groves process in not practical. Operations Research **55**, 191–197 (2007)
48. Steiglitz, K.: Snipers, Shills and Sharks: eBay and Human Behavior. Princeton University Press, Princeton, New Jersey (2007). Page 298.
49. Vickrey, W.: Counter-speculation, auctions and competitive sealed tenders. Journal of Finance **XVI**, 8–37 (1961)
50. Wilkie, S.: Spectrum auctions are not a panacea: Theory and evidence of anti-competitive and rent-seeking behavior in FCC rulemakings and auction design (2007). Working Paper

Chapter 8
The Design of Partially Survivable Networks

Syam Menon

Abstract In cellular telecommunications networks, a mobile telephone switching office (MTSO) co-ordinates the activities of the mobile units either directly or through intermediate hubs. In order to increase reliability, cells often split their traffic via multiple hubs — this ensures partial survivability in the face of equipment failures. This chapter examines three related problems in the context of partial survivability. The first is a single-period model that addresses the question of assigning cells to hubs at minimum cost, while incorporating diversity requirements. The second extends the problem into a multi-period context, while the third incorporates capacity requirements at the hubs. Results from extensive computational tests are presented for the first two versions of the problem, while limited results are presented for the third.

8.1 Introduction

Cellular telecommunications systems have three basic components — mobile units, base transceivers or hubs at the cell sites and the mobile telephone switching office (MTSO). The MTSO co-ordinates the activities of the base transceivers in addition to connecting calls to the wired telephone network. Calls from (or to) mobile units are routed via the base transceivers to the MTSO for processing. The mobile units connect to the transceivers via radio links while the transceivers connect to the MTSO either via microwave or via terrestrial digital lines. Readers interested in details of the components of cellular networks are referred to books in this area (e.g. Lee [11]).

The number of simultaneous calls that can be handled by a cell can be expressed in terms of the number of 64 kbps DS-0 channels it needs in order to communicate

Syam Menon
School of Management, University of Texas at Dallas, Richardson, TX 75083, USA, e-mail: syam@utdallas.edu

J. Kennington et al. (eds.), *Wireless Network Design: Optimization Models and Solution Procedures*, International Series in Operations Research and Management Science 158,
DOI 10.1007/978-1-4419-6111-2_8,

with the MTSO. Every cell has to be connected to the MTSO, either directly or through hubs that aggregate traffic from multiple cells and connect to the MTSO via higher capacity links. With increasing size, the latter approach becomes more attractive as it enables the owner of the network to reduce costs by taking advantage of economies of scale.

The increased efficiency comes with reliability concerns however, as the failure of even one high capacity line could have a significant impact on the entire network. Common approaches to tackling this issue include the use of multiple paths (Cardwell et. al. [3]) and self-healing SONET rings (e.g., Sosonsky and Wu [17]). SONET rings are particularly convenient as they provide high capacity at a reasonable cost. In addition, if the cable is severed at some point in the ring, data can re-route itself with little to no loss in information. The increased flexibility and bandwidth availability of SONET provides significant advantages over older systems. These include reduced equipment requirements, increased network reliability, and a synchronous structure which greatly simplifies the interface to digital switches and digital add-drop multiplexers.

Most large companies lease transmission capacity on their SONET rings to other organizations. Dutta and Kubat [4] introduced the problem of designing partially survivable interconnect networks for a cellular system that has access to this leased capacity. They represented the problem as an integer program that minimizes the cost of transmission subject to reliability requirements and assumed a self-healing ring topology for the backbone network. They showed that the problem was NP-hard and developed an efficient heuristic based on Lagrangian relaxation. This was further addressed in Menon and Amiri [13], who showed that the problem could be solved efficiently via dynamic programming. This single-period problem is the first one addressed in this chapter.

Cellular telecommunications systems tend to be more flexible than traditional ones. As a result, traditional approaches to telecommunications network design are often inappropriate for the design of cellular networks, and approaches which explicitly incorporate the increased flexibility into the design process are necessary. The second problem we consider in this chapter is a multi-period model that was addressed in Kubat and Smith [8] and Menon and Amiri [14]. It exploits the flexibility of cellular systems by designing a multi-period cell-MTSO interconnection network. The objective is to minimize the total interconnection costs over the planning horizon, while determining homing decisions for the cells in each planning period.

The third problem considered in this chapter was introduced by Kubat et al. [9]. In this problem, reliability requirements are incorporated in the face of capacity constraints. They represent the problem as an integer program that minimizes the cost of transmission subject to reliability requirements while assuming a tree topology for the underlying network. We introduce a new formulation for this problem and present a procedure based on column generation to solve it.

Each problem is discussed in a separate section of this chapter. The single period problem is addressed in Section 8.2, while the multi-period and capacity constrained versions are developed in Sections 8.3 and 8.4, respectively. Conclusions are in Section 8.5.

8.2 The Single Period Problem

All cells in a cellular network must be able to communicate with each other and with the MTSO. This communication is achieved by electronic connection of the base stations to the MTSO through a fixed network, which may use copper cable, fiber optics, microwave radio, or (less frequently) satellite links. Transmission rates for this fixed part of the network tend to be high — for example, as early as 2000, AirTouch Cellular had built its own OC-12 and OC-48 optical rings to backhaul cellular traffic among hubs and MTSOs in the Los Angeles metropolitan area.

In the single period problem we are given a backbone network comprised of a self-healing SONET ring of known capacity. The locations of the nodes (i.e., the hubs and the MTSO) on the ring are assumed to be known. The traffic on the ring is assumed to be fully protected as the ring is self-healing. The demand at a cell is the number of DS-0 channels needed during "busy hour". In order to increase reliability, cells are often connected to multiple nodes on the ring; the number of nodes to which a cell is to be connected is its *diversity requirement*. It is assumed that a cell's traffic is split equally among the nodes it is connected to — for instance, a cell with a diversity requirement value of 2 would have half its traffic routed through one node and the remaining half through another node. This ensures partial survivability in case of call-node link failure, as only 50% of the traffic from this cell will be lost. The objective is to find a least-cost cell-to-hub interconnection network that satisfies the diversity requirements.

Figure 8.1 shows an example network. Nodes **a**–**e** are hubs which are connected by a high capacity ring while the cells **u**–**z** need to be assigned. In this example, the diversity requirements of nodes **v**, **x** and **y** are 1, 2 and 3, respectively. The arrows denote one possible assignment of cells to nodes.

8.2.1 Problem Formulation

Dutta and Kubat [4] formulated the problem as $(P1)$ below, after defining the following. V is the set of cells to be connected to the ring, while H is the set of hubs (excluding MTSO). R is the set of nodes including the MTSO (i.e., $R = H \cup \{\text{MTSO}\}$. The capacity of the ring is K, while the demand from cell i is D_i; both are in numbers of DS-0 circuits. S_i is the diversity requirement of cell i, and c_{ij} is the cost of connecting cell i to hub j, in dollars per year. The decision variable x_{ij} is 1 if there is a link from cell i to hub j, and 0 otherwise.

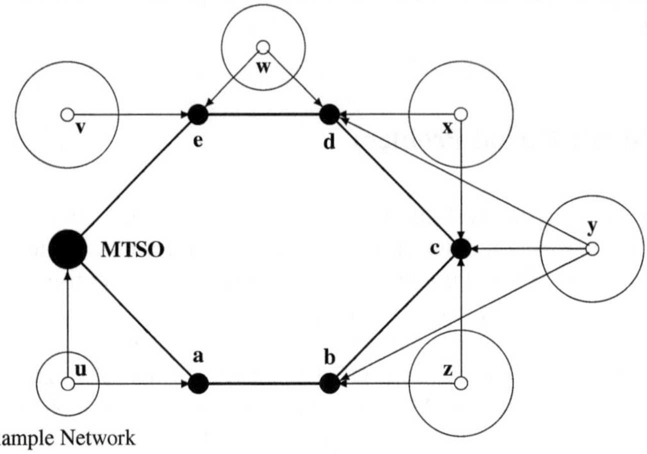

Fig. 8.1 Example Network

$$
(P1) \quad
\begin{aligned}
\min & \sum_{i \in V} \sum_{j \in R} c_{ij} x_{ij} \\
\text{st} & \sum_{j \in R} x_{ij} = S_i \quad \forall i \in V \qquad (8.1) \\
& \sum_{i \in V} \sum_{j \in H} \frac{D_i}{S_i} x_{ij} \le 2K \qquad (8.2) \\
& x_{ij} \in \{0,\ 1\} \; \forall i \in V,\ \forall j \in R \qquad (8.3)
\end{aligned}
$$

The objective function minimizes the total cost of connecting cells to nodes on the ring. Constraint set 8.1 ensures that the diversity requirement of every cell is met, while constraint 8.2 ensures that the ring capacity is not exceeded. All cell demands will ultimately terminate at the MTSO, and therefore these flows must enter the MTSO in either the clockwise or the counter-clockwise direction. Consequently, the total traffic in the ring cannot exceed $2K$ when the ring capacity is K. Note that if traffic is allowed to flow just in one direction, the only alteration to the formulation would be to limit the total traffic on the ring to K. As already mentioned, Dutta and Kubat [4] solve problem $(P1)$ via a heuristic technique based on Lagrangian relaxation.

8.2.2 Reduction to a Binary Knapsack Problem

Formulation $(P1)$ involves $|V| \times |R|$ variables and $|V|+1$ constraints. In this section, we show how $(P1)$ can be reduced to a significantly smaller binary knapsack problem (with at most $|V|$ variables). This is convenient, as the size of the binary knapsack problem to be solved is independent of the values of $|H|$ and S_i. A variety of specialized solution approaches are available to solve the binary knapsack problem (e.g., Balas and Zemel [1], Pisinger [15]). Most of these approaches are based on dynamic programming and are pseudo-polynomial in the capacity of the knapsack. As we shall see, the capacity of the binary knapsack problem to be solved is very small, and this makes its solution through one of these specialized approaches almost instantaneous. The solution technique presented here relies on the following result.

Lemma 8.1. *Every cell $i \in V$ will be connected to the cheapest $(S_i - 1)$ hubs in the optimal solution.*

Proof: At most one connection can be to the MTSO. Therefore, at least $(S_i - 1)$ connections have to be to the hubs. There are no capacity restrictions specific to the hubs. Therefore, these $(S_i - 1)$ connections to the ring can be to *any* of the hubs. The cheapest way to achieve this is to connect to the $(S_i - 1)$ cheapest hubs. $\triangleleft$

For each cell i, let $\bar{c}^i_j$ represent the costs c_{ij} in ascending order (i.e., $\bar{c}^i_1 \leq \bar{c}^i_2 \leq \cdots \leq \bar{c}^i_{|V|}$). Based on Lemma 8.1, all the variables associated with $\bar{c}^i_1, \bar{c}^i_2, \cdots, \bar{c}^i_{(S_i-1)}$ will be set to 1. Similarly, all the variables associated with $\bar{c}^i_{(S_i+1)}, \bar{c}^i_{(S_i+2)}, \cdots, \bar{c}^i_{|V|}$ will be set to 0.

So, the only *decision* to be made concerns whether to connect each cell i to the S_i-th cheapest hub, or to the MTSO directly. Let j_i be the index of the S_i-th cheapest hub for cell i. Incorporating this into $(P1)$, we get

$$
(P2)\qquad
\begin{aligned}
\min \sum_{i\in V}(c_{i0}x_{i0} + c_{ij_i}x_{ij_i}) &+ \sum_{i\in V}\sum_{j<j_i}\bar{c}^i_j \\
\text{s.t. } x_{i0} + x_{ij_i} &= 1 && \forall i \in V && (8.4)\\
\sum_{i\in V}\frac{D_i}{S_i}x_{ij_i} &\leq 2K - \sum_{i\in V}(S_i - 1)\times\frac{D_i}{S_i} && && (8.5)\\
x_{i0}, x_{ij_i} &\in \{0,\ 1\} && \forall i \in V && (8.6)
\end{aligned}
$$

The second term $(\sum_{i\in V}\sum_{j<j_i}\bar{c}^i_j)$ in the objective function represents the cost of assigning each cell i to its cheapest $(S_i - 1)$ hubs. The right hand side of constraint 8.5 is the remaining capacity of the ring after these assignments have been made. This value has to be non-negative in order for a feasible solution to exist — i.e., the problem is feasible only if $\sum_{i\in V}(S_i - 1)\times\frac{D_i}{S_i} \leq 2K$.

Constraints 8.4 imply that $x_{ij_i} = (1 - x_{i0})$. Incorporating this into the objective function of $(P2)$ and re-arranging the terms, we get the equivalent formulation $(P3)$ below.

$$
(P3)\qquad
\begin{aligned}
\sum_{i\in V}c_{i0} + \sum_{i\in V}\sum_{j<j_i}\bar{c}^i_j &- \max\sum_{i\in V}(c_{i0} - c_{ij_i})x_{ij_i} \\
\text{s.t. } \sum_{i\in V}\frac{D_i}{S_i}x_{ij_i} &\leq 2K - \sum_{i\in V}D_i + \sum_{i\in V}\frac{D_i}{S_i} && && (8.7)\\
x_{ij_i} &\in \{0,\ 1\} && \forall i \in V && (8.8)
\end{aligned}
$$

This is a binary knapsack problem involving at most $|V|$ variables. The terms $(\sum_{i\in V} c_{i0})$ and $(\sum_{i\in V}\sum_{j<j_i} \overline{c}^i_j)$ are constants which can be dropped when solving the knapsack problem $(P3)$. Note that the knapsack capacity has reduced from $2K$ to $2K - \sum_{i\in V}(S_i - 1) \times \frac{D_i}{S_i}$, thereby making the knapsack problem easier to solve. Some of the objective function coefficients of $(P3)$ could be negative, in which case the associated variables can be set to zero. As a result, there may be fewer than $|V|$ relevant variables in $(P3)$. The binary knapsack problem is relatively easy to solve, taking only milliseconds even on problems involving OC-192 (9,953.28 Mbps) capacities. Consequently, we have not presented computational results here; interested readers are referred to Menon and Amiri [13] for detailed results.

8.3 The Multi-period Problem

As mentioned earlier, cellular networks are more flexible than traditional ones. In this section, we adapt the problem presented in Section 8.2 into a multi-period framework. It was introduced by Kubat and Smith [8] and studied further by Menon and Amiri [14]. In order to incorporate the multi-period requirements, the connection-related decisions were limited to determining whether a cell connected to the MTSO or the ring in each period. In the context of the problem studied in Section 8.2, we can see that if the decision to connect (or not connect) cell i to the MTSO has already been made, the resulting single period problem is trivial — the optimal solution is to connect cell i to the cheapest $(S_i - 1)$ (or S_i) hubs, where S_i is the diversity requirement.

In this section, we present a column generation based solution procedure for the problem introduced in Kubat and Smith [8]; this approach solves problems much larger than those that could be solved using previous approaches.

8.3.1 Problem Description and Formulations

As in Section 8.2, we assume a high capacity backbone network with a known capacity. This capacity could vary over time depending on leasing decisions made by the cellular company. Cell demands are assumed known for the entire planning horizon, as are backbone capacities and the location of nodes on the backbone. The entire backbone network is viewed as a single hub which is connected to the MTSO. The problem is to determine the least-cost assignments of each cell in each time period. The problem becomes difficult as there are costs associated with connecting to the hub and the MTSO in each period and also with setting up or terminating a link from a cell to the hub in any period. Readers are referred to Kubat and Smith [8] for additional details.

The multi-period cellular design problem can be formally described as $(M1)$ below, after defining parameters N, the set of cells; T, the number of time periods in the planning horizon; d_{ij}, the demand from cell i in period j; r_{ij}, the cost of connecting cell i to the root (MTSO) in period j; h_{ij}, the cost of connecting cell i to the hub (backbone) in period j; s_{ij}, the cost of setting up a new link between cell i and the hub in period j; t_{ij}, the cost of terminating an existing link between cell i and the hub in period j; k_j, the capacity available at the hub in period j, and variables x_{ij}, which is 1 if cell i is assigned to the hub in period j and, 0 otherwise; u_{ij} which is 1 if a new link is established between cell i and the hub in period j; and v_{ij} which is 1 if an existing link between cell i and the hub is terminated in period j. For notational convenience, x_{i0} is defined to be 0.

$$
(M1) \quad
\begin{array}{llll}
\min & \sum_{i\in N}\sum_{j=1}^{T} h_{ij}x_{ij} + \sum_{i\in N}\sum_{j=1}^{T} r_{ij}(1-x_{ij}) & & \\
+ & \sum_{i\in N}\sum_{j=1}^{T} s_{ij}u_{ij} + \sum_{i\in N}\sum_{j=1}^{T} t_{ij}v_{ij} & & \\
\text{s.t.} & \sum_{i\in N} d_{ij}x_{ij} & \le k_j & j=1,\ldots,T \quad (8.9)\\
& u_{ij} - x_{ij} + x_{i(j-1)} & \ge 0 & \forall i \in N,\ j=1,\ldots,T \quad (8.10)\\
& v_{ij} - x_{i(j-1)} + x_{ij} & \ge 0 & \forall i \in N,\ j=1,\ldots,T \quad (8.11)\\
& x_{ij} & \in \{0,\ 1\} & \forall i \in N,\ j=1,\ldots,T \quad (8.12)\\
& u_{ij}, v_{ij} & \ge 0 & \forall i \in N,\ j=1,\ldots,T \quad (8.13)
\end{array}
$$

The objective is to minimize total cost, comprising the cost of connecting cells to the hub or the root in each period and the cost of establishing or terminating links from the cells to the hub in each period. Constraints 8.9 ensure that hub capacity is not exceeded in any period; constraints 8.10 and 8.11 ensure that the set-up costs and termination costs, respectively are captured; constraints 8.12 enforce integrality of the assignment variables. Integrality of the set-up and termination forcing variables u_{ij} and v_{ij} need not be explicitly enforced — if the x_{ij} are integral, these variables will be integral as well.

Another formulation for this problem was introduced in Kubat and Smith [8]. This formulation is amenable to column generation, and is presented as $(M2)$ below.

$$
(M2) \quad
\begin{array}{llll}
\min & \sum_{g\in G} f_g\rho_g & & \\
\text{s.t.} & \sum_{g\in G} a_{ig}\rho_g = 1 & \forall i \in N & (8.14)\\
& \sum_{g\in G} b_{jg}\rho_g \le k_j & \forall j=1,\ldots,T & (8.15)\\
& \rho_g \in \{0,\ 1\} & \forall g \in G & (8.16)
\end{array}
$$

Variable ρ_g represents a feasible schedule for a particular cell — for example, in a three period model, a schedule $[0,\ 1,\ 1]$ for cell i indicates that cell i is connected to the MTSO in the first period and to the hub in the next two periods. G is the set of all feasible schedules. Parameter a_{ig} is 1 if cell i is represented in variable $g \in G$, and 0 otherwise. Parameter b_{jg} is d_{ij} if cell i is connected to the hub in period j, and 0 otherwise. The coefficient f_g in the objective function is the cost associated with the schedule; it can be calculated as $f_g = \left(\sum_{j=1}^{T} h_{ij}x_{ij} + \sum_{j=1}^{T} r_{ij}(1-x_{ij}) + \sum_{j=1}^{T} s_{ij}u_{ij} + \sum_{j=1}^{T} t_{ij}v_{ij} \right)$, where cell i is the cell represented by variable ρ_g. The objective function minimizes total cost while constraints 8.14 ensure a valid schedule for each cell. Constraints 8.15 require the capacity limitations at the hub to be adhered to in each period.

We assume that all schedules are feasible for every cell. Under this assumption, there are $|N| \times 2^{|T|}$ feasible schedules; this number grows exponentially with the number of time periods. Kubat and Smith [8] solve the problem by enumerating all feasible schedules for small problems (where $N = 5$ and $T = 3$), or via heuristic approaches based on Lagrangian relaxation and cut generation applied to formulation $(M2)$. While explicit enumeration is reasonable for small problems, this results in intractable formulations as the size of the problem (in particular, the number of time periods) increases. In this section, we present a column generation based approach to solving this problem, which is observed to give gaps of less than 1% even on problems involving 300 cells and 15 time periods. While we have terminated column generation when the linear programming optimal solution is identified, we also outline two valid branching schemes for branch and price.

8.3.2 *The Solution Procedure*

The problem represented by $(M2)$ is NP-complete as even the single period case is a binary knapsack problem. Therefore, specialized solution approaches need to be developed if large instances of the problem are to be solved. Here, we present a description of the key features of the column generation based algorithm, along with details on the subproblem formulation used to generate attractive columns.

8.3.2.1 Linear Programming Column Generation

As already discussed, enumerating all feasible schedules is not a reasonable task. We start with a subset $G' \subseteq G$ and solve the linear programming relaxation of the resulting *restricted master* problem. The dual prices obtained from solving the restricted master problem are then used to implicitly price out columns that might be beneficial to add to the master problem.

A common approach to solving linear programs with a large number of columns is implicit enumeration (Gilmore and Gomory [5, 6]). We use the standard approach, and initialize the master problem with every cell being connected to the MTSO. Instead, one could initialize using a feasible solution obtained via a heuristic procedure. Solving the master problem provides dual prices which are used to generate columns by solving a subproblem for each of the cells. We need not go over all the cells in every iteration — we can switch to the master problem as soon as we have columns to add to it; the actual number of columns to add at each iteration can be determined empirically. We iterate between the master and subproblems until no more columns price out, at which point we have the optimal solution to the linear programming relaxation of $(M2)$.

8.3.2.2 Representing the Subproblem

One important factor in the success of a column generation based solution procedure is the ease with which the subproblem can be represented and solved. The subproblem gets solved many times in the solution process, and it is critical that the subproblem be easy to solve. A column is attractive to the linear programming relaxation of the master problem if the associated reduced cost is negative. If we associate dual variables $\pi \in R^{|N|}$ with the constraints 8.14 and $\lambda \in R^T_-$ with the constraints 8.15, a column s could (potentially) improve our objective function value if and only if $\left(f_g - \pi_i - \sum_{j=1}^{T} \lambda_j b_{jg} \right) < 0$. Therefore, the key issue becomes one of identifying a schedule with this feature if it exists.

At any iteration of the column generation process, we have a subproblem for each cell i that is best described by Figure 8.2. In this figure, every time period j is represented by two sets of nodes R_j and H_j, corresponding to the cell being connected to the MTSO and the hub, respectively. In addition, a source node P and a sink node Q are added, with the source node being connected to the nodes associated with time period 1 (edges $P \rightarrow R_1$ and $P \rightarrow H_1$) and the sink node being connected to the final period (edges $R_j \rightarrow Q$ and $H_j \rightarrow Q$). There are edges from every MTSO node R_j, $j = 1, \ldots, (T-1)$ indicating that the cell is connected to the root ($R_j \rightarrow R_{(j+1)}$) or the hub ($R_j \rightarrow H_{(j+1)}$) in the next period. Similarly, there are edges from every hub node H_j, $j = 1, \ldots, (T-1)$ indicating that the cell is connected to the root ($H_j \rightarrow R_{(j+1)}$) or the hub ($H_j \rightarrow H_{(j+1)}$) in the next period. The $P \rightarrow R_1$ edge and the $R_j \rightarrow R_{(j+1)}$ edges have cost r_{i1} and $r_{i(j+1)}$ associated with them, respectively; the $P \rightarrow H_1$ edge and the $R_j \rightarrow H_{(j+1)}$ edges have cost $(h_{i1} + s_{i1})$ and $(h_{i(j+1)} + s_{i(j+1)})$ associated with them, respectively; the $H_t \rightarrow R_{(j+1)}$ edges have cost $(r_{i(j+1)} + t_{i(j+1)})$ associated with them, and the cost on the edges $R_j \rightarrow Q$ and $H_j \rightarrow Q$ are 0. In addition, there are costs of $(-\lambda_j \times d_{ij})$ associated with each of the H_j nodes as a result of the dual multipliers. Given this description, the subproblem becomes one of identifying the lowest cost to send one unit of flow from P to Q.

In the absence of the node weights $(-\lambda_j \times d_{ij})$, this problem is equivalent to one of finding the shortest path from P to Q, and any shortest path algorithm (e.g., Dijkstra's) will provide the solution. In the case of the special network of the sub-

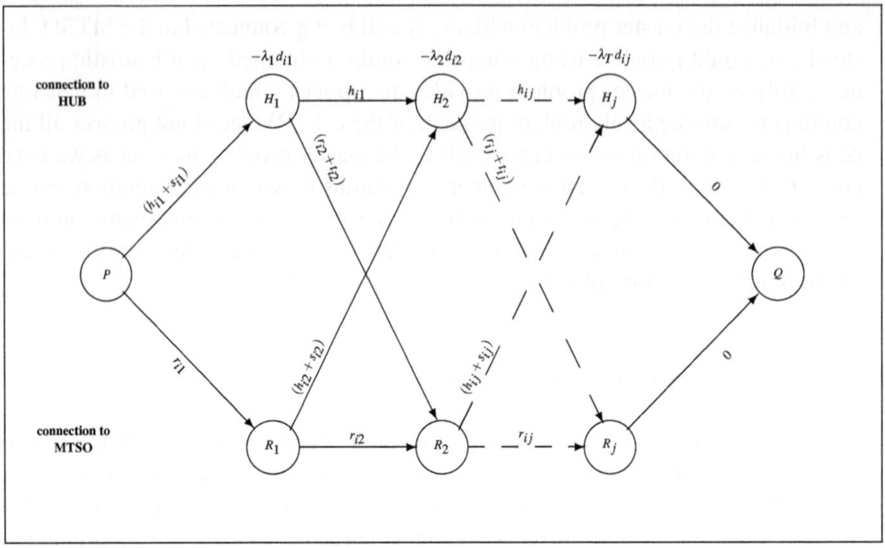

Fig. 8.2 Representing the Subproblem

problem, since we have a graph without cycles and since every period has to be visited exactly once, we can add the node weights to the weights on the edges entering the nodes H_j without loss of generality. Note that this problem can also be represented as a min-cost-flow problem where one unit of flow needs to be sent from P to Q. In our context, the min-cost-flow formulation (without the node weights) is (S_1) below.

$$
(S_1)\quad
\begin{aligned}
\min \quad & \sum_{(i,\,k)\in E} c_{ik} z_{ik} & & \\
\text{s.t.} \quad & z_{PR_1} + z_{PH_1} = 1 & & (8.17)\\
& z_{PR_1} - z_{R_1R_2} - z_{R_1H_2} = 0 & & (8.18)\\
& z_{PH_1} - z_{H_1R_2} - z_{H_1H_2} = 0 & & (8.19)\\
& z_{R_{(j-1)}R_j} + z_{H_{(j-1)}R_j} - z_{R_jR_{(j+1)}} - z_{R_jH_{(j+1)}} = 0 \; j=2,\ldots,(T-1) & & (8.20)\\
& z_{R_{(j-1)}H_j} + z_{H_{(j-1)}H_j} - z_{H_jR_{(j+1)}} - z_{H_jH_{(j+1)}} = 0 \; j=2,\ldots,(T-1) & & (8.21)\\
& z_{R_{(T-1)}R_T} + z_{H_{(T-1)}R_T} - z_{R_TQ} = 0 & & (8.22)\\
& z_{R_{(T-1)}H_T} + z_{H_{(T-1)}H_T} - z_{H_TQ} = 0 & & (8.23)\\
& z_{R_TQ} + z_{H_TQ} = 1 & & (8.24)\\
& z_{ik} \in \{0,1\} \forall (i,k) \in E & & (8.25)
\end{aligned}
$$

The c_{ik} are the costs on the edges as identified earlier; E is the set of edges in the network; z_{ik} is a variable indicating flow between nodes i and k. Constraint 8.17 indicates flow out of P; constraints 8.18 and 8.19 are flow balance constraints on the R_1 and H_1 nodes, respectively; constraints 8.20 and 8.21 are flow balance constraints on the intermediate R_j and H_j $(t = 2, \ldots, (T-1))$ nodes, respectively; constraints 8.22 and 8.23 are flow balance constraints on the R_T and H_T nodes, respectively; constraint 8.25 indicates flow into Q.

One way in which the node weights could be incorporated into this formulation is by adding constraints $y_1 \geq z_{PH_1}$ for time period 1 and $y_j \geq z_{R_{(j-1)}H_j} + z_{H_{(j-1)}H_j}$ for the other time periods, and adding the term $(+\sum_{j=1}^{T} e_j y_j)$ into the objective function (e_j is the weight at node H_j). However, since at most one of the edges coming into an H_j node will be set to 1, we can add the node weights to the cost of every edge coming into H_j and solve (S_1) with these modified edge weights to get the optimal solution to the problem. This implies that we can solve a shortest path problem to solve the subproblem. The reduced cost can then be obtained by subtracting the dual price π_i associated with cell i from the subproblem objective function value.

8.3.2.3 Integer Programming and Column Generation

The most common technique for solving integer programs is branch and bound (Land and Doig [10]). In a column generation framework, switching on the integrality requirements after solving the LP relaxation usually does not guarantee the optimal integral solution. However in many cases, if enough good quality columns have been generated, switching on the integrality requirements and solving the resulting integer program gets us very close to the integer optimum. We have used this approach in this paper, as the gaps are observed to be very small.

Exact column generation procedures for integer programs are relatively more difficult. The difficulty usually lies in the development of an effective way to incorporate the linear programming column generation into a branch and bound scheme. However, a considerable volume of work has focused on this general branch-and-price approach (e.g., Barnhart et al. [2], Vance et al. [18]). Here, we briefly outline two valid branching schemes for branch and price.

Branching Scheme I

Identify a fractional variable v_{s1}; let i be the cell associated with this variable. From constraint set 8.14, we can infer that there is at least one other variable involving this cell that is fractional. Find one such variable, say v_{g2}. As there are no identical columns, there is at least one period where these columns differ. Find one such period, say t. A valid branching scheme is one which branches on cell i being connected to the hub in period j, or not. Under this scheme, the subproblems at the branches would be solved as follows. If on the branch where i is connected to the hub in j, the edges of the subproblem which allow for connecting this cell to the MTSO (i.e., $z_{R_{(j-1)}R_j}$ and $z_{H_{(j-1)}R_j}$) are set to 0. For all cells $k \neq i$, no alteration is needed in the subproblem. If on the branch where i is connected to the MTSO in

j, the edges of the subproblem which allow for connecting this cell to the hub (i.e., $z_{R_{(j-1)}H_j}$ and $z_{H_{(j-1)}H_j}$) are set to 0. For all cells $k \neq i$, no alteration is needed in the subproblem. These alterations do not destroy the structure of the subproblem.

Branching Scheme II

Another branching strategy involves finding a fractional variable ρ_g and setting it to 0 or 1. Let i be the cell associated with ρ_g. Under this scheme, column generation within branch and bound would proceed as follows. If on the branch where ρ_g is set to 1, no subproblem need be solved for cell i; no alterations need be made in the subproblem for the other cells. If on the branch where ρ_g is set to 0, no alterations need be made in the subproblem for cells $k \neq i$. For cell i, a constraint needs to be added to the subproblem which will prevent column ρ_g from being generated again. In any feasible subproblem solution, exactly $(T+1)$ variables will be set to 1. A valid constraint that can be added to (S_1) would simply add up the subproblem variables that were set to 1 in column ρ_g and say that this sum has to be less than or equal to T. This constraint does destroy subproblem structure.

It seems likely that the first branching scheme would require a large number of branches relative to the second one. However, the second scheme requires the solution of a problem which is a network problem with side constraints, while the first one does not. It is also worth noting that the second scheme is similar to the scheme followed by a generic branch and bound approach when the integer program is solved after switching on the integrality requirements. The only difference is that columns are being generated in the branch and bound process.

8.3.3 Computational Experiments

The column generation approach was tested extensively on a large number of data sets generated at random based on cost parameters suggested in Kubat and Smith [8]. Results based on the column generation implementation are presented here, along with basic information on the data sets used.

Three small categories ($|N|$=10, 20, 50) and five large categories ($|N|$=100, 150, 200, 250, 300) were created. For each category, four time periods (T=3, 5, 10, 15) were used. Three levels (high, medium, low) were used for the capacity of the backbone (hub). These capacities were drawn from uniform distributions; $U(15,35)$, $U(10,20)$ and $U(5,10)$, respectively for the small categories and $U(100,200)$, $U(75,150)$ and $U(50,100)$, respectively for the large categories. Cell demand was generated from a uniform distribution, $U(2,10)$.

The cost of connecting to the hub (h_{ij}) is generated from a uniform distribution $U(100,1000)$; the cost of connecting to the MTSO (r_{ij}) is generated from a uniform distribution $U(2000,5000)$; the cost of setting up a link from a cell to the hub (s_{ij}) is generated from a uniform distribution $U(500,1000)$ and the cost of terminating an existing link between a cell and the hub (t_{ij}) is generated from a uniform distribution $U(100,500)$. The random number generator used is from Knuth [7]. Ten problem instances were generated in each of the $(8 \times 4 \times 3) = 96$ groups.

Table 8.1 provides the results from applying the solution procedure, for the high, medium and low capacity levels. $|N|$ is the number of cells and T is the number of time periods. The columns under each of the capacity categories represent the average number of variables generated as a fraction of the total (in %) and the average gaps (in %), where the gaps are obtained as $(\frac{UB-LB}{LB}) \times 100$, with UB being the objective function value corresponding to the best integer feasible solution obtained after spending a maximum of two minutes on branch and bound, and LB being the linear programming relaxation value of $(M2)$. The variables and gaps reported are averages over the 10 data sets of each type. The code was implemented using the LINDO Callable Library [12] on an Intel Pentium II running at 450 MHz.

IP branch and bound was terminated after a maximum of two minutes of execution. The gaps between the column generation LP and the integer solution obtained by switching on the integrality requirements are low in general and decrease with problem size. The low gaps suggest that the linear programming relaxation of formulation $(M2)$ is reasonably tight. In most of the realistic cases ($|N| > 20$), the gaps are extremely low (well below 1%). This suggests that a branch and price approach is not necessary to obtain near-optimal solutions. Even in cases where the LP-IP gaps were high (essentially when $|N| = 10$), it was noted that this was a result of the LP bound being relatively weak for the small problems.

In general, as the problem size got larger, the performance of the column generation approach improved. This could be partly due to the fact that more time is spent in linear programming column generation, which increases the number of good quality columns generated. The fraction of variables generated decreased with problem size, as would be expected. Conditioning on the number of cells, the gaps increased marginally with the number of time periods.

8.4 The Single Period Problem with Capacity Restrictions

When capacities are introduced at the hubs, the problem becomes more difficult. Kubat et al. [9] present a problem of designing interconnect networks for a cellular system that incorporate reliability requirements in the face of capacity constraints, where the backbone has a tree topology with the MTSO at the root. The links in the tree have known capacities, and it is possible for two hubs with distinct logical channels to the MTSO to share the same physical link. In order to provide increased reliability, it is also desirable to have a cell connect not just to multiple hubs, but to multiple subtrees that do not have any physical links in common between the hubs and the MTSO. This is particularly relevant when the backbone topology is a tree, as the failure of even one high capacity line could affect the entire network. As before, the number of nodes (hubs + MTSO) on distinct subtrees to which a cell is required to be connected is its *diversity requirement*. The objective is to find a least-cost cell-

|N|	T	High Capacity		Medium Capacity		Low Capacity	
		Var (%)	Gap (%)	Var (%)	Gap (%)	Var (%)	Gap (%)
10	3	43.33	3.919	45.00	3.374	45.28	3.031
10	5	17.26	5.493	17.67	3.506	19.31	3.144
10	10	1.05	5.247	1.15	3.749	1.21	2.753
10	15	0.05	7.314	0.06	4.732	0.06	3.640
20	3	43.40	1.342	43.89	1.260	43.82	1.118
20	5	15.57	1.235	15.10	0.996	15.26	0.882
20	10	0.93	1.781	0.97	1.356	1.00	1.102
20	15	0.04	2.552	0.04	2.048	0.05	1.012
50	3	39.22	0.107	42.06	0.266	40.69	0.308
50	5	14.44	0.183	13.38	0.187	13.78	0.304
50	10	0.76	0.285	0.75	0.347	0.79	0.265
50	15	0.03	0.524	0.03	0.381	0.04	0.260
100	3	40.07	0.039	39.49	0.047	40.50	0.073
100	5	13.57	0.103	13.44	0.076	13.24	0.045
100	10	0.66	0.354	0.69	0.384	0.67	0.252
100	15	0.03	0.634	0.03	0.553	0.02	0.583
150	3	39.70	0.024	39.11	0.020	37.42	0.026
150	5	13.27	0.044	12.43	0.034	12.35	0.022
150	10	0.63	0.257	0.62	0.230	0.63	0.157
150	15	0.02	0.510	0.03	0.545	0.03	0.314
200	3	36.62	0.012	35.44	0.009	34.69	0.014
200	5	12.89	0.022	12.65	0.011	12.04	0.019
200	10	0.62	0.188	0.63	0.164	0.59	0.063
200	15	0.02	0.404	0.02	0.304	0.02	0.251
250	3	36.97	0.007	36.81	0.004	36.15	0.008
250	5	12.19	0.020	12.03	0.011	11.43	0.015
250	10	0.59	0.176	0.57	0.097	0.57	0.051
250	15	0.02	0.367	0.02	0.370	0.02	0.287
300	3	32.69	0.005	32.00	0.006	31.63	0.006
300	5	12.20	0.008	11.94	0.010	10.98	0.006
300	10	0.56	0.188	0.56	0.161	0.56	0.033
300	15	0.02	0.304	0.02	0.216	0.02	0.168

Table 8.1 Multi-period Column Generation Results

to-hub interconnection network that satisfies the diversity requirements and capacity restrictions.

Figure 8.3 shows an example network. Hubs **a** – **c** are connected by a high capacity line to the MTSO while **v** – **z** are the cells which need to be connected. Hub **b** is on the path connecting hub **c** to the MTSO and as a consequence, no cell can

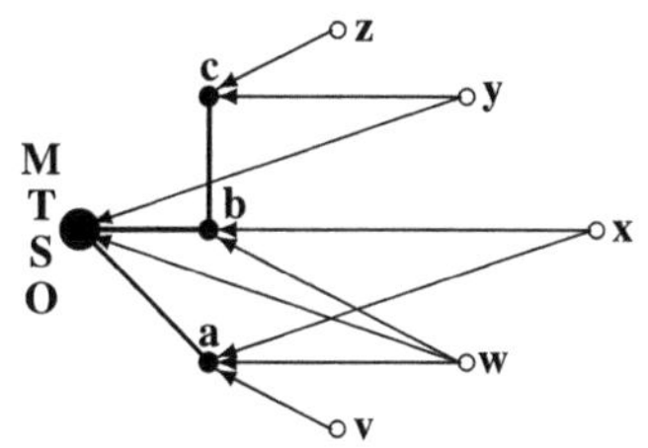

Fig. 8.3 Example Network

home to both of these hubs. The diversity requirements of nodes **v**, **x** and **w** are 1, 2 and 3, respectively. The arrows denote one possible assignment of cells to nodes.

The formulation in Kubat et al. [9] can be expressed as $(CP1)$ below, after defining the following. H, R, V, c_{ij}, d_i, and s_i are as defined in Section 8.2. k_j is the capacity of the path connecting hub j to the MTSO, in number of DS-0 circuits. a_{jl} is 1 if traffic from hub j flows via hub l, and 0 otherwise. Q is a set of trees such that $T_s \cap T_t = \emptyset \; \forall T_s,\ T_t \in Q$. The decision variable x_{ij} is 1 if traffic from cell i is routed through node j, and 0 otherwise.

$$(CP1)\qquad \begin{aligned} \min \sum_{i\in V}\sum_{j\in R} c_{ij}x_{ij} & & & \\ \text{s.t.} \sum_{j\in R} x_{ij} &= s_i & \forall i\in V & \quad (8.26)\\ \sum_{i\in V}\sum_{j\in R}\frac{d_i}{s_i}a_{jl}x_{ij} &\le k_l & \forall l\in H & \quad (8.27)\\ \sum_{j\in T_q} x_{ij} &\le 1 & \forall i\in V,\ \forall T_q\in Q & \quad (8.28)\\ x_{ij} &\in \{0,\ 1\} & \forall i\in V,\ \forall j\in R & \quad (8.29) \end{aligned}$$

The objective function minimizes the total cost of connecting cells to hubs or the MTSO. Constraints 8.26 ensure that the diversity requirement of each cell is met by the assignments, while constraints 8.27 ensure that the capacities of the path connecting the hubs to the MTSO are not exceeded. Constraints 8.28 ensure that any given cell connects to at most one hub across all the hubs in any given tree. Kubat et al. [9] solve problem $(CP1)$ via three heuristic techniques. Readers are referred to Kubat et al. [9] for additional details.

8.4.1 A Partitioning Formulation

The cell assignment problem described in this paper can be viewed as a partitioning problem where the cells are partitioned into distinct sets and assigned to various trees (or individual hubs, if there is only one hub in a given tree) in such a way that exactly s_i of the sets selected contain cell i. In addition, any feasible assignment has to ensure that the link capacity k_j is not exceeded for hub j and that cell i does not home to multiple hubs in the same tree.

A few additional definitions are needed before presenting the formulation. W is the set of all feasible entities to which cells can be assigned. Therefore, $W = \{\text{MTSO}\} \cup Q$, as cells can be assigned to the MTSO and to any of the subtrees $T_w \in Q$. For notational convenience, we refer to the MTSO as T_0 in the context of set W. Variable θ_g represents a feasible sub-assignment of cells to elements in W, while parameter r_p represents the total cost associated with this sub-assignment. Here, a feasible sub-assignment is simply a subset of cells that can be assigned to an element in W without exceeding its capacity. If a tree $T_w \in W$ contains multiple hubs, the capacity of the link associated with each hub cannot be exceeded. P is the set of all feasible sub-assignments. Parameter α_{ip} is 1 if cell i is represented in sub-assignment $p \in P$, and 0 otherwise. Similarly, parameter β_{wp} is 1 if $T_w \in W$ is involved in sub-assignment $p \in P$, and 0 otherwise. The partitioning based formulation is $(CP2)$ below.

$$
(CP2) \qquad
\begin{aligned}
\min \; & \sum_{p \in P} r_p \theta_g \\
\text{s.t.} \; & \sum_{p \in P} \alpha_{ip} \theta_g = s_i && \forall i \in V && (8.30) \\
& \sum_{p \in P} \beta_{wp} \theta_g \leq 1 && \forall T_w \in W && (8.31) \\
& \theta_g \in \{0, 1\} \; \forall p \in P &&&& (8.32)
\end{aligned}
$$

The objective function minimizes cost while constraints 8.30 require that each cell i be in exactly s_i of the sub-assignments selected. Constraints 8.31 ensure that any sub-assignment involving a tree $T_w \in W$ can be selected at most once. A sub-assignment has to satisfy capacity restrictions, and a column is feasible if and only if it satisfies $\sum_{i \in V} \sum_{j \in R} \frac{d_i}{s_i} a_{jl} \alpha_{ip} \leq k_l$ for all hubs l that are a part of $T_w \in W$. That is, if $T_w \in W$ is the MTSO, there are no capacity restrictions, if $T_w \in W$ is an individual hub j, there is a single constraint to be met, and if $T_w \in W$ has multiple hubs in it, this requirement has to be satisfied for each hub in T_w.

8.4.1.1 Representing the Subproblem

Enumerating all feasible sub-assignments is not a reasonable task. We start with a subset $P' \subseteq P$ and solve the linear programming relaxation of the resulting *restricted master* problem. The dual prices obtained from solving the restricted master problem are then used to implicitly price out columns that might be beneficial to add to the master problem.

A column is attractive to the linear programming relaxation of the master problem if the associated reduced cost is negative. If we associate dual variables $\pi \in R^{|V|}$

with the constraints 8.4, $\lambda \in R_{-}^{|W|}$ with the constraints 8.5, a column p could (potentially) improve our objective function value if and only if $(r_p - \sum_{i \in V} \pi_i \alpha_{ip} - \lambda_w) < 0$, where $T_w \in W$ is the subtree involved with this sub-assignment. r_p, the cost associated with column p, is the total connection cost $\sum_{i \in V} c_{ij} \alpha_{ip}$. Any subset of cells form a feasible column so long as their combined demand does not exceed the capacities of the hub(s) involved. Therefore, an attractive column exists if and only if the problem (S_w) has a negative objective function value for some $T_w \in W$.

$$
(S_w) \qquad \begin{aligned}
\min & \sum_{i \in V} \sum_{j \in w} (c_{ij} - \pi_i) z_{ij} - \lambda_w \\
\text{s.t.} & \sum_{i \in V} \sum_{j \in R} \frac{d_i}{s_i} a_{jl} z_{ij} \le k_l \quad \forall l \in T_w \qquad (8.33) \\
& \sum_{j \in w} z_{ij} \le 1 \quad \forall i \in V \qquad (8.34) \\
& z_{ij} \in \{0,\ 1\} \ \forall i \in V,\ j \in w \qquad (8.35)
\end{aligned}
$$

This is a binary multiple knapsack problem, with the objective function minimizing the reduced cost and constraint 8.33 ensuring that we will remain within capacity for each hub l contained in T_w. Constraints 8.34 enforces the requirement that a cell i can home at most once to T_w. Even though it is NP-complete, it can be solved relatively efficiently using either an integer programming package like LINDO [12] or a specialized solution approach (Pisinger [16]). When the $T_w \in W$ is a single hub, (S_w) simplifies further, to a binary knapsack problem. For the most part, a greedy heuristic can be used to solve (S_w), where the knapsacks are packed in greedy fashion, with an optimal approach being resorted to only if the heuristic fails to generate a column with negative reduced cost. As in Section 8.3, we have resorted to the easier approach of switching on the integrality requirements after solving the LP relaxation.

8.4.1.2 Reducing the Impact of Degeneracy

In the face of degeneracy, basic solutions tend to designate a small subset of the constraints without slack as "binding", and assign all the importance to these constraints. As a result, degenerate basic solutions are often accompanied by extreme dual prices. This is undesirable when doing column generation, as this could increase solution times considerably through the generation of a large number of unnecessary variables. Note that the equality constraints 8.4 can be replaced by inequality ($\geq$) constraints without loss of generality. If a cell is covered by more than s_i columns, a feasible solution to $(P2)$ that is at least as cheap can be obtained by introducing a new column that is identical to one of the selected columns featuring cell i, except for the absence of the cell in question. This new column is both feasible and cheaper (if the connection cost of the discarded cell is positive). Therefore, we can replace the equalities (=) in constraints 8.4 with inequalities (≥). This alteration reduces the need for other techniques to combat degeneracy.

8.4.1.3 A Greedy Heuristic

After the linear programming solution has been obtained, and before branching is initiated, a greedy column generation approach can be used to probe for better integer solutions. The logic behind this heuristic is as follows. After linear programming optimality has been obtained, the columns are scanned to identify "good" columns to fix at 1. All columns set to 1 in the linear programming solution are fixed at 1, as is one additional column identified as a good one to fix, based on some criterion (e.g., the variable value, the ratio of the objective function value divided by the number of cells covered by this column, to name two). Column generation is continued until the linear programming optimal solution to this restricted problem is obtained. This process is repeated until all cells have been assigned. At this point, all the variables which were fixed earlier to 1 are reset. This greedy approach can be repeated until a termination criterion is reached (e.g. no improvement in the integer solution obtained, no more than k columns added). This is a convenient feature to add, as it could improve the quality of solutions obtained via branch and bound.

In addition, any time an integral objective function value is encountered during column generation, the corresponding solution can be scanned for integrality. If the solution is integral, the incumbent integral solution and objective function value can be updated.

8.4.2 Computational Experiments

Limited computational experiments suggest that this approach performs well. The data for these experiments were generated as in Kubat et al. [9]. The parameters for the data in the first set of experiments are $31 \leq |V| \leq 50$, $4 \leq |H| \leq 6$, and $|Q| = 2$, while those for the second set are $81 \leq |V| \leq 50$, $4 \leq |H| \leq 6$, and $|Q| = 2$. Table 8.2 and Table 8.3 present the results of these experiments. The last column in these tables provide the results of the column generation approach, while the three columns before it present the results from the three heuristics from Kubat et al. [9]. For this limited set of experiments, the column generation approach seems to be doing well.

8.5 Conclusions

This chapter studied three problems encountered in the design of partially survivable cellular networks. The first involved the assignment of cells to nodes on a SONET

LP	Heuristic 1 Gap (abs)	Heuristic 1 Time (sec)	Heuristic 2 Gap (abs)	Heuristic 2 Time (sec)	Heuristic 3 Gap (abs)	Heuristic 3 Time (sec)	Col. Gen. Gap (abs)	Col. Gen. Time (sec)
123397.81	2383.19	3.37	1018.19	17.60	1018.19	6.40	577.19	0.23
48404.98	1736.02	1.38	3868.02	8.02	2152.02	2.24	475.02	0.17
69759.00	441.00	3.45	441.00	3.37	441.00	3.49	111.00	0.17
113425.00	857.00	3.26	1146.00	52.61	1146.00	5.92	687.00	0.63
106138.00	671.00	2.67	7610.00	24.71	671.00	5.14	563.00	0.17
98469.34	1902.66	1.58	5106.66	22.98	1264.66	2.50	1200.66	0.63
182132.00	0.00	8.64	0.00	8.64	0.00	8.64	0.00	0.17
119684.00	3393.00	1.45	2189.00	2.11	2189.00	2.18	453.00	0.00
99243.35	2085.65	3.32	2981.65	52.03	2085.65	5.78	453.65	0.86
68149.79	766.21	5.05	1326.21	28.10	766.21	9.49	150.21	0.17

Table 8.2 Results from Data Set I

LP	Heuristic 1 Gap (abs)	Heuristic 1 Time (sec)	Heuristic 2 Gap (abs)	Heuristic 2 Time (sec)	Heuristic 3 Gap (abs)	Heuristic 3 Time (sec)	Col. Gen. Gap (abs)	Col. Gen. Time (sec)
236426.14		23.50	147.86	335.23		41.30	147.86	0.75
346227.56	1229.44	24.53	1229.44	866.85	709.44	47.48	189.44	1.21
151202.63		23.76	870.38	511.84		43.44	270.38	0.46
245311.00	0.00	23.29	0.00	23.29	0.00	23.29	0.00	0.23
414656.88	975.13	33.74	1623.13	153.13	921.13	56.62	273.13	3.51
225457.56	4435.44	28.13	1495.44	79.70	1936.44	54.21	256.44	2.53
206971.95	2656.05	20.18	4084.05	484.31	3489.05	36.50	361.05	1.09
352435.78	10590.22	27.31	11319.22	657.53	10590.22	48.75	573.22	7.26
204829.08	3214.92	21.91	4222.92	452.49	3739.92	42.18	295.92	11.00
288682.34	3591.66	17.92			5791.66	33.01	466.66	5.12

Table 8.3 Results from Data Set II

ring while considering diversity requirements. It was reduced to a relatively small binary knapsack problem, and even the largest problem tried solved in a few milliseconds. The second problem extended the problem to a multi-period setting, and a column generation approach was used for solution. It solved very large problems efficiently, both in terms of time and solution quality. The third problem introduced capacity limitations at the hubs. A column generation procedure was proposed and the results from preliminary experiments look encouraging.

Acknowledgements Portions of this chapter have appeared in print previously.
I thank J. MacGregor Smith for providing the Mathematica code for data generation and heuristics used in Kubat et al. [9].

References

1. Balas, E., Zemel, E.: An Algorithm for Large Zero–One Knapsack Problems. Operations Research **28**, 1130–1154 (1980)
2. Barnhart, C., Johnson, E., Nemhauser, G., Savelsbergh, M., Vance, P.: Branch and Price : Column Generation for Solving Huge Integer Programs. Operations Research **46**, 316–329 (1998)
3. Cardwell, R., Monma, C., Wu, T.: Computer Aided Design Procedures for Survivable Fiber Optic Telephone Networks. IEEE Journal on Selective Areas in Communications **7**(8), 1188–1197 (1989)
4. Dutta, A., Kubat, P.: Design of Partially Survivable Networks for Cellular Telecommunications Systems. European Journal of Operations Research **118**, 52–64 (1999)
5. Gilmore, P., Gomory, R.: A Linear Programming Approach to the Cutting Stock Problem. Operations Research **9**, 849–859 (1961)
6. Gilmore, P., Gomory, R.: A Linear Programming Approach to the Cutting Stock Problem–Part II. Operations Research **11**, 863–888 (1963)
7. Knuth, D.: The Art of Computer Programming, Volume 2: Seminumerical Algorithms. Third edn. Addison-Wesley Longman, Reading, MA (1997)
8. Kubat, P., Smith, J.: A Multi-period Network Design Problem for Cellular Telecommunication Systems. European Journal of Operations Research **134**, 439–456 (2001)
9. Kubat, P., Smith, J., Yum, C.: Design of Cellular Networks with Diversity and Capacity Constraints. IEEE Transactions on Reliability **49**(2), 165–175 (2000)
10. Land, A., Doig, A.: An Automatic Method for Solving Discrete Programming Problems. Econometrica **28**(2), 497–520 (1960)
11. Lee, W.: Mobile Cellular Telecommunications: Analog and Digital Systems. Second edn. McGraw-Hill, New York (1995)
12. LINDO Systems, Inc.: LINDO Callable Library User's Manual (2001)
13. Menon, S., Amiri, A.: Designing Partially Survivable Cellular Telecommunications Networks. Information Technology and Management **6**(2/3), 293–303 (2005)
14. Menon, S., Amiri, A.: Multi-period Cellular Network Design via Price Influenced Simulated Annealing (PISA). IEEE Transactions on Systems, Man and Cybernetics – Part B **36**(3), 600—610 (2006)
15. Pisinger, D.: A Minimal Algorithm for the 0-1 Knapsack Problem. Operations Research **45**, 758–767 (1997)
16. Pisinger, D.: An Exact Algorithm for Large Multiple Knapsack Problems. European Journal of Operations Research **114**, 528–541 (1999)
17. Sosonsky, J., Wu, T.: SONET Ring Applications for Survivable Fiber Loop Networks. IEEE Communications Magazine pp. 51–58 (1991)
18. Vance, P., Barnhart, C., Johnson, E., Nemhauser, G.: Airline Crew Scheduling : A New Formulation and Decomposition Algorithm. Operations Research **45**, 188–200 (1997)

Part III
Optimization Problems in Ad Hoc Networks

Chapter 9
Routing in Mobile Ad Hoc Networks

Khaldoun Al Agha and Steven Martin

Abstract Ad hoc networks were created to provide communication between peers without any network infrastructure. They could help in a hostile environment for military and rescue entities, and also for commercial applications such as gaming or facilities for networking. To improve the provided services on those networks, many quality of service (QoS) frameworks have been proposed to improve the performance of ad hoc networks, and to offer many possibilities for important applications to use priorities, admission control, or quasi guarantees. The dynamicity of ad hoc networks introduces complexity for such QoS schemes. In this chapter, we offer a summary of all existing ad hoc protocols at different levels of the network architecture and the proposed QoS frameworks.

9.1 Ad Hoc Networks

In a wireless network, a node can successfully communicate with nodes within its transmission range. Nodes outside this range, but close enough to detect the signal, are in the carrier sensing range. In a wireless ad hoc network, no infrastructure or centralized administration is needed to establish or manage data exchanges. Such a network is formed dynamically and nodes organize themselves. The main restriction of an ad hoc network is its area, since any node has to be within the transmission range of any other node it communicates with. To extend the area of the network, that is to say to enable communication between two distant nodes, intermediate nodes must relay messages. This is the characterization of a wireless multi-hop ad hoc network. Moreover, nodes are free to move independently and in any direction.

Khaldoun Al Agha
LRI, Bâtiment 490, Université Paris-Sud, 91405 Orsay cedex, France, e-mail: alagha@lri.fr

Steven Martin
LRI, Bâtiment 490, Université Paris-Sud, 91405 Orsay cedex, France, e-mail: smartin@lri.fr

J. Kennington et al. (eds.), *Wireless Network Design: Optimization Models and Solution Procedures*, International Series in Operations Research and Management Science 158,
DOI 10.1007/978-1-4419-6111-2_9, © Springer Science+Business Media, LLC 2011

Thus, delivering packets to a specific destination becomes a complex task due to frequent and unpredictable topology changes.

The development of wireless ad hoc networks was initially motivated by military applications, but nowadays, these networks are involved in many civil applications. Therefore, these networks have to offer differentiated services, according to the application's requirements. This also implies admission control mechanisms. All these topics have constituted hot research issues for the last decade. Obviously, solutions leading to single-hop and multi-hop ad hoc networks are complex. They are all the more complex when considering specific problems such as connectivity, security, quality of service, energy consumption, and routing optimization. Indeed, not only do ad hoc networks inherit the classical problems of wireless networks (e.g. air medium propagation, unreliability, and hidden nodes), but also generate new problems and complexities (e.g. autonomy, lack of infrastructure, dynamicity, and scalability).

The remainder of this chapter is structured as follows. First, in Section 9.2, we present routing issues, the constraints to take into account and the characteristics of routing protocols. Then, we describe in Section 9.3 the two main link layer solutions which enable single-hop ad hoc networks: the IEEE 802.11 standard and the Bluetooth specification. In Section 9.4, the existing ad hoc routing protocols are presented: ad hoc on-demand distance vector (AODV), dynamic source routing (DSR), optimized link state routing (OLSR), and topology broadcast based on reverse path forwarding (TBRPF). Finally, in Section 9.5 we present several methods that provide quality of service (QoS) in mobile ad hoc networks: flexible QoS model for MANETs (FQMM), core extraction distributed ad hoc routing (CEDAR), ticket based probing (TBP), INSIGNIA, stateless wireless ad hoc network (SWAN), and QoS for OLSR (QOLSR). Section 9.6 concludes the chapter.

9.2 What is Routing?

The main idea of an ad hoc network is that each wireless node can communicate directly with any other wireless node, without the use of any existing network infrastructure. Thus, the architecture is totally decentralized. Such networks are limited by the fact that each wireless node has to be within the transmission range of all the others; this leads to small networks as illustrated in Figure 9.1. The main problems in ad hoc networks are how to transmit in, and share ,the air medium. There exist two major link layer technologies used in ad hoc networks: the IEEE 802.11 standard and the Bluetooth specifications (see Section 9.3).

The architecture of an ad hoc network applies to small networks, due to the limited transmission range of wireless network interfaces. As shown in Figure 9.1, extending connectivity to a larger area requires a multi-hop ad hoc network architecture in which each node assumes the role of a router and relays packets toward their final destinations. Thus, network nodes cooperate to forward packets for each other to allow communication over multiple "hops" between nodes not directly within wireless transmission range of one another.

However, this characteristic is not sufficient to ensure network connectivity. Furthermore, packets must be directed through the network. In other words, a path has to be determined from the source to its destination. This is so-called routing. A simple way to perform routing would be to enter static routes by integrating a fixed routing table in each node according to the topology. Obviously, as nodes in a multi-hop ad hoc network move about or join/leave the network, and as wireless transmission conditions such as sources of interference change, this technique is not practical. Routing must result from a specific algorithm which calculates routes automatically and dynamically, that is to say from a multi-hop wireless ad hoc network routing protocol.

Before presenting the standardized routing protocols in Section 9.4, we remind the reader that multi-hop ad hoc networks have the following characteristics:

- Dynamic topologies. Nodes, which are able to forward packets, can move arbitrarily. This mobility results in the topology changing randomly and rapidly at unpredictable times.
- Limited physical security. Attacks, such as denial-of-service or eavesdropping, are often easier to launch in wireless networks than in wired ones. Ad hoc networks at least have the advantage of being totally decentralized.
- Omnidirectional radio transmission. Since packets are broadcast to all neighboring nodes, simultaneous data exchange between nodes close to each other is not possible.
- Possible unidirectional links. Because of interference, fading, or limited transmission range of some wireless network interfaces, links can be unidirectional.
- Constraints on bandwidth and energy. These resources are quite limited. The former due to the intrinsic properties of radio interfaces and the effects of multiple access, noise, etc.; the latter due to the characteristics of wireless mobile nodes forming the multi-hop ad hoc network, which often rely on batteries.

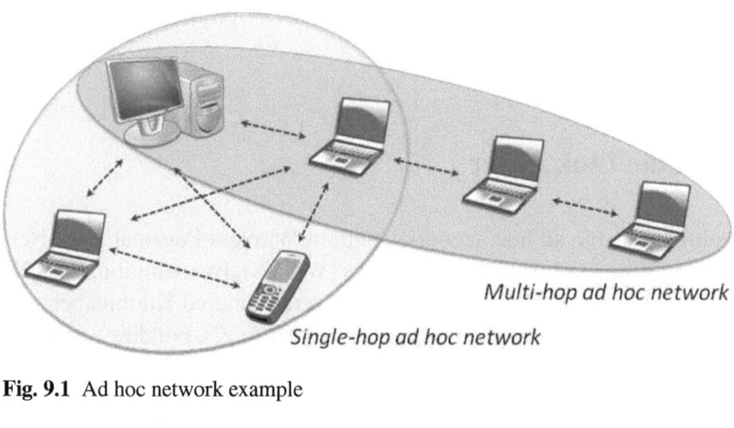

Fig. 9.1 Ad hoc network example

Routing does not only consist of finding a route between two nodes, the goal is to calculate an optimized route according to one or several constraints (number of hops, maximum bandwidth, minimum end-to-end delay, etc.). Not designed specifically for ad hoc networks, classical routing protocols (e.g. RIP and OSPF) are inapplicable. They must be upgraded in order to be efficient and reliable, taking into account the wireless and decentralized architecture, reducing control traffic and managing mobility. Therefore, multi-hop ad hoc routing protocols present much originality, since the mobile nodes in the network dynamically establish routing among themselves to form their own network. Nevertheless, these protocols apply properties of classical routing schemes. Indeed, we can distinguish two techniques to calculate the best route:

- Distance-vector routing. Each node informs (periodically and after topology changes) its neighbors of the estimated direction and distance to any other node. The cost of reaching a destination is computed using various route metrics. The Bellman-Ford algorithm is then used to calculate routes. Distance-vector routing protocols have low computational complexity and message overhead. On the other hand, the algorithm may struggle to converge on the new topology when network conditions change.
- Link-state routing. Each node broadcasts its links' states with its neighbors in the network. Being acquainted with which nodes are connected to which other nodes, every node attempts to construct a map of the connectivity of the network. Then, the Dijkstra algorithm enables calculation of optimal routes. Instead of sharing routing tables with the neighborhood, nodes applying a link-state routing protocol flood the network only with their connectivity information.

These two techniques, well known in wired networks with RIP and OSPF, are used in multi-hop ad hoc routing protocols (e.g., AODV and OLSR). Many other differences appear between routing protocols. For instance, a packet may contain in its header the complete path to follow in order to reach the destination (source routing) or only the next node to its destination (hop-by-hop). We will see in Section 9.4 that a major distinction in multi-hop ad hoc routing protocols concerns the time when a route is calculated. This defines two kinds of protocols: table driven and on demand protocols.

9.3 Ad Hoc in the Link Layer

Several technologies enable ad hoc networks, both in Wireless Personal Area Networks (WPAN) and Wireless Local Area Networks (WLAN). By definition, WPANs are limited to a few meters, with data rates up to several hundred Kilobits per second. As for WLANs, their communicating range is larger (e.g. a building) and they can achieve data rates of tens of Megabits per second. A WLAN is obviously, and above all, a LAN. Thus, in spite of wireless characteristics (e.g. bandwidth and en-

ergy constraints), the requirements of LANs and WLANs are the same: high data rates, full connectivity, broadcast capability, etc.
We present in the following the two main link layer technologies for ad hoc networks: the IEEE 802.11 standard for WLANs and the Bluetooth specifications for WPANs. As detailed in the previous section, a "single-hop" ad hoc network implies that all the stations must be within the same transmission area to be able to communicate. The range of the network can be extended by adding network layer protocols, including routing mechanisms. Solutions leading to multi-hop ad hoc networks are described in the next section.

9.3.1 Ad Hoc Mode in Wi-Fi

Proposed in 1997, the IEEE 802.11 standard [2] (commercially called Wi-Fi, the acronym of Wireless Fidelity) defines the two first layers of the OSI model for a wireless local area network: the physical layer and the link layer. While three different physical layers are proposed, the link layer is unified. Indeed, to transmit data between nodes, it can use either: infrared (IF) pulse position modulation, direct sequence spread spectrum (DSSS), or frequency-hopping spread spectrum (FHSS). Frequencies are in the Industrial, Scientific and Medical (ISM) band, ranging from 2.4000 GHz to 2.4835 GHz. This standard allows data rates equal to 1 or 2 Mbit/s. Since then, it has been extended by several task groups, which are designated by letters. In 1999, the extensions IEEE 802.11a and IEEE 802.11b were proposed in order to improve data rates: up to 54 Mbit/s in the 5 GHz band for the former and up to 5.5 or 11 Mbit/s in the 2.4 GHz band for the latter (with backward compatibility). In 2003, the IEEE 802.11g extended the IEEE 802.11b to support 54 Mbit/s in the 2.4 GHz band, by applying the same OFDM based transmission scheme as IEEE 802.11a. Nowadays, there exist around thirty amendments, offering higher throughput, security services, handover capabilities, etc. [1]. Two operational modes are specified in the standard:

- Infrastructure-based mode. In this mode, all the communications between nodes have to go via a specific node: the Access Point (AP). The network area is then determined by the transmission range of the AP. The set of nodes around the AP is called Basic Service Set (BSS). To extend the network, it is possible to merge several BSSs. For this, the AP can be connected to another AP, generally with a wired network such as Ethernet.
- Ad hoc mode. No AP is required in this mode, the network being totally distributed. Nodes can communicate if they are within their respective transmission range. They constitute an Independent Basic Service Set (IBSS).

Moreover, the Medium Access Control (MAC) layer of the IEEE 802.11 standard defines a contention-free service, called the Distributed Coordination Function (DCF) based on a multiple access method. As an optional access method, the standard defines the Point Coordination Function (PCF), implemented on top of the

DCF and based on a polling scheme. Since there is no central authority to act as a point coordinator in ad hoc networks, the PCF cannot be readily applied in such environments.

In the DCF mode, a Carrier Sense Multiple Access with Collision Avoidance (CSMA/CA) is used. Hence, to transmit a packet, a node has to first listen to the channel. If it is idle, then the node can send its packet. Otherwise, that is to say if the node senses energy above a specific threshold on the medium, the transmission is deferred for a random interval, reducing the probability of collisions on the channel. More precisely, when the current transmission ends, the node wanting access waits a Distributed Inter-Frame Spacing (DIFS) before choosing a random back-off value. This value is then decremented only when the channel is idle. As soon as the value reaches zero, the node sends its packet.

Despite this mechanism, collisions may occur. Contrary to the CSMA/CD used in the IEEE 802.3 standard, collision detection is not feasible in the IEEE 802.11 standard, due to the use of half duplex antennas. This hinders the node from sending and receiving simultaneously. Collisions are then detected indirectly; to confirm the correct reception of information, acknowledgment (ACK) packets are sent from the receiver to the sender. In order to send the ACK just after the received packet, another inter-frame spacing is defined: the Short Inter-Frame Spacing (SIFS). This one being shorter than the DIFS, ACKs are sent as a priority, preceding any other packet transmission.

Another problem occurs in wireless networks: the hidden node problem. Indeed, consider Figure 9.2 where node *C* wants to transmit a packet to node *B*, which is already receiving a packet from node *A*. As *C* and *A* are not within the transmission range of each other, when node *C* applies the CSMA protocol and listens to the channel, it cannot detect the transmission between *A* and *B*. This creates a collision, scrambling data.

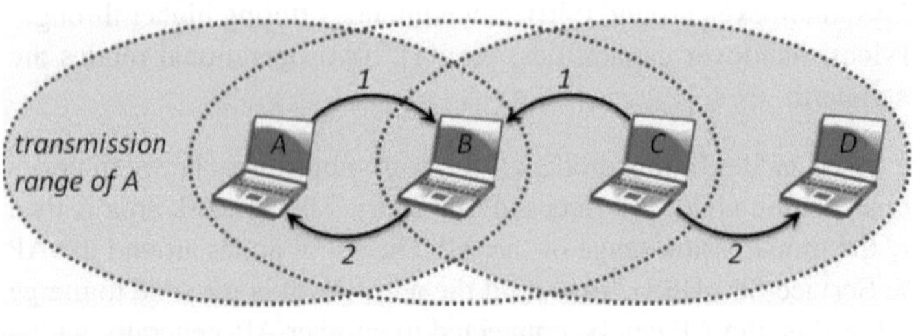

Fig. 9.2 Hidden node (1) and exposed node (2) problems

To solve this problem, the physical carrier sensing mechanism CSMA can optionally be supplemented by an additional mechanism, a virtual carrier-sensing mechanism based on the exchange of Request To Send (RTS) and Clear To Send (CTS) packets between the sender and the intended receiver. The IEEE 802.11 RTS/CTS exchange is defined as follows. Before sending packets, the source sends to the intended receiver a short packet, named RTS. If the RTS packet is well transmitted, the receiver replies (after a SIFS) by another short packet, named CTS. Then, the source knows that the destination is ready to receive the data frame. Information contained in both the RTS and CTS packets is the identity of the requesting node (the intended receiver), the total duration of the transmission (time interval needed to transmit the RTS and CTS packets), the data frame, and the related acknowledgment. Therefore, all the nodes within the transmission range of either the source or the destination know the overall time interval of the coming transmission. It is important to note that this mechanism partly overcomes the hidden node problem. Moreover, RTS/CTS may decrease throughput [19]. For this reason, an RTS threshold can be adjusted to apply this technique only for large data frame transmissions that considerably reduce network performance in the presence of collisions [14].

Another technique to increase the transmitting power of the nodes could solve the problem. Indeed, this enables the nodes to hear the hidden ones, which are no longer hidden. Unfortunately, this technique amplifies another well-known problem in wireless networks: the exposed node problem. As shown in Figure 9.2, when B is transmitting to node A, even if C wants to send packets to D (a transmission that would not cause a collision at A), the CSMA protocol hinders node C from transmitting, thus reducing throughput.

9.3.2 Bluetooth Link Layer

Bluetooth technology was introduced in 1998 by a specific working group, the Bluetooth SIG (Special Interest Group), including industry leading companies such as Ericsson, Intel, Nokia, Toshiba, 3Com, and Motorola. These specifications have been extended by the IEEE 802.15 working group, in charge of developing a family of communications standards for WPANs. The goal of this technology was a wireless communication system based on short-range radio at low-cost. The application field was clearly the PANs by making feasible wireless ad hoc transmissions between different equipment such as laptops, printers, keyboards, cellular phones, wireless handhelds, car radios, etc. Several Bluetooth specification versions have been released since the original one (version 1.0A) in 1999.

The Bluetooth specification defines the lowest layers, especially the radio and the baseband layers [5]. The radio layer defines how to transmit data over the air medium. Bluetooth technology operates in the 2.4 GHz ISM band, employs a fast frequency-hopping spread spectrum (FHSS) radio technique, and uses a binary Gaussian frequency-shift keying (GFSK) modulation. These techniques enable very low-cost implementations. The baseband defines how Bluetooth devices can communicate with each other. A Bluetooth network is formed in an ad hoc manner (without any infrastructure assistance). The basic single-hop ad hoc network is named piconet in the Bluetooth parlance. The transmission range of a specific Bluetooth device depends on its transmit power. Three classes are defined: class 1 (100 mW), class 2 (2.5 mW) and class 3 (1 mW) radios. The network area spreads from one meter to one hundred meters, according to the transmit power class.

During the piconet establishment, one node is designated master and the other nodes slaves. A piconet contains at most seven active slaves (actively involved in communications). A node that is not the master or an active slave is either parked, if it is registered with the master, or in stand-by mode if it does not belong to a piconet. The role of the master is to arrange transmissions in the piconet, by polling the slaves. Hence, a slave can deliver a packet only if it has received the authorization from the master. Moreover, all transmissions have to go by the master, even if the data exchange is between two slaves. For identification, a locally unique Active Member Address (AM_ADDR) is assigned to each active slave by the master.

To communicate, all the nodes belonging to a given piconet are synchronized to the same channel, which is defined as the followed sequence of the frequency hops (up to 1,600 per second). Indeed, transmissions are made on a slotted scheme. The duration of a basic slot is equal to 625 μs. During a slot, the carrier frequency remains unchanged. Frequency hopping occurs at the end of the slot. However, when a packet occupies multiple slots (three or five), no frequency change is applied. When hopping resumes, the next frequency is the same as the one that would have been chosen in single slot transmission. According to the Bluetooth clock of the master, slots in a piconet are even or odd. The master and the slaves alternating transmissions in a Time Division Duplex (TDD) manner, the even slots are used by the master and the odd ones are used by the active slaves, knowing that a slave can transmit only if it has just received a poll packet from the master. Several studies have investigated the performance of a piconet by considering, for example, specific scheduling algorithms [6] or finite device buffer sizes [13].

A slave may belong to several piconets. In this way, multi-hop ad hoc networks can be created. Such a Bluetooth network, interconnecting several piconets, is called a scatternet. More precisely, a slave can be temporarily separated from a piconet to become the slave of another master, and then come back to the initial piconet. A slave can also become the master of another piconet. These situations are illustrated in Figure 9.3. Several solutions are proposed in the literature in order to form a scatternet [4].

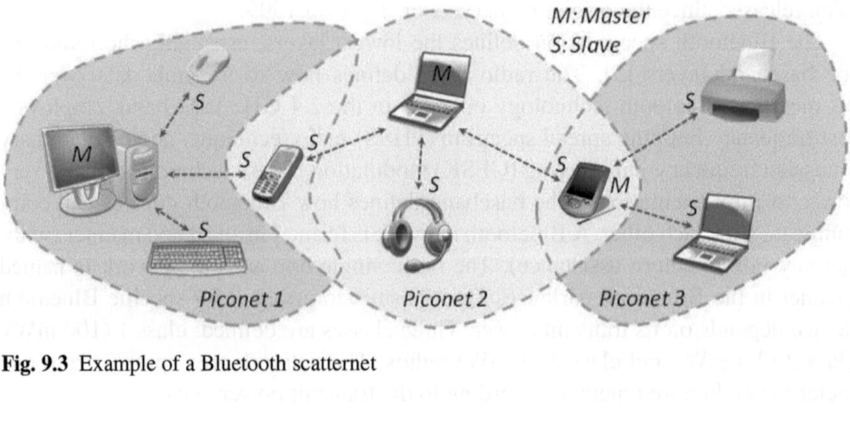

Fig. 9.3 Example of a Bluetooth scatternet

9.4 Ad Hoc Routing Protocols

Routing consists of finding a set of links which allow the transfer of information from a source point to a destination. In ad hoc networks, links are mostly radio interfaces connecting two mobile nodes. The use of radio interface and mobile nodes makes the routing harder than in fixed networks because of many problems such as interference, dynamicity, non-symmetric links, narrow band resources, and non-deterministic access. For those reasons, applying the same routing protocols created for the IP network is not possible and may create hard problems such as delivery divergence, high delay, and small delivery rate of information.

A huge number of studies were done for creating or adapting protocols for the ad hoc environment. Those protocols are divided into three sets: reactive, proactive, and hybrid. The reactive protocols minimize the number of exchanged control information messages when nodes are idle and no traffic is active. They react only when a node decides to send information, at which time the network starts a route discovery process that could be very expensive in terms of resources. Proactive protocols, on the other hand, exchange control information periodically in order to maintain information on the network topology and have information on routes for all the destinations. Finally, the hybrid protocols are a mixed of reactive and proactive.

At the IETF standardization process, the MANET (Mobile Ad hoc NETwork) working group was created in order to provide RFC documents on reference protocols and the result was four experimental standards: two reactives (DSR and AODV) and two proactives (OLSR and TBRPF). Phase 2 was initiated and work is in progress to merge the reactive protocols towards a single protocol, named DYMO, and the proactive protocols towards a single protocol named OLSRv2.

9.4.1 Introduction to Ad Hoc Routing Protocols

The military was probably the first to require the use of ad hoc networks. Military operations are often very dynamic in nature and cannot be in most cases based on a fixed and pre-existing infrastructure. Communications between the various entities of a transaction will often be on a wireless network without knowledge of the network formed by these entities.

The US Department of Defense has been involved in projects of advanced research and development in areas such as communication networks required by military applications. In 1973, DARPA (Defense Advanced Research Agency), established by the U.S. Department of Defense, initiated a research project on the technique of packet switching in wireless environments. This request was motivated by the need to provide network access to mobile entities and to establish communications in a mobile environment. The project was based on the packet radio transmission developed at the University of Hawaii in the ALOHA project in the 1970s.

Initially, the system could send data packets using radio to reach destinations located at one hop. The development of a new network called PRNet (Packet Radio

Network) has enabled radio multi-hop communications on an extensive geographic area. In Figure 9.4, the PRs represent nodes which are connected by radio links. All PRs can communicate directly by radio (left side of the figure) or by using intermediate PRs (right side). The later is called multi-hop communications. If, for example, a communication must pass by three radio links, one speaks of three hops. The PR-Net is the basis of ad hoc networks. The reader will find a complete description of packet radio networks in [11].

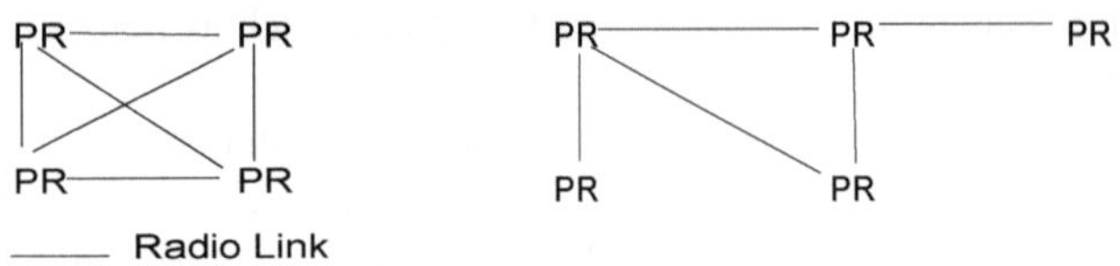

Fig. 9.4 Examples of PR-network

MANET was created in 1995 in order to provide RFC to propose routing protocols for ad hoc networks. At that time, only Hiperlan1 could provide information forwarding in an ad hoc environment. However, Hiperlan1 was creating a layer 2 routing which was not compliant with other standards. The objective of MANET is to adapt IP routing to work in ad hoc networks. We summarize four protocols which belong to two different concepts: reactive and proactive.

9.4.2 Reactive Protocols

Reactive protocols offer routes for information flows only on demand. It means that the protocol will react to find a route when a source node needs to send information to a destination node. During the remaining time, the node is idle or active to participate in the forwarding process to serve other source nodes. The reactive protocols do not consume many resources when idle, however to discover a route (since they don't have any topology information), they flood requests among the network and in this case consume a huge amount of the network resources. Examples of reactive protocols are AODV [16] and DSR [10].

9.4.2.1 AODV (Ad hoc On Demand Distance Vector)

AODV today has many successes and is used in many systems. The Zigbee and the Wi-Fi mesh networks use an adapted version of AODV. AODV belongs to the class of distance vector routing protocols. In these protocols, to reach a destination, the mobile node uses the next hop allowing the smallest distance in number of hops between it and the destination. AODV is a reactive protocol; hence mobile nodes do not retain any information for nodes not concerned by the active traffic information flows. After traffic forwarding is achieved, the routing information is maintained for a fixed time before it is purged from the routing table.

When a mobile node wishes to send packets to a destination, it starts checking if it has an available route in its routing table. If this is not the case, it will initiate a route discovery by sending a Route Request packet (RREQ) to locate the mobile recipient. This packet is distributed to its neighbors. The mobiles receiving the RREQ will look to determine if they have a route available to the recipient in their routing tables. If this is not the case, they in turn rebroadcast the RREQ packet to their neighbors and keep a trace[1]. RREQ packets are flooded throughout the network. When such a packet reaches the destination, or a mobile node that has a route to this recipient, a response packet (RREP) is generated by the mobile and sent back again, thanks to information stored in the caches traversed by the RREQ. During the flooding of the RREP packet, each mobile node will update its routing table by keeping the identity of the destination according to the selected route.

For example, in Figure 9.5, the mobile node X, which has no route to mobile node Y, initiates a route discovery process. As we can see, the RREQ generates many redundancies and uses a high number of resources. From the MAC layer point of view, many nodes will compete in order to win a slot within the conflict resolution protocol. When the number of requests is very high, this protocol might result in very poor network performance. In a dense network (high number of neighbors per node), the uncontrolled flooding is a disaster for network performance.

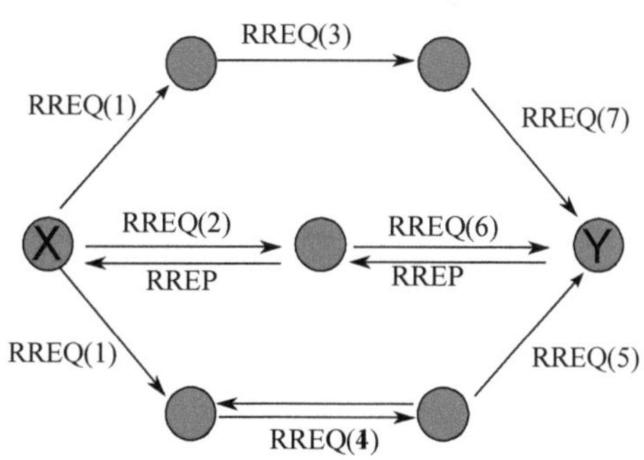

Fig. 9.5 Route discovery flooding

9.4.2.2 DSR (Dynamic Source Routing)

DSR is also a reactive protocol and it works in a very similar way to AODV. The main difference is in the routing procedure. AODV is table driven and DSR uses source routing. Nodes in DSR are idle and when a node decides to send traffic to a destination, it will initiate a route discovery in an identical procedure to AODV. Then to indicate the route, instead of creating local routing tables, each intermediate node

[1] Using sequence numbers on packets avoids infinite loops.

prints adds its identity to the packet header. The drawback of the source routing is of course the additional overhead in the packets. The advantage of source routing is that the route information is retained in the packets and nodes could at any moment be aware of the route from the source to the destination. This is not the case with routing tables because the unique information that they keep is the next hop. Knowing the complete route could facilitate the reaction of mechanisms for quality of service, security, energy, etc.

9.4.3 Proactive Protocols

A proactive routing protocol maintains regularly updated information on the network at each node. This information is obtained through messages sent periodically or triggered in response to important events that occur in the network. This information allows each mobile to keep topology knowledge offering the possibility to calculate a route for each destination in the network. The routing policy adopted in packet radio networks has a proactive approach. Examples of proactive routing protocols are DSDV [17] (Destination-Sequenced Distance-Vector), OLSR [8] (Optimized Link State Routing), and TBRPF [15] (Topology Broadcast Based on Reverse-Path Forwarding). OLSR and TBRPF became Experimental RFC. DSDV is, however, completely obsolete.

9.4.3.1 OLSR (Optimized Link State Routing Protocol)

OLSR attempts to maximize the dissemination of information packets in the network and reduce the overhead due to the flooding procedure. For this, the protocol is based on the concept of multipoint relays. A multipoint relay (MPR) for a node is one of its neighbors which is authorized to retransmit the information generated by the mobile. Each node selects its MPR according the following rule: choose a minimal subset of neighboring nodes that could cover all two-hop neighbors with symmetric links. Figure 9.6 displays a comparison between classical flooding and MPR. The denser the network, the more efficient is MPR. The MPR selection is related to the dominating set problem which is NP-complete making it necessary to use heuristics to reduce the time needed for the selection.

To establish its list of multipoint relays, each mobile regularly sends its identity and its list of neighbors. With this information, each mobile will know the network topology in its two-hop neighborhood. Thus it will be able to determine a list of multipoint relays which will cover all the neighbors in a two-hop distance. Each terminal informs its multipoint relay of its choice, which will act as a router for this terminal. The second step consists of sending the list of selected MPRs to all the nodes of the network (MPR flooding) and thus, each node which receives this information will be able to build a topology map and calculate the best route for each destination.

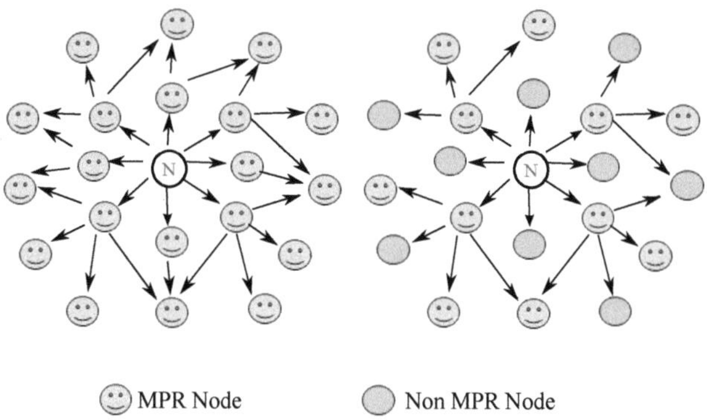

Fig. 9.6 Route discovery flooding

9.4.3.2 TBRPF (Topology Dissemination Based on Reverse-Path Forwarding)

As with OLSR, TBRPF is a proactive protocol that seeks to reduce the use of bandwidth by control messages. As opposed to OLSR, which maintains a partial topology of the network (using MPR), all mobiles in TBRPF have a knowledge of the complete topology of the network. Such knowledge is more expensive but allows the use of multiple paths or paths with a certain quality of service. For the network topology, each node will maintain a tree of shortest paths (it may be possible to use a different metric) to all other nodes. The tree is unique to each mobile. When a mobile detects a state change at one of its neighboring links, it sends an update on its associated tree. Each internal node in the tree forwards the update message. The trees themselves are constructed according to the update messages: when a node receives such a message, it computes the tree related message initiator. The update messages are sent only when detecting a status change (appearance or disappearance of a link). Only the neighbor discovery messages are sent periodically.

9.5 Quality of Service in Ad Hoc Networks

For the last several years, there has been a significant increase of interest in supporting QoS constraints in multi-hop mobile ad hoc networks (MANET). The specificity of MANETs make existing solutions for wireline networks unsuitable and a broad range of novel approaches have been studied. When considering QoS support, there are several aspects that must be taken care of:

1. One must choose a QoS model, i.e., the eventual goals one wants to achieve. The goals are tightly related to the applications that are to take advantage of the QoS support. Real-time audio/video applications require constant bandwidth and little delay, whereas FTP transfers can go with no more than classical best-effort.

Moreover, the former can adapt its throughput to available bandwidth and sustain data loss to some extent, while the latter requires a lossless channel. Most of the time, the model must comprise bandwidth and delay constraints.
2. A QoS-aware routing protocol should be used in order to find a suitable multi-hop path given some link constraints supported by the model.

3. If reservation is to be made, which is the case for most applications with QoS requirements, a signaling system has to be chosen to communicate between the source node and all the intermediate nodes. In many cases a strict reservation is not possibly because of the dynamic topology variations. Likewise, in case of constraint breakage, the signaling is used by the intermediate nodes either to repair the link—by reestablishing a bypassing path—or to report the failure to the source node. Lastly, when the connection ends, the reservation has to be torn down. Once more, because of the varying topology, the reservation is maintained in a soft state, i.e., it is never definitive and has to be renewed on a regular basis. This is due to the fact that a link failure may cause a partition of the network that would no longer allow contact between intermediate nodes. Signaling generally comes in two flavors: in-band and out-of-band. The first consists of control information piggybacked to regular data packets (e.g. as options in the IP header) and has the advantage of not transmitting additional packets that would contend with data packets and potentially waste bandwidth or delay. The second is a special kind of control packet that can be transmitted independently of any data flow.
4. Nodes must support some packet scheduling (fair queuing) algorithms in order to prioritize some packets at the expense of others as well as shape or police some flows. Optionally, if the MAC layer provides some kind of support for QoS, it can be taken advantage of in upper layers. Support can exist in direct low level delay respect or bandwidth allocation (e.g. in TDMA-based MAC protocols) or solely in link quality accounting capability.

In remainder of this section we present several methods that provide QoS in MANETs.

9.5.1 FQMM (Flexible QoS Model for MANETs)

FQMM [20] is an attempt to build a model explicitly for MANETs. It can be seen as a mix of IntServ and DiffServ in the way that it supports both per-flow and per-class granularity. Flows with borderline QoS requirements are granted per-flow processing while the others are aggregated in classes. As in DiffServ, there are three types of nodes: the sender (ingress node), the routers (interior nodes) and the receiver (egress node). The ingress node is required to police its outgoing traffic to meet a given traffic profile. Traffic profiles are expressed in terms of percentage of available bandwidth to cope with changing link conditions. The routing algorithm is left unspecified and is assumed to be of multi-path kind (i.e., all routes to a target are known, allowing for multiple choices with respect to some bandwidth requirements), though the use of some QoS-aware routing protocol is encouraged.

9.5.2 CEDAR (Core-Extraction Distributed Ad hoc Routing)

CEDAR [18] is a reactive protocol which optimizes routing requests by using core nodes, i.e., nodes that belong to the dominating set of the network. Each node periodically broadcasts BEACON messages which are used for one-hop neighborhood discovery. In a first phase of the protocol, each node must elect its core node. The set of nodes that are the core node of a neighbor is called the core of the network. Then, local topology information is disseminated by means of piggybacking link information to BEACON messages to the three-hop neighborhood of the nodes, allowing each core node to maintain virtual links to its neighboring core nodes. Thus, core nodes maintain information about the local core graph.

When a node s needs a route to d, it requests a route calculation to its core node, dom(s). By means of some chosen reactive routing protocols, dom(s) finds a core path to dom(d). After having determined the route to dom(d), s sends packets to d by means of source-routing. Now if QoS constraints are to be honored for a given route calculation, each core node must find a virtual link to the next core node towards dom(d) that satisfies the constraints. The use of core nodes for route calculation assumes that the core path is a good hint towards the optimal route. But it does not ensure that the best route can be found.

9.5.3 TBP (Ticket-based Probing)

TBP [7] attempts to achieve near optimal performance while avoiding flooding, and accounting for link information imprecision. QoS-satisfying routes are found by sending probes carrying some limited total amount of tickets towards the destination using distance vector information. At each intermediate node, a probe may be split up with the tickets distributed between the probes resulting from the split. The total number of tickets must remain the same along the propagation. The goal is to maintain a compromise between the odds of finding an optimal route (more tickets sent) and reduced control overhead (fewer tickets sent). If only one ticket is sent, then the plain, old shortest path is searched.

TBP relies on the assumption that stable links tend to remain stable, as opposed to so called transient links. Each node i collects statistical information about delay $D_i(t)$, bandwidth $B_i(t)$ and cost $C_i(t)$ to each other node t in the network. It is thus not meant to be scalable, but is rather designed for moderate scale networks. Along with $D_i(t)$ and $B_i(t)$, each node maintains the association variation $\Delta D_i(t)$ and $\Delta B_i(t)$ by which the next reported value will differ with the current one.

The protocol makes a distinction between delay and bandwidth constrained paths. Thus, it either searches a route for least-cost delay-constrained connections, or least-cost bandwidth-constrained connections. For each purpose, it defines two kinds of tickets—yellow ones and green ones—that make the probes either look for feasible paths or least-cost paths. The relative quantity of each kind governs which kind of path is going to be searched.

9.5.4 INSIGNIA

INSIGNIA [12] is meant to provide a complete framework for QoS support in MANET. It is based on the assumption that applications requiring QoS support also provide a way to modulate this requirement and can sustain service degradation by the use of adaptive services. The model thus defines three operating modes for a flow: best effort (BE), base QoS (BQ) constraints with minimum bandwidth, and enhanced QoS (EQ) with maximum bandwidth. It also assumes, quite rightfully, that the destination node is the best place to measure the QoS of a flow and make decisions about service degradation.

The framework provides a form of in-band signaling between source and destination by the addition of an option in the IP header of packets, the purpose of which is to attach QoS requirements to a flow as well as a means to alert the destination in case the requirements are not fulfilled at a bottleneck node. It is not specifically designed to operate with a given routing protocol and is rather meant to support any MANET routing protocol. Thus, it does not take any advantage of QoS-aware routing but rather relies on service adaptation on the application's side. Moreover, the framework remains independent of any MAC layer specificities which should ensure easy implementation with existing networking equipment. The destination is required to report the QoS level to the source in the feedback in order to modulate the flow to suit QoS conditions. Rerouting is provided by the routing protocol and quality is subject to degradation in case of failure to reserve resources along the new path. Old reservations disappear if they are not regularly refreshed.

9.5.5 SWAN (Stateless Wireless Ad hoc Network)

SWAN [3] provides QoS support through service differentiation. It is another framework independent of the routing protocol and the MAC layer, though it relies on the latter for bandwidth and delay measurement. The novelty is that routing nodes do not maintain any state information about the flows they forward, hence the term "stateless". Each node differentiates two kinds of traffic through DSCP flags in the IP header: real time (RT) and best effort (BE). RT flows are forwarded normally, whereas BE are fed to a traffic conditioner. The node actively monitors RT bandwidth usage and sets the conditioner's parameters accordingly to shape BE traffic to the residual bandwidth so that packets queues are short to avoid delay increase.

When a node intends to open a QoS-constrained connection, it sends a probe to measure bottleneck bandwidth and available delay. Here it seems that any routing protocol may be used, one that is QoS aware comes in especially handy. While traversing the network towards the destination, the probe is marked by the routing nodes with the measurement of the actual link delay and its minimum bandwidth. The destination copies this gathered information in the reply packet and sends it back to the source which can then decide whether to admit the flow or not. Once the flow is admitted, it has to be routed through the chosen path.

9.5.6 QOLSR (Quality of Service for OLSR)

QOLSR [9] is an extension of the OLSR protocol which is capable of providing multiple-metric routing. This extension can be added to OLSR functioning. No additional control traffic is generated (only augmented HELLO and TC messages). QOLSR protocol uses standard multipoint relays (MPRs) to ensure that the overhead is as low as possible for forwarding control traffic. Local QoS metrics on links are used to elect the quality of service MPRs (QMPRS). This information is then flooded in the network by TC messages to calculate routing tables. QOLSR can find optimal paths on the known partial topology having the same bandwidth performances as those on the whole network. These paths contain only QMPRs as intermediate nodes between a source destination pair. QOLSR carries out different functions which are required to perform the task of routing:

- Neighbor sensing allows, by means of hellos messages, each node to detect the neighbor nodes with which it has a direct and bi-directional link.
- Neighbor QoS measurement is done by the node to estimate the QoS conditions (available bandwidth, delay, loss probability, cost, security, power consumption, etc.) on links to each neighbor having direct and symmetric links.
- Quality of service multipoint relay selection is made by the node to design its own set of QMPRs. This set is calculated to contain a subset of the 1-hop neighbors which provides maximum bandwidth and minimum delay from each 2-hop neighbor to the given node. The QMPR set needs to be optimal, however it should be small enough to minimize the number of generated TC messages in the network. Topology Control extension messages are sent to all the nodes by each QMPR node at regular intervals to declare its QMPR selector set and QoS conditions.
- Each node maintains a routing table which allows it to route packets for the other destinations in the network with optimal metrics respecting QoS constraints.

9.6 Conclusion

In this chapter, we summarized the main issues in ad hoc networks. We started from the first ad hoc network created in the 70's and then we presented the current ad hoc standards. Those standards could offer communication on a single hop as in Bluetooth or Wi-Fi and on a multi-hop such as MANET. The chapter ends by introducing the existing frameworks that propose mechanisms to provide quality of service on the different existing MANET protocols. Nowadays, the tendencies are oriented to make those proposals implementable on existing materials. Proposed implementations still suffer from the lack of physical standards to be deterministic and to have the ability to guaranty the quality or to provide an accurate reference of this quality.

References

1. http://www.ieee802.org/11/: Web site of the IEEE 802.11 working group
2. IEEE Std 802.11-1997 Part11: Wireless LAN Medium Access Control (MAC) and Physical Layer (PHY) Specifications
3. Ahn, G., Campbell, A., Veres, A., Sun, L.: SWAN: Service differentiation in Stateless Wireless Ad hoc Networks. IEEE INFOCOM 2002 Proceedings **2**, 457–466 (2002)
4. Basagni, S., Bruno, R., Petrioli, C.: A performance comparison of scatternet formation protocols for networks of Bluetooth devices. First IEEE International Conference on Pervasive Computing and Communications (2003)
5. Bisdikian, C.: An overview of the Bluetooth wireless technology. IEEE Communication Magazine **39**, 86–94 (2001)
6. Bruno, R., Conti, M., Gregori, E.: Bluetooth: Architecture, protocols and scheduling algorithms. Cluster Computing Journal **5**, 117–131 (2002)
7. Chen, S., Nahrstedt, K.: Distributed quality-of-service routing in ad-hoc networks. IEEE Journal on Selected Areas in Communications **17**, 1488–1505 (1999)
8. Clausen, T., Jacquet, P.: Optimized Link State Routing (OLSR) protocol. RFC 3626, IETF (2003)
9. H. Badis and K. Al Agha: QOLSR, QoS routing for ad hoc wireless networks using OLSR. European Transactions on Telecommunications Journal, Wiley **15**, 427–442 (2005)
10. Johnson, D., Hu, Y., Maltz, D.: The Dynamic Source Routing protocol (DSR). RFC 4728, IETF (2007)
11. Jubin, J., Tornow, J.: The DARPA packet radio network protocols. Proceedings of the IEEE **75**, 21–32 (1987)
12. Lee, S., Ahn, G., Zhang, X., Campbell, A.: INSIGNIA: An IP-based quality of service framework for mobile ad hoc networks. Journal of Parallel and Distributed Computing **60**, 374–406 (2000)
13. Misie, J., Misic, V.: Performance analysis of Bluetooth piconets with finite baseband buffers. Wireless communications and mobile computing **5**, 917–925 (2005)
14. Mjidi, M., Chakraborty, D., Nakamura, N., Shiratori, N.: The impact of dynamic RTS threshold adjustment for IEEE 802.11 MAC protocol. Mobile Information Systems **5**, 5–20 (2009)
15. Ogier, R., Templin, F., Lewis, M.: Topology dissemination Based on Reverse-Path Forwarding (TBRPF). RFC 3684, IETF (2004)
16. Perkins, C., Belding-Royer, E., Das, S.: Ad hoc On-demand Distance Vector (AODV) routing. RFC 3561, IETF (2003)

17. Perkins, C., Bhagwat, P.: Highly dynamic Destination-Sequenced Distance-Vector routing (DSDV) for mobile computers. ACM SIGCOMM Computer Communication Review **24**, 234–244 (1994)
18. Sivakumar, R., Sinha, P., Bharghavan, V.: CEDAR: a Core-Extraction Distributed Ad hoc Routing algorithm. IEEE Journal on Selected Areas in Communications **17**, 1454–1465 (1999)
19. Wu, C., Hou, T.: The impact of RTS/CTS on performance of wireless multihop ad hoc networks using IEEE 802.11 protocol. IEEE International Conference on Systems, Man and Cybernetics **4**, 3558–3562 (2005)
20. Xiao, H., Seah, W., Lo, A., Chua, K.: A flexible quality of service model for mobile ad-hoc networks. IEEE VTC-spring **1**, 445–449 (2000)

Chapter 10
Compact Integer Programming Models for Power-optimal Trees in Ad Hoc Wireless Networks

Dag Haugland and Di Yuan

Abstract This chapter examines two types of optimization problems of minimizing the total transmission power required to satisfy some connectivity requirement for a group of nodes in ad hoc wireless networks. The first problem type is broadcast and multicast of messages from a source node to the rest of the group. The second problem type, also known as range assignment, amounts to forming a strongly connected subgraph containing all nodes of the group by bi-directional links. Optimal solutions to these problems are characterized by trees or arborescences. As a consequence of power minimization in a wireless transmission environment, the structural properties of these optimal trees and arborescences differ significantly from the classical ones. We discuss compact integer programming models for both problem types. In addition to reviewing models based on network flows, we develop a new compact model for range assignment. We provide theoretical analysis of how the models relate to each other in the strengths of their linear programming (LP) relaxations. Experimental results are provided to illustrate the performance of the models in bounding and in approaching integer optimality. The new compact model for range assignment turns out to be very competitive in both aspects.

10.1 Introduction

Power efficiency is a key aspect in many applications of ad hoc wireless networks, where the networking devices set up a self-organizing communication environment without pre-installed infrastructure (cf. mobile cellular networks). Two devices in ad

---

Dag Haugland
Department of Informatics, University of Bergen, PB. 7800, N-5020 Bergen, Norway, e-mail: Dag.Haugland@ii.uib.no

Di Yuan
Department of Science and Technology, Linköping University, SE-601 74 Norrköping, Sweden, e-mail: diyua@itn.liu.se

J. Kennington et al. (eds.), *Wireless Network Design: Optimization Models and Solution Procedures*, International Series in Operations Research and Management Science 158,
DOI 10.1007/978-1-4419-6111-2_10,

hoc networks can set up a communication link if the transmission power used allows them to directly reach each other. Devices can act as relaying nodes to provide multi-hop communication for others, if one-hop communication is not possible.

There is a wide range of current and potential applications of ad hoc networks. Applications in conferencing, home networking, emergency disaster relief, and survivable radio networks for military use are discussed by, for example, Perkins [44]. In [23], Conti and Giordano give a number of examples of using the ad hoc networking paradigm in wireless mesh, opportunistic, vehicular, and sensor networks. For application scenarios in wireless mesh and sensor networks, two excellent surveys are provided by Akyildiz et al. in [3] and [4], respectively. A platform of integrating ad hoc networking into GSM in order to extend service coverage is presented by Aggélou [1]. Experimental studies and real-world operation of ad hoc networking for group and peer-to-peer communication, radio dispatch systems, and sensor deployment for environment observation are reported in [24].

The devices in ad hoc networks are typically heavily energy-constrained (e.g., having battery as the only energy source). Hence, accomplishing communication tasks using a minimum amount of power is of high relevance. In addition to energy conservation, power minimization is of significance for reducing interference and thereby enhancing performance in, for example, wireless mesh systems. Unlike wired networks, where a transmission typically involves one sender and one receiver, multicast is an inherent transmission characteristic of the wireless medium, as a message transmitted by a node is received by all the nodes within the range defined by the transmission power. This is referred to as the multicast advantage [49]. Consequently, the power necessary to send a message to a set of nodes is not the sum, but the maximum of the powers required to reach each of the nodes in the set, i.e., the cost occurs at nodes instead of links.

In this chapter, we examine optimization problems of minimizing the total power required to satisfy some connectivity requirement for a group of nodes in ad hoc wireless networks. Two types of connectivity requirement are considered. The first is to connect a source node to the rest of the group nodes by directed links, and the second is to provide strong connectivity between all group members by bi-directional links. Optimal solutions to the two problem types are characterized by arborescences and trees, respectively. Interestingly, they can be viewed as the node-cost peers of several classical optimal tree and arborescence problems. The structural properties of our problems differ significantly from the classical ones, however, as a result of the multicast advantage. In particular, the problems considered in the chapter are all $\mathscr{N}\mathscr{P}$-hard. We refer to [16] for a compact survey of the approximability aspect.

Integer linear programming is one approach for the tree and arborescence problems considered in this chapter. A problem solution can be computed by integer programming and used by the network, however, only if centralized computation is feasible (e.g., the nodes are static and some central unit is powerful enough to afford the computational effort). Although this may be a realistic assumption in mesh networks, it is certainly not the case in many ad hoc networking scenarios, for which solutions have to rely on fast but heuristic algorithms. Yet, integer programming is very important for the purpose of evaluating heuristic algorithms in terms of optimality, which is not possible by comparing heuristics among themselves.

We are interested in not only using the models to approach integer solutions, but also the bounds provided by the linear programming (LP) relaxation of the models. The strength of LP relaxation is of importance for two reasons. First, the LP bound typically plays a critical role in the scalability of a model in solving the problem. Second, for instances where integer optimum is out of reach because of instance size, a strong model remains useful for providing a good lower bound for performance evaluation of heuristics, whereas LP-weak ones do not.

The chapter focuses on compact integer programming models. A model is compact if the numbers of variables and constraints are polynomial in instance size. Because the problems concern power minimization subject to a connectivity requirement, most of the compact models are, as can be expected, based on the use of network flows. For the second problem type, we will, in addition to reviewing some flow models in the literature, give a new compact model. In total, seven models are presented. For each problem type, we give theoretical analysis of how the models relate to each other in the strengths of their LP relaxations. Experimental results are provided to illustrate the performance of the models in bounding and in approaching integer optimality.

10.1.1 Problem Overview

The first problem type we consider is the minimization of the total power for broadcast and multicast communications. These types of communications are used not only for data of broadcast or multicast nature, but also for various kinds of control information. Multicast is defined by a group of nodes, among which one is the message source and the others are the destinations. The messages generated by the source are to be routed to the destinations, either directly or via relaying. The power assignment at the nodes must directly or indirectly connect the source to the destinations. Note that message relaying may involve some nodes outside the group, if doing so can reduce the total power consumption. Broadcast is the special case of multicast where the group consists of all nodes in the network.

Following the problem names introduced in [49], throughout the chapter we use minimum energy broadcast tree (MEBT) and minimum energy multicast tree (MEMT) to refer to the broadcast and multicast cases, respectively. Abbreviation MET is used as a general notation to denote both problems. It can be easily proved that an optimal structure to MET is a directed tree rooted at the source node, i.e., an arborescence. Indeed, MEBT and MEMT resemble the classical minimum cost arborescence and the minimum Steiner arborescence problems, respectively.

The second problem type arises in topology control in ad hoc wireless networks. Given a group of nodes, the problem amounts to minimizing the power needed to provide strong connectivity between the group members, that is, all nodes in the group can reach each other in single or multiple hops. One additional requirement is that all links used in the solution must be symmetric, i.e., every link in the solution is bi-directional, in order to ease link-protocol operations. Similar to MEMT, some nodes outside the group may be used as relay points. A special case of the problem is that the group spans all nodes in the network.

Since the power determines the transmission range of a node, the term range assignment has been used in several references [6, 19] to refer to the problem of allocating power to connect nodes in an ad hoc wireless network by symmetric links. In this chapter, we will use the term Steiner range assignment problem (SRAP) if the group is formed by a proper subset of the nodes. The case where all nodes are in the group will be referred to as the all-to-all range assignment problem (ARAP). Moreover, RAP refers to the union of SRAP and ARAP. We remark that in the literature, the definition of RAP is narrower and refers to ARAP only. It is easy to see that the optimal solution to RAP is a tree, and SRAP and ARAP are the node-cost peers of the classical Steiner tree and minimum spanning tree (MST), respectively.

10.1.2 Outline of the Chapter

The remainder of the chapter is organized as follows: In Section 10.2, we provide formal definitions of MET and RAP. In Section 10.3, we review models based on network flows for MET, and discuss the strengths of the LP relaxations. Network flow models for RAP and the relationships between their LP bounds are presented in Section 10.4. In Section 10.5, we present a new compact model, which we refer to as the multi-tree model, for RAP, and prove that its LP is stronger than all the flow models in Section 10.4. Experimental comparison is presented in Section 10.6. Finally, concluding remarks are given in Section 10.7.

10.2 Problem Definitions

In this section, we define some mathematical notation, and give formal definitions of MET and RAP. Related works in the literature are reviewed. The section also uses an example to illustrate optimal trees for MET and RAP, and compare them to some classical optimal trees in graphs.

10.2.1 Notation and Assumptions

We model an ad hoc wireless network by a graph $G = (V, E)$, where V and E denote the sets of nodes (network units) and potential links between the nodes, respectively. For convenience, we let $N = |V|$. The elements of E are edges (i.e., not directed).

Without any loss of generality, we assume $i < j$ for all $(i, j) \in E$. Let A denote the set of arcs derived from E, i.e., $A = \{(i, j) : i, j \in V, (i, j) \text{ or } (j, i) \in E\}$. We use D to denote the group of nodes for which the connectivity requirement of MET or RAP is defined. For MET, one node in D, denoted by s, is the source, and the remaining $|D| - 1$ nodes are the destinations.

The minimum transmission power required at node i to reach node j is denoted by p_{ij}. We assume that the power parameter is symmetric, i.e., $p_{ij} = p_{ji}, (i, j) \in E$. Following most of the references on MET and RAP, we assume omni-directional antenna, and use the formula $p_{ij} = \kappa d_{ij}^{\alpha}$, where d_{ij} is the distance between i and j, α is an environment-dependent parameter (typically $2 \leq \alpha \leq 4$), and κ is a constant, to generate instances of MET and RAP, although this power formula is not a necessary assumption in the integer programming models. Merely for the sake of simplifying notation, we do not impose a maximum power limit at the nodes, and as a consequence G is a complete graph. In the presence of power limits, the problem of finding a feasible solution becomes $\mathscr{NP}$-complete, but all integer programming models suggested in this chapter remain valid after a minor change of notation. We use a vector $\mathbf{P} = (P_1, \ldots, P_N)$ to denote the power assignment. The vector elements will be used as variables in some of the models.

To present and analyze the models, it is often necessary to treat the power levels of a node in non-decreasing order. For any $i \in V$, let $\pi_i : \{1, \ldots, N-1\} \mapsto V \setminus \{i\}$ be a bijection such that $p_{i,\pi_i(1)}, \ldots, p_{i,\pi_i(N-1)}$ is monotonously non-decreasing. Under the aforementioned power formula, $\pi_i(k)$ is the kth closest node to i. To simplify the notation, we use $p_{(ik)}$ as a short-hand notation for $p_{i,\pi_i(k)}$. This notation style will be used as well for other entities indexed by pairs of nodes. Thus p_{ik} and $p_{(ik)}$ denote different entities at node i. The former is the power required to reach node k, whereas the latter is the power for reaching the kth closest node (i.e., node $\pi_i(k)$).

To ease the discussion of LP bounds, we use LP-M to denote, for an integer linear programming model M, the LP relaxation, and LP*(M) the optimal value of LP-M. As most of the models in the chapter use network flows, we introduce a compact notation for flow balance equations. Let $\mathscr{F}(G, \mathbf{b})$ denote the set of feasible flows in G with respect to vector $\mathbf{b} \in \Re^N$, where $\mathbf{b}$ contains the nodes' supply or demand. In other words, $\mathscr{F}(G, \mathbf{b}) = \{\mathbf{x} \in \Re_+^{|A|} : \sum_{j \in V:(i,j) \in A} x_{ij} - \sum_{j \in V:(j,i) \in A} x_{ji} = b_i, i \in V\}$. Two instances of vector $\mathbf{b}$ will be used: $\mathbf{b}_{sD}$ and $\mathbf{b}_{st}$. The former refers to the vector where $b_s = |D| - 1$, $b_i = -1$ for all $i \in D \setminus \{s\}$, and $b_i = 0$ otherwise. The definition of $\mathbf{b}_{st}$, for any $t \in D \setminus \{s\}$, is $b_s = 1$, $b_t = -1$, and $b_i = 0$ otherwise.

Our network flow models should hence be understood in the following way: The statement $\mathbf{x} \in \mathscr{F}(G, \mathbf{b}_{sD})$ means that $\mathbf{x}$ is a vector of flow variables in G such that the supply at the source node s is $|D| - 1$ and the demand at each of the destinations is 1. For all other nodes, conservation of flow is satisfied, which means that both supply and demand are zero. If $\mathbf{x} \in \mathscr{F}(G, \mathbf{b}_{st})$ then the supply at s and the demand at destination t are both equal to 1, whereas conservation of flow is satisfied at all other nodes, including the destinations $D \setminus \{t\}$.

The flow sets $\mathscr{F}(G, \mathbf{b}_{sD})$ and $\mathscr{F}(G, \mathbf{b}_{st})$ are used respectively in single-commodity and multi-commodity flow models. Network models of this kind are frequently used in the literature in order to formulate connectivity constraints, especially in applications of network design problems. For further reading on network flow models, see the textbook of Ahuja et al. [2].

10.2.2 Minimum Energy Broadcast and Multicast

By the notation in Section 10.2.1, MET can be formally defined as follows:

[MET] Find a power vector $(P_1, P_2, \ldots, P_N) \in \Re_+^N$ of minimum sum, such that the induced graph (V, E^P), where $E^P = \{(i,j) \in A : p_{ij} \leq P_i\}$, has a path from s to each $t \in D \setminus \{s\}$.

If $D = V$, we arrive at the broadcast case MEBT, otherwise it is MEMT. In an optimal solution to MET, (V, E^P) contains an arborescence having s as the root. For MEMT, the arborescence may use some nodes in $V \setminus D$.

Formal analysis of the $\mathcal{NP}$-hardness of MET is provided in [14, 15]. The problem class was first introduced by Wieselthier et al. [49, 50, 51, 52]. For MEBT, the authors developed a solution approach, referred to as the broadcast incremental power (BIP) algorithm, by an adaptation of Prim's algorithm for MST. For MEMT, the authors proposed a multicast incremental power (MIP) algorithm that combines BIP and a pruning phase.

Given that MET is $\mathcal{NP}$-hard, we know that there exists no polynomial algorithm for MET unless such an algorithm exists for all $\mathcal{NP}$-hard problems, which in its turn contradicts the well-known '$\mathcal{P}$ is not $\mathcal{NP}$' conjecture (see e.g. Fortnow [31] for a discussion of the $\mathcal{P}$ versus $\mathcal{NP}$ problem, including failed attempts to prove the conjecture). Discouraging results on the computational tractability of a problem often trigger the interest in fast solution approaches with a lower ambition than finding an optimal solution. If no bound on the distance from optimality can be found for the output of the method, it is referred to as a heuristic method. Otherwise, the method is called an approximation algorithm.

More precisely, an approximation algorithm for MET is a polynomial algorithm that, for all instances of the problem, produces a solution with no more total power consumption than some constant times the smallest achievable total power consumption. The constant in question must be common for all instances of the problem, and the smallest valid constant is called the approximation ratio of the algorithm. For a more detailed introduction to approximation algorithms, we recommend Chapter 35 of the textbook of Cormen et al. [25].

It has been proved that for geometric instances, BIP is an approximation algorithm for MEBT, and theoretical performance analysis has been conducted in several works. Wan et al. [47] gave $[13/3, 12]$ as a range of the approximation ratio of BIP. In [11], the lower bound is strengthened from 13/3 to approximately 4.6. In [36], Klasing et al. corrected the upper bound in [47] to 12.15. Navarra [41] improved the upper bound to 6.33. Ambühl [8] proved that a better upper bound is 6. We note that the bounds in [8, 36] are derived for the MST heuristic (i.e., use the MST as the solution of MEBT), but these results apply to BIP due to a lemma in [47]. For the MST heuristic, the bound in [8] is in fact tight. For MIP, Wan et al. [48] showed that the algorithm does not have a constant approximation ratio, but algorithms having this property can be obtained by adapting several algorithms developed for the Steiner tree problem.

Since the introduction of MET and the first algorithms BIP and MIP, many approximation algorithms and heuristics have been proposed in the literature. We refer to [12, 14, 15, 18, 30, 33, 36, 37, 42, 48, 53, 55], and the references therein, for additional details. Some of the algorithms [42, 48, 55] solve MET; others are specifically developed for the broadcast case MEBT. The algorithms differ in how they construct an arborescence and/or how to manipulate a given arborescence to seek for improvement.

Integer programming for MET has been investigated in [7, 10, 28, 38, 54]. The models in the references use network flow, sub-tour elimination, graph cuts, or constraints of set-covering type with an implicit tree representation, to enforce connectivity from the source to the destinations. Some comparative studies are reported in [10, 38]. The integer programming approaches in [7, 54] are specifically designed for MEBT; the other references deal with the general problem MET. In Section 10.3, we review and compare some compact MET models presented in [10, 28].

10.2.3 Range Assignment

In range assignment, the power vector **P** must enable a subgraph containing sufficiently many bi-directional links to strongly connect all nodes in D. Below we give a formal definition.

[RAP] Find a power vector $(P_1, P_2, \ldots, P_N) \in \Re^N_+$ of minimum sum, such that the induced graph (V, E^P), where $E^P = \{(i,j) \in E : p_{ij} \leq P_i \text{ and } p_{ji} \leq P_j\}$, connects the nodes in D.

Problems ARAP and SRAP correspond to $D = V$ and $D \subset V$, respectively. There always exists an optimal solution, in which (V, E^P) contains a tree. For ARAP, this is a spanning tree of graph G. For SRAP, the tree spans all nodes in D and possibly some Steiner nodes as well.

RAP arises in the context of optimal topology control [46] of wireless networks by means of power adjustment. Up to date, the literature has focused on the ARAP problem. The $\mathcal{NP}$-hardness of ARAP in a 3D Euclidean space (with the aforementioned power formula) was proved by Kirousis et al. [35]. Later, Clementi et al. [19] provided the 2D-space extension of the $\mathcal{NP}$-hardness result. Polynomial-time algorithms for the 1D case of ARAP are provided in [21, 26]. An extension of ARAP, where the power formula includes a node-specific scaling factor $c_i, i \in V$, i.e., $p_{ij} = c_i \kappa d_{ij}^{\alpha}$, is studied in [9]. Approximation algorithms and heuristics for ARAP with a hop limit constraint are provided in [17, 20, 22, 27].

In [5], Althaus et al. presented an integer programming model of ARAP using sub-tour elimination constraints. The authors presented a constraint generation scheme for solving the model, and provided an analysis of two approximation algorithms. In [6], the authors give, in addition to a more extensive presentation of their work in [5], a polynomial time approximation scheme. They also commented on the approximability of SRAP. Das et al. [29] presented an integer programming model based on network flow. Although the problem addressed in [29] assumes that nodes have sectored antennas instead of omni-directional antennas, the application of the model to RAP is straightforward. Using sub-tour elimination for ensuring connectivity, the authors of [13] computationally compared, for relatively small networks, three alternatives of variable definition to represent the total power consumption: a set of binary variables, one continuous power variable per node, and a single continuous variable for the entire network. The first two alternatives will be examined in more detail in Sections 10.3.1 and 10.4. In [39], Montemanni and Gambardella proposed two integer programming models based on network flow and graph cuts, respectively, and compare solution approaches using these models to that in [5]. An overview of the integer programming models and solution procedures in [5, 29, 39] and comparative numerical results are provided in [40]. Note that, in [40] and some other references, ARAP is referred to as the minimum power symmetric connectivity problem. In Sections 10.4 and 10.5, we review some flow-based models in [29, 40], and present new compact models.

10.2.4 Optimal Trees and Arborescence: Illustrative Example

In this section, we give an example illustrating optimal trees and arborescences as defined by problems MET and RAP, respectively, and compare them with optimal solutions to related well-studied problems. Since the related problems are defined in terms of cost minimization, we let cost and power be synonymous in this context.

If the cost parameters were associated with links instead of nodes, MET and RAP become classical optimal arborescences and trees. Specifically, MEBT, MEMT, ARAP, and SRAP are peers of minimum cost arborescence, Steiner arborescence, minimum spanning tree, and Steiner tree, respectively. It is instructive to use an example to illustrate the optimal solutions to MET and RAP, and compare them to the classical optimal tree and arborescence solutions. Figure 1(a) shows the topology of a network of 10 nodes. The power parameters are displayed in Figure 1(b). Note that, under the assumption of symmetric link costs, minimum cost arborescence and MST are equivalent (except the specification of link direction in the former), and have the same optimal value. The observation holds also for optimal Steiner arborescence and Steiner tree.

Figure 10.2 shows the optimal arborescence of MEBT and that of a group of 5 nodes for MEMT, as well as the corresponding optimal solutions of the classical minimum arborescence problem and the Steiner arborescence problem. There are clear differences between the minimum cost arborescence and the optimal MEBT arborescence. Once arc $(4,8)$ is used in the MEBT solution, the additional powers needed at node 4 to reach nodes 2,6 and 7 are less than those required by the min-

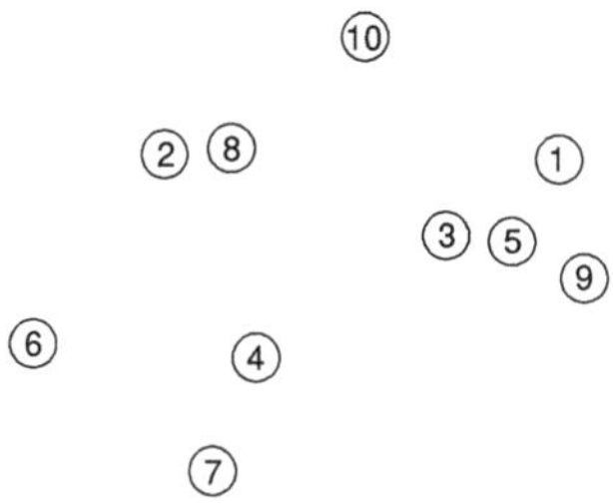

(a) Network topology

	1	2	3	4	5	6	7	8	9	10
1	-	103.08	12.00	87.12	6.42	217.10	145.65	84.20	7.70	44.14
2		-	55.88	26.63	76.70	32.37	60.44	1.00	118.87	33.93
3			-	34.51	1.70	131.40	74.70	43.46	11.75	37.09
4				-	9.11	37.92	10.54	25.43	75.34	75.38
5					-	162.02	91.96	61.77	4.69	46.61
6						-	37.71	42.87	213.99	132.11
7							-	62.84	118.32	142.03
8								-	100.31	24.51
9									-	76.08
10										-

(b) Power matrix

Fig. 10.1 A network example.

imum cost arborescence. For the same reason, arc $(3,10)$ is used to reach node 10 in Figure 2(b) instead of $(8,10)$. The difference between Figures 2(c) and 2(d) also originates from the multicast nature of wireless transmission.

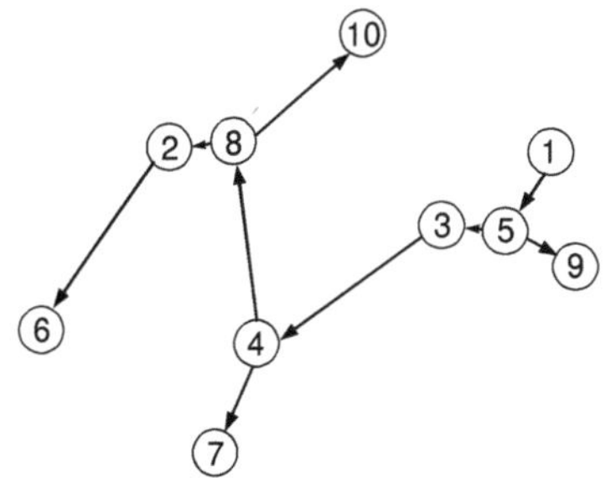

(a) Minimum cost arborescence (root = 1)

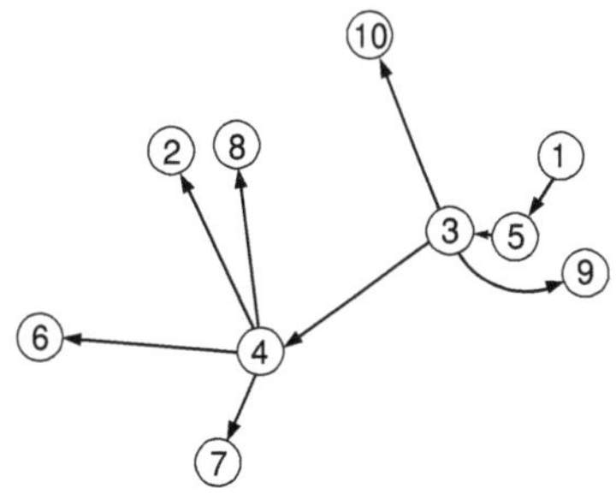

(b) Optimal MEBT tree

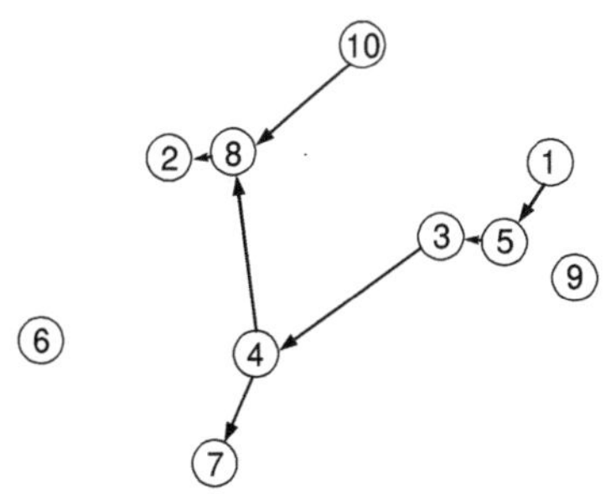

(c) Optimal Steiner arborescence for group {1, 2, 3, 7, 10} with root = 1

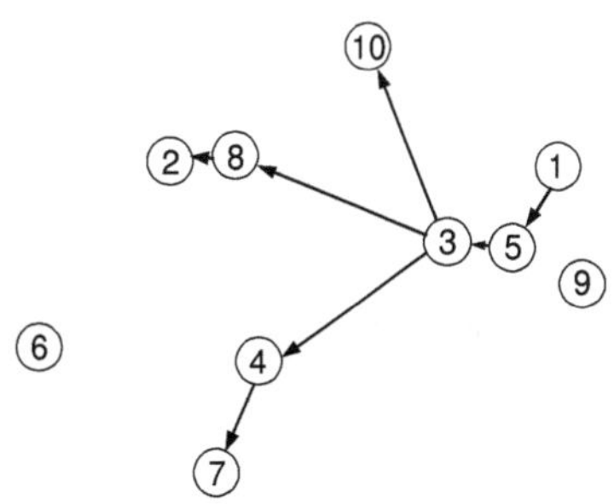

(d) Optimal MEMT tree for group {1, 2, 3, 7, 10} with root = 1

Fig. 10.2 Optimal arborescence for the network example.

In Figure 10.3, we illustrate, for the same network example and node group, the optimal trees for ARAP and SRAP. The figure shows also two classical optimal trees: MST and Steiner tree. The optimal ARAP tree differs from MST in the use of edge $(4,6)$ instead of $(2,6)$. Although edge $(4,6)$ is more costly than $(2,6)$, the former is preferable to the latter, since node 4 has links to nodes 3 and 8, and the additional power to reach node 6 is marginal. The difference between the optimal Steiner tree and the optimal SRAP tree is, again, a result of the node-oriented cost in SRAP. In the optimal SRAP tree, node 3 can reach node 1 for free, because of the existence of edge $(3,4)$. Thus the power required to connect nodes 1 and 2 equals p_{13} (i.e., the power assigned to node 1). This gives lower power consumption than using the Steiner node 5, as $p_{13} \leq 2p_{15}$.

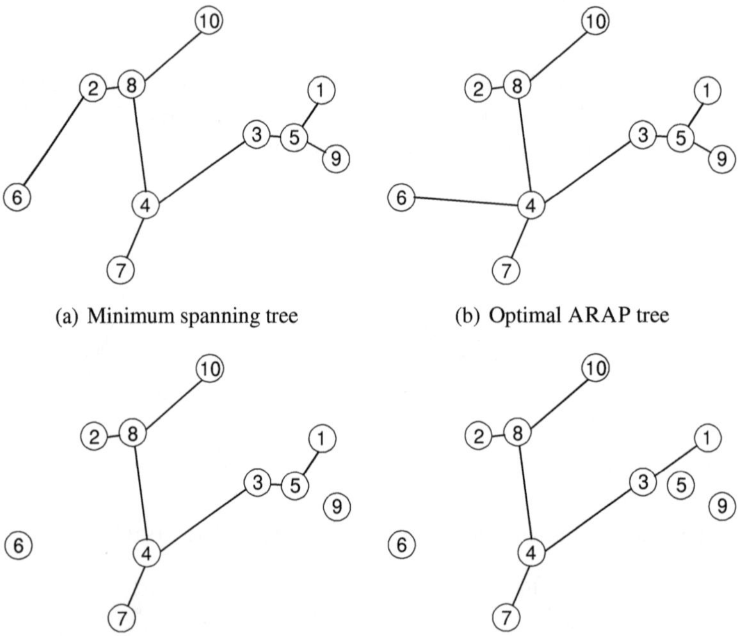

(a) Minimum spanning tree

(b) Optimal ARAP tree

(c) Optimal Steiner tree for group {1, 2, 3, 7, 10}

(d) Optimal SRAP tree for group {1, 2, 3, 7, 10}

Fig. 10.3 Optimal trees for the network example.

10.3 Network Flow Models for MET

In this section, we review some integer programming models for MET. In these models, connectivity is ensured by sending some flow from the source node to the destinations. The models differ in size and in the strength of the LP relaxations.

10.3.1 Models and Analysis

The first flow model is presented by Das et al. in [28]. Below, we give the variable definitions and the model.

$$P_i = \text{Power of node } i \in V,$$

$$x_{ij} = \text{Flow on arc } (i,j) \in A,$$

$$z_{ij} = \begin{cases} 1 & \text{if arc } (i,j) \text{ is used in the solution,} \\ 0 & \text{otherwise.} \end{cases}$$

$$\begin{aligned}
\text{[MET-Das]} \quad \min \quad & \sum_{i \in V} P_i \\
\text{s. t.} \quad & \mathbf{x} \in \mathscr{F}(G, \mathbf{b}_{sD}), && (10.1)\\
& x_{ij} \leq (|D|-1) z_{ij}, (i,j) \in A, && (10.2)\\
& p_{ij} z_{ij} \leq P_i, (i,j) \in A, && (10.3)\\
& \mathbf{z} \in \{0,1\}^{|A|}. && (10.4)
\end{aligned}$$

By constraint (10.2), we can have flow only on arcs $(i,j) \in A$ for which $z_{ij} = 1$. The flow conservation constraints (10.1) hence imply that the graph induced by $\mathbf{z}$ contains paths from s to each $t \in D$. In this induced graph, the p_{ij}-value of any arc leaving node i is by (10.3) a lower bound on P_i, and thus P_i becomes no smaller than the power required at i. Finally, minimizing the objective function ensures that P_i attains exactly this power level.

The model MET-Das has $|V| + 2|A|$ constraints, $|V| + |A|$ continuous variables, $|A|$ binary variables, and a very sparse constraint matrix. Explicit representation of power in dedicated variables is however not needed, and the number of continuous variables can be reduced if we define the binary variables:

$$y_{ij} = \begin{cases} 1 & \text{if the power of node } i \text{ equals } p_{ij}, \\ 0 & \text{otherwise.} \end{cases}$$

Using the y-variables to replace the representation of node power in MET-Das, we obtain a second model, which we denote by MET-F1. The model no longer requires the use of z-variables.

$$\begin{aligned}
\text{[MET-F1]} \quad \min \quad & \sum_{(i,j) \in A} p_{ij} y_{ij} \\
\text{s. t.} \quad & (10.1), \\
& x_{(ik)} \leq (|D|-1) \sum_{\ell=k}^{N-1} y_{(i\ell)}, i \in V, k \in \{1, \ldots, N-1\}, && (10.5)\\
& \mathbf{y} \in \{0,1\}^{|A|}. && (10.6)
\end{aligned}$$

As stated by constraint (10.5), flow can be assigned to arc $(i,\pi_i(k))$ only if at least one of $y_{(ik)},\ldots,y_{(i,N-1)}$ equals one, i.e., only if the power of node i equals at least $p_{(ik)}$. Constraints (10.1) ensure again that there is a path of such arcs from the source to each destination. In comparison to MET-Das, the numbers of variables and constraints have gone down by $|V|$ and $|A|$, respectively, but the coefficient matrix is denser.

Proposition 10.1. *LP*(*MET-*Das)* $\leq$ *LP*(*MET-*F1).*

Proof. Assume $(\mathbf{x},\mathbf{y})$ is a feasible solution to LP-MET-F1, and define $z_{(ik)} = \min\left\{\sum_{\ell=k}^{N-1} y_{(i\ell)}, 1\right\}$ and $P_i = \max_{j:(i,j)\in A} p_{ij}z_{ij}$. If G has a directed cycle such that $\mathbf{x}$ is positive on each arc in the cycle, we can, without violating the constraints or altering the objective function value of LP-MET-F1, send flow in the reverse direction until at least one arc gets zero flow and the flow on all other arcs remains non-negative. We can thus assume that the flow vector $\mathbf{x}$ is acyclic, which means that (10.1) implies $x_{ij} \leq |D|-1 \ \forall (i,j) \in A$. Hence (10.2) is satisfied for all i,k for which $z_{(ik)} = 1$, and if $z_{(ik)} = \sum_{\ell=k}^{N-1} y_{(i\ell)}$, (10.2) follows from (10.5). Therefore, $(\mathbf{x},\mathbf{y},\mathbf{P})$ is feasible in LP-MET-Das, and by the definition of $\mathbf{P}$ and $\mathbf{z}$, we get $\sum_{i\in V} P_i$
$= \sum_{i\in V} \max_{j\in V:(i,j)\in A} p_{ij}z_{ij}$
$\leq \sum_{i\in V} \max_{k=1,\ldots,N-1} p_{(ik)} \sum_{\ell=k}^{N-1} y_{(i\ell)} \leq \sum_{i\in V} \max_{k=1,\ldots,N-1} \sum_{\ell=k}^{N-1} p_{(i\ell)}y_{(i\ell)}$
$= \sum_{i\in V} \sum_{\ell=1}^{N-1} p_{(i\ell)}y_{(i\ell)} = \sum_{(i,j)\in A} p_{ij}y_{ij}$, which completes the proof.

Introducing multi-commodity flow variables, where each commodity corresponds to a unique destination and vice versa, often leads to stronger formulations for network design problems. We now go on to demonstrate how this can be accomplished in the case of MET.

For consistency, we will reuse the variable notation $\mathbf{x}$ to denote flows in a multi-commodity model of MET. Doing so will not cause ambiguity, as the precise meaning of $\mathbf{x}$ will be clear from the context.

$$x^t_{ij} = \text{Flow to destination } t \in D\setminus\{s\} \text{ on arc } (i,j).$$

In compact notation, we have defined $\mathbf{x} \in \Re_+^{|A|(|D|-1)}$. For each destination t, we denote its components in $\mathbf{x}$, i.e., the flow vector of t, by $\mathbf{x}^t$. The multi-commodity flow model for MET is as follows:

$$\text{[MET-F2]} \quad \min \sum_{(i,j)\in A} p_{ij}y_{ij}$$
$$\text{s. t. } (10.6),$$
$$\mathbf{x}^t \in \mathscr{F}(G,\mathbf{b}_{st}), t \in D\setminus\{s\}, \tag{10.7}$$
$$\sum_{\ell=k}^{N-1} x^t_{(i\ell)} \leq \sum_{\ell=k}^{N-1} y_{(i\ell)}, i \in V, k \in \{1,\ldots,N-1\}, t \in D\setminus\{s\} \tag{10.8}$$

Constraint (10.8) is a strengthened version of the more intuitive constraint $x_{(ik)} \leq \sum_{\ell=k}^{N-1} y_{(i\ell)}$, stating that arc $(i,\pi_i(k))$ can carry flow only if the power at i is at least

$p_{(ik)}$. But in any optimal integer solution, if there is any flow leaving node i and arriving at destination t, then the flow amount is one unit, and it is carried by exactly one of the outgoing arcs of node i. Thus, $\sum_{\ell=k}^{N-1} x^t_{(i\ell)} \leq 1$, and the same sum is zero if $\sum_{\ell=k}^{N-1} y_{(i\ell)} = 0$, which proves the validity of (10.8).

In models MET-F1 and MET-F2, a node will use at most one of its candidate power levels. This corresponds to the following inequalities:

$$\sum_{j \in V \setminus \{i\}} y_{ij} \leq 1, i \in V. \tag{10.9}$$

Although (10.9) is valid, it is redundant for defining the integer optimum. A simple proof based on contradiction is to assume that, at integer optimum, there are multiple y-variables being equal to one at some node. Setting these y-variables other than the one corresponding to the highest power to zero, it is easily verified that (10.5) and (10.8) remain satisfied, and the total power decreases. Hence constraints (10.9) are not needed for the correctness of the models. In fact, they have no impact on the LP optimum either, as stated in the proposition below.

Proposition 10.2. *The LP relaxations of models* MET-*F1 and* MET-*F2 have an optimal solution satisfying* (10.9).

Proof. Consider MET-F2. Assume $(\mathbf{x}, \mathbf{y})$ satisfies (10.7)-(10.8), and that $\mathbf{y} \in [0,1]^{|A|}$. We can assume that $\mathbf{x}^t$ is acyclic for all t, which by (10.7) implies that $\sum_{\ell=k}^{N-1} x^t_{(i\ell)} \leq 1$. Define $\hat{\mathbf{y}} \in [0,1]^{|A|}$ by

$$\hat{y}_{(ik)} = \begin{cases} y_{(ik)}, & \text{if } \sum_{\ell=k}^{N-1} y_{(i\ell)} \leq 1, \\ \left(1 - \sum_{\ell=k+1}^{N-1} y_{(i\ell)}\right)^+, & \text{otherwise.} \end{cases}$$

Then $\sum_{\ell=k}^{N-1} \hat{y}_{(i\ell)} = \min\left\{1, \sum_{\ell=k}^{N-1} y_{(i\ell)}\right\} \leq 1$. It follows that $(\mathbf{x}, \hat{\mathbf{y}})$ satisfies (10.8), and is hence feasible in the LP-relaxation of MET-F2. Since $\hat{\mathbf{y}} \leq \mathbf{y}$, we also have $\sum_{(i,j)\in A} p_{ij}\hat{y}_{ij} \leq \sum_{(i,j)\in A} p_{ij} y_{ij}$, which completes the proof. For MET-F1 the proof is similar.

The following proposition states the relation between the LP relaxations of MEP-F1 and MEP-F2. By this proposition and Proposition 10.1, the LP relaxations of the models in this section, following the order in which the models are presented, become increasingly stronger.

Proposition 10.3. *LP**(MET-*F1*) $\leq$ *LP**(MET-*F2*).

Proof. The inequality is obtained by observing that any feasible multi-commodity flow in MET-F2 can be transformed into a feasible single-commodity flow in MET-F1 by defining $x_{ij} = \sum_{t \in D \setminus \{s\}} x^t_{ij} \; \forall (i,j) \in A$.

In addition to the flow-based models, two cut-based models for MET were proposed in [10]. Their underlying idea is that for any node set S containing the source but not all destinations, some arc in the corresponding cut set must be used for transmission. Expressed in terms of $\mathbf{y}$, this means that $y_{ij} = 1$ for some $i \in S$ and some $j \in V$ for which there exists some $k \in \bar{S} = V \setminus S$ such that $p_{ij} \geq p_{ik}$. Note that, since i can send directly to k if it is assigned power p_{ij}, node j does not have to be found in $\bar{S}$.

The number of constraints in the cut-based models is exponential in $|V|$, and therefore constraint generation schemes to solve them were proposed in [10]. The strongest cut-based model was shown to be exactly as strong as MET-F2; however, their overall performance turned out to be inferior to that of MET-F2. Since the cut-based models are non-compact, they will not be further discussed in the chapter.

10.3.2 Models Based on Incremental Power

The power of the nodes can be represented in an incremental way. This amounts to introducing the following variables:

$$y'_{ij} = \begin{cases} 1 & \text{if the power of node } i \text{ is at least } p_{ij}, \\ 0 & \text{otherwise.} \end{cases}$$

At each node $i \in V$, a necessary condition for feasibility is $y'_{(ik)} \leq y'_{(i,k-1)}, k = 2,\ldots,N$, that is, if the power of node i is greater than or equal to that of a link, then the y'-variables of all links requiring less power must be equal to one as well. Introducing a notation convention $p_{(i0)} = 0, i \in V$, the objective function is $\sum_{i \in V} \sum_{k=1}^{N-1} (p_{(ik)} - p_{(i,k-1)}) y'_{(ik)}$. By utilizing these results, we can reformulate MET-F1 and MET-F2 using $\mathbf{y}'$ and thus treating the power levels incrementally.

The relationship between the y- and y'-variables is defined by the following equations. Note that the mapping defined by the equations is unique.

$$y'_{(ik)} = \sum_{\ell=k}^{N-1} y_{(il)}, \; k = 1,\ldots,N-1, \tag{10.10}$$

$$y_{(i,N-1)} = y'_{(i,N-1)}, y_{(il)} = y'_{(i\ell)} - y'_{(i,\ell+1)}, \; \ell = 1,\ldots,N-2. \tag{10.11}$$

As results of (10.10) and (10.11), a solution defined in the y-variables corresponds to a solution in the y'-variables, and vice versa. This holds also for the LP relaxations of the models, provided that (10.9) is present. Because of this type of equivalence, and the observation that reformulation by incremental power does not bring noticeable computational benefit [10], we do not discuss the models using incremental power in any further detail.

10.4 Network Flow Models for RAP

The models in Section 10.3 can be adapted to RAP, by including additional constraints to ensure strong connectivity between nodes in D by bi-directional links. For the resulting models, propositions corresponding to those in Section 10.3 apply.

The first flow model for RAP is due to Das et al. [29]. Among the nodes in D, one is selected as the source node. As with MET, we denote it by s. Utilizing the underlying idea of MET-Das leads to the following model:

$$\begin{aligned} \text{[RAP-Das]} \quad \min \quad & \sum_{i \in V} P_i \\ \text{s. t.} \quad & (10.1), (10.2), (10.3), (10.4), && (10.12) \\ & z_{ij} = z_{ji}, (i,j) \in E. && (10.13) \end{aligned}$$

In RAP-Das, the last set of constraints (10.13) ensures symmetry. As a result of the constraints, a link can be used if and only if the power values of both nodes are high enough to establish bi-directional communication.

A second flow model, obtained by adapting MET-F1 to RAP, is among the models proposed by Montemanni and Gambardella in [39]. We denote the model by RAP-MG to indicate the two authors. In [40], RAP-MG has been shown to be competitive in comparison to RAP-Das [29] and a model using sub-tour elimination constraints by Althaus et al. [5, 6].

$$\begin{aligned} \text{[RAP-MG]} \quad \min \quad & \sum_{(i,j) \in A} p_{ij} y_{ij} \\ \text{s. t.} \quad & (10.1), (10.5), (10.6), \\ & x_{\pi_i(k),i} \leq (|D|-1) \sum_{\ell=k}^{N-1} y_{(i\ell)}, i \in V, k \in \{1, \ldots, N-1\} && (10.14) \end{aligned}$$

The last set of constraints in RAP-MG plays the same role as (10.13). We remark that the flow model in [39] is in fact constructed using incremental power variables. The equivalence to RAP-MG follows from the discussion in Section 10.3.2. In [40], some valid inequalities for RAP-Das and RAP-MG are presented and used. We do not discuss these inequalities here, but refer to [40] for details. The inequalities will however be used for some of the numerical illustrations in Section 10.6.

By introducing symmetry-enforcing constraints to MET-F2, a multi-commodity flow model for RAP is obtained. We denote it by RAP-F2.

$$
\begin{aligned}
\text{[RAP-F2]} \quad \min \quad & \sum_{(i,j)\in A} p_{ij} y_{ij} \\
\text{s. t.} \quad & (10.6), (10.7), (10.8), \\
& \sum_{\ell=k}^{N-1} x^t_{\pi_i(\ell),i} \leq \sum_{\ell=k}^{N-1} y_{(i\ell)}, i \in V, k \in \{1,\ldots,N-1\}, t \in D \setminus \{s\}.
\end{aligned} \tag{10.15}
$$

As with the models in the previous section, constraints (10.9) are valid but redundant for defining the integer optimum of RAP, as well as at LP optimum. We formulate the latter by a proposition, and omit the proof as it is very similar to the one for Proposition 10.2.

Proposition 10.4. *The LP relaxations of models* RAP-*MG and* RAP-*F2 have an optimal solution satisfying* (10.9).

In terms of LP strength, we have observations similar to those in Section 10.3, leading to the following proposition. We omit the proofs as they can be easily obtained by adapting those of Propositions 10.1 and 10.3.

Proposition 10.5. *LP*(*RAP-*Das)* $\leq$ *LP*(*RAP-*MG)* $\leq$ *LP*(*RAP-*F2).*

10.5 A Strong Multi-tree Model for RAP

Khoury, Pardalos and Hearn [34] have formulated the minimum Steiner tree problem by exploiting the idea that any Steiner tree can be considered as $|D|$ Steiner arborescences rooted at distinct destinations. One can use a binary variable to indicate whether or not an arc in A is in the arborescence rooted at $t \in D$. This leads to the multi-tree formulation for the problem, the strength of which has been analyzed in depth by Polzin and Daneshmand [45].

In this section, we apply the multi-tree idea, and introduce the variables

$$
q^t_{ij} = \begin{cases} 1 & \text{if arc } (i,j) \text{ is used by node } i \text{ to reach node } t \in D \text{ in the tree} \\ 0 & \text{otherwise.} \end{cases}
$$

10.5.1 The Model

In addition to the q- and y-variables, the latter being defined in Section 10.3.1, the multi-tree model uses two more sets of binary variables. Note the reuse of notation z. The reason is that, except for being defined for edges instead of arcs, its meaning is identical to the z-variables in the previous sections. Let

$$z_{ij} = \begin{cases} 1 & \text{if edge } (i,j) \text{ is included in the tree,} \\ 0 & \text{otherwise.} \end{cases}$$

$$w_i = \begin{cases} 1 & \text{if node } i \text{ is included in the tree,} \\ 0 & \text{otherwise.} \end{cases}$$

The multi-tree model for RAP, denoted by RAP-MT, is defined by (10.16)-(10.24).

The first set of constraints (10.17) states the cardinality property of a tree: The number of edges selected equals the number of tree nodes minus one. Equations (10.18) and (10.19) state that w_i is one if and only if any outgoing arc of node i is used to reach any node $t \in D$, and that w_i must be one for all nodes in D, respectively. In the latter case, exactly one q-variable of the outgoing arcs is one for each of the other members of D, that is, node $i \in D$ must use exactly one arc to reach any other node in D. Equations (10.20) form a core set of constraints in the model. They connect the q- and z-variables. Note that, as a result of the tree structure, if $z_{ij} = 1, (i,j) \in E$, then exactly one of the two directions (i,j) and (j,i) will be used to reach a destination $t \in D$. If $z_{ij} = 0$, then clearly all q-variables must be zero in both directions of the edge. Therefore, we have an equality in (10.20). The next two sets of constraints, (10.21) and (10.22), are special cases of (10.20): For edge (i,j), arc (i,j) is used to reach $j \in D$ if and only if $z_{ij} = 1$. The same applies to the opposite direction. Node power is defined by (10.23), of which the construction is similar to that of (10.8).

$$[\text{RAP-MT}] \quad \min \sum_{(i,j)\in A} p_{ij} y_{ij} \tag{10.16}$$

$$\text{s. t.} \quad \sum_{(i,j)\in E} z_{ij} = \sum_{i\in V} w_i - 1, \tag{10.17}$$

$$\sum_{j:(i,j)\in A} q^t_{ij} = w_i, t \in D, i \in V \setminus \{t\}, \tag{10.18}$$

$$w_i = 1, i \in D, \tag{10.19}$$

$$q^t_{ij} + q^t_{ji} = z_{ij}, (i,j) \in E, t \in D \setminus \{i,j\}, \tag{10.20}$$

$$q^j_{ij} = z_{ij}, (i,j) \in E : j \in D, \tag{10.21}$$

$$q^i_{ji} = z_{ij}, (i,j) \in E : i \in D, \tag{10.22}$$

$$\sum_{\ell=k}^{N-1} q^t_{(i\ell)} \leq \sum_{\ell=k}^{N-1} y_{(i\ell)}, i \in V, k \in \{1,\ldots,N-1\}, t \in D, \tag{10.23}$$

$$\mathbf{q} \in \{0,1\}^{|A|\times|D|}, \mathbf{w} \in \{0,1\}^{|V|}, \mathbf{y} \in \{0,1\}^{|A|}, \mathbf{z} \in \{0,1\}^{|E|} \tag{10.24}$$

That RAP-MT is a correct model for RAP may not be very obvious. The model is correct, if the z-variables form a tree by the constraints of the model. Consider ARAP (i.e., $D = V$). In this case, all w-variables and consequently the right-hand sides of (10.18) are fixed to one. Moreover, the right-hand side of (10.17) becomes $|V| - 1$. Thus exactly $|V| - 1$ edges must be selected. Suppose that these edges do not form a tree. Then the graph defined by the edges has at least two disconnected components. Each of these graph components must have at least three edges, because constraints (10.20)–(10.22) cannot be satisfied otherwise. In addition, at least one of the components is a tree (i.e., contains no cycles), because there are $|V|-1$ edges in total. Consider any such tree component, and any one of its leaves, say m. Select arbitrarily a node n in another component. By (10.18), at node m, exactly one outgoing q-variable to destination n equals one. Thus if we start at m and follow this q-variable, we arrive at another node, say j. If j is not a leaf, another outgoing q-variable will lead to a new node, which cannot be i because of (10.20). Repeating the argument, we eventually arrive at a leaf node. At this point, (10.18) is violated since the node has only one edge (used by the q-variable in the incoming direction). We obtain a contradiction, and therefore RAP-MT is correct for ARAP.

We leave out the proof for SRAP, as the proof is quite lengthy and technical. However, it follows from Section 5 in [45] that constraints (10.17)-(10.22) and (10.24) define the set of feasible solutions to the Steiner tree problem. The additional constraints (10.23) ensure that $\mathbf{y}$ is assigned values according to its definition in Section 10.3.1, and hence RAP-MT is a correct model also for SRAP. Note that Proposition 10.2 as well as the proof given for MET-F2 apply also to RAP-MT.

10.5.2 LP Strength

For the minimum Steiner tree problem, it was shown in [45] that a multi-tree formulation is stronger than a multi-commodity flow formulation. We demonstrate that this property is inherited by RAP.

Proposition 10.6. *LP^*(RAP-F2) $\leq$ LP^*(RAP-MT).*

Proof. Consider any feasible solution $(\mathbf{q},\mathbf{y},\mathbf{z},\mathbf{w})$ to LP-RAP-MT. Define $\mathbf{x}^t = (\mathbf{q}^t - \mathbf{q}^s)^+ \; \forall t \in D \setminus \{s\}$. By (10.20), we get for all $i \in V$ and $t \in D \setminus \{i,s\}$ that

$$\sum_{j:(i,j)\in A} x^t_{ij} - \sum_{j:(j,i)\in A} x^t_{ji} =$$

$$\sum_{j:(i,j)\in A, q^t_{ij}\geq q^s_{ij}} \left(q^t_{ij} - q^s_{ij}\right) - \sum_{j:(j,i)\in A, q^t_{ji}\geq q^s_{ji}} \left(q^t_{ji} - q^s_{ji}\right) =$$

$$\sum_{j:(i,j)\in A, q^t_{ij}\geq q^s_{ij}} \left(q^t_{ij} - q^s_{ij}\right) - \sum_{j:(j,i)\in A, q^t_{ij}< q^s_{ij}} \left(z_{ij} - q^t_{ij} - z_{ij} + q^s_{ij}\right) =$$

$$\sum_{j:(i,j)\in A} \left(q^t_{ij} - q^s_{ij}\right) = 0,$$

where (10.18) has been used to obtain the last equality. That $\mathbf{x}^t$ satisfies conservation of flow also when $t = i$ follows from

$$\sum_{j:(t,j)\in A} x^t_{tj} - \sum_{j:(j,t)\in A} x^t_{jt} =$$
$$\sum_{j:(t,j)\in A} \left(q^t_{tj} - q^s_{tj}\right)^+ - \sum_{j:(j,t)\in A} \left(q^t_{jt} - q^s_{jt}\right)^+ =$$
$$-\sum_{j:(j,t)\in A} \left(z_{jt} - q^s_{jt}\right)^+ = -\sum_{j:(t,j)\in A} q^s_{tj} = -w_t = -1.$$

The equalities follow from applying the definition of $\mathbf{x}^t$, the facts that $q^t_{tj} = 0$ and $q^t_{jt} = z_{jt}$, (10.20), (10.18), and (10.19), respectively. Thus, $\mathbf{x}^t \in \mathscr{F}(G, \mathbf{b}_{st})$. Furthermore, (10.23) implies (10.8) since $\mathbf{x}^t \leq \mathbf{q}^t$. Utilizing (10.20) and (10.23), we get $\sum_{\ell=k}^{N-1} x^t_{\pi_i(\ell),i} = \sum_{\ell=k}^{N-1} \left(q^t_{\pi_i(\ell),i} - q^s_{\pi_i(\ell),i}\right)^+ = \sum_{\ell=k}^{N-1} \left(-q^t_{(i\ell)} + q^s_{(i\ell)}\right)^+ \leq \sum_{\ell=k}^{N-1} q^s_{(i\ell)} \leq \sum_{\ell=k}^{N-1} y_{(i\ell)}$, which implies (10.15). The proof is complete by observing that RAP-F2 and RAP-MT have identical objective functions.

The inequality in Proposition 10.6 may be strict. Some instances where this occurs are reported in the next section.

10.6 Experimental Results

In this section, we report some experimental results of solving MET and RAP using the models discussed in the previous sections. Networks of sizes 10, 20, 30, and 40 nodes are used in the experiments. They are obtained by following the instance generation procedure in [49] with $\alpha = 2$ in the power formula (Section 10.2.1). For each network size N, two groups of instances with $|D| = N/2$ and $|D| = N$, respectively, are generated. There are hence eight network groups in total. The number of instances in each group is 20. For MET, a node in D is randomly chosen as the source s.

We have applied CPLEX [32] (version 10.1) to all the models. Throughout the experiments, the CPLEX network optimizer is switched on, and all other solver parameters follow their default values. All experiments have been conducted on a server with an Opteron processor at 2.4 GHz and 7 GB RAM. For all models, we set a time limit of one hour for solving the LP relaxation, and one additional hour for approaching integer optimum.

For making comparisons among the models, solution time will be used as one of the performance indicators. However, the aim of the experiments is not to declare the best model for MET or RAP in terms of solution time. For each of the models, the time can be indeed improved by, for example, pre-processing, tweaking the solver parameters, or implementing methods other than using a standard solver. Instead of focusing on time only, the experiments are intended to also shed light on other aspects such as LP strength, size of the branch and bound (B&B) tree, and scalability.

Table 10.1 Results of solving the LP relaxations of the MET models.

Network $N/\|D\|$	MET-Das Max/min/**ave.**	MET-F1 Max/min/**ave.**	MET-F2 Max/min/**ave.**
10/5			
Gap (%)	86.9/74.8/**80.9**	78.8/47.6/**66.9**	0.0/0.0/**0.0**
Time	<0.1/<0.1/<**0.1**	<0.1/<0.1/<**0.1**	<0.1/<0.1/<**0.1**
10/10			
Gap (%)	91.8/85.4/**89.4**	88.9/69.4/**81.2**	2.1/0.0/**0.1**
Time	<0.1/<0.1/<**0.1**	<0.1/<0.1/<**0.1**	<0.1/<0.1/<**0.1**
20/10			
Gap (%)	94.0/88.8/**92.3**	88.4/78.0/**83.6**	3.8/0.0/**0.3**
Time	<0.1/<0.1/<**0.1**	<0.1/<0.1/<**0.1**	0.2/<0.1/<**0.1**
20/20			
Gap (%)	96.5/93.9/**94.9**	93.3/86.0/**89.0**	8.3/0.0/**0.9**
Time	<0.1/<0.1/<**0.1**	<0.1/<0.1/<**0.1**	1.1/0.3/**0.6**
30/15			
Gap (%)	96.6/93.9/**95.3**	91.7/84.9/**89.3**	6.1/0.0/**1.3**
Time	0.2/<0.1/**0.1**	0.1/<0.1/<**0.1**	6.2/1.3/**3.0**
30/30			
Gap (%)	97.6/95.8/**96.8**	95.7/89.3/**92.7**	9.2/0.0/**3.5**
Time	0.1/<0.1/**0.1**	0.1/<0.1/<**0.1**	28.0/7.4/**16.8**
40/20			
Gap (%)	97.4/96.1/**96.8**	94.3/90.2/**92.4**	12.4/0.0/**4.5**
Time	0.6/0.3/**0.5**	0.3/0.2/**0.2**	74.5/19.3/**28.5**
40/40			
Gap (%)	98.2/97.1/**97.7**	96.1/92.3/**94.7**	5.7/0.0/**2.9**
Time	0.7/0.4/**0.5**	0.2/0.1/**0.2**	157.5/46.7/**101.1**

10.6.1 Results for MET

In Table 10.1, we report results of solving the LP relaxations of the three models for MET. For each network group, the table contains the maximum, minimum, and average values of two entities over the 20 instances: the LP gap in relation to integer optimum and the LP solution time. The solution times are in seconds. In Table 10.1 as well as subsequent tables, average values are shown in bold face, and fractional values below 0.1 are not displayed to save space.

From the table, it is apparent that both LP-MET-Das and LP-MET-F1 can be solved rapidly for all network groups. In addition, LP solution time for these two models is insensitive to the group size $|D|$. Model MET-F2 is significantly larger than MET-Das and MET-F1, and consequently the solution time of LP-MET-F2 is magnitudes higher than those of LP-MET-Das and LP-MET-F1. Also, the group size has a clear impact on the solution time of LP-MET-F2. From a bounding viewpoint, LP-MET-F2 is superior in giving sharp bounds on the integer optimum. The gap of LP-MET-Das is huge – The average value is above 90% in most cases. The bound of LP-MET-F1 is only slightly better. For LP-MET-F2, the gap is no more than a few percent in average. One can also observe that the LP gap tends to grow with instance size for all the three models.

Table 10.2 summarizes the results of approaching integer solutions of MET. For each network group, the first row shows, within brackets, the numbers of instances (among a total of 20) solved to integer optimality by the models within the time limit. Then, the maximum, minimum, and average values of three performance indicators are displayed. The first indicator is the final gap. If an instance is solved to optimality, the gap is zero. Otherwise it is the difference between the best known lower and upper bounds reported by the solver when B&B terminates by time limit. If all 20 instances are solved to optimality, we use '–' to compactly denote zero gap. The second and third performance indicators are the solution time spent and the size of the B&B tree, respectively, for reaching the reported gap values.

From Table 10.2, we observe that, for networks of up to 20 nodes, integer optimum can be computed in a short amount of time by any of the models. The number of instances solved to optimality drops quickly for MET-Das and MET-F1, when $N \geq 30$. For $N = 40$, integer optimum becomes out of reach for the two models, and the remaining gap after running B&B for one hour is large. For MET-F2, integer optimum is guaranteed for size up to $N = 40$ and $|D| = 20$. For the largest network group with $|D| = 40$, more than half of the instances are solved to optimality by MET-F2, and the average remaining gap for the rest of the instances is quite small. MET-F2 also requires less computing time than the other two models. Thus MET-F2 scales better than MET-Das and MET-F1 in approaching integer solutions.

What has been observed above is a typical scenario for integer programming models that differ in size as well as the LP bound. MET-Das and MET-F1 are much smaller in size than MET-F2, thus solving LP-MET-Das and LP-MET-F1 is very fast. On the other hand, LP-MET-Das and LP-MET-F1 give significantly weaker bounds than LP-MET-F2. As a result, applying B&B to MET-Das and MET-F1 requires magnitudes more nodes for reaching optimum in comparison to MET-F2. For MET, a model with tight LP bound, although requires substantially more computation per node in the B&B tree, translates into moderate tree size and better overall scalability.

10.6.2 Results for RAP

In the computational experiments of the RAP models, we have chosen to enhance the two models RAP-Das and RAP-MG by means of valid inequalities. Without any valid inequality, the two models behave similarly to MET-Das and MET-F1, respectively. The valid inequalities originate from the work by Montemanni et al. [40]. In the reference, three and six sets of valid inequalities are presented for RAP-Das and RAP-MG, respectively. We refer to [40] for details. All these inequalities have been included in the experiments. We denote the resulting models by RAP-Das+ and RAP-MG+. As [40] deals with ARAP (i.e., $N = |D|$), we have adapted some of the inequalities in order to generalize them to SRAP.

Table 10.2 Results of computing integer solutions of MET.

Network $N/\|D\|$	MET-Das [Solved] Max/min/**ave.**	MET-F1 [Solved] Max/min/**ave.**	MET-F2 [Solved] Max/min/**ave.**
10/5	[20]	[20]	[20]
Gap (%)	–	–	–
Time	<1/<1/**<1**	<1/<1/**<1**	<1/<1/**<1**
B&B	32/0/**8.4**	0/0/**0**	0/0/**0**
10/10	[20]	[20]	[20]
Gap (%)	–	–	–
Time	< 1/<1/**<1**	<1/<1/**<1**	<1/<1/**<1**
B&B	47/0/**14.5**	0/0/**0**	0/0/**0**
20/10	[20]	[20]	[20]
Gap (%)	–	–	–
Time	20/<1/**6**	2/<1/**<1**	<1/<1/**<1**
B&B	6117/37/**1624.1**	2/0/**0.1**	0/0/**0**
20/20	[20]	[20]	[20]
Gap (%)	–	–	–
Time	49/2/**14**	2/<1/**<1**	4/<1/**2**
B&B	6809/110/**1828.6**	20/0/**2.8**	4/0/**0.3**
30/15	[16]	[13]	[20]
Gap (%)	23.2/0.0/**2.2**	21.8/0.0/**4.6**	–
Time	3669/27/**1292**	3626/2/**1531**	25/<1/**6**
B&B	169039/1214/**53054.6**	1218231/0/**453479.8**	29/0/**4.2**
30/30	[9]	[8]	[20]
Gap (%)	15.4/0.0/**5.0**	48.8/0.00/**10.4**	–
Time	3645/286/**2602**	3624/2/**2693**	282/5/**81**
B&B	66730/4720/**35022.4**	2077864/0/**850846.9**	207/0/**47.1**
40/20	[0]	[0]	[20]
Gap (%)	70.5/21.8/**39.7**	70.3/16.8/**40.8**	–
Time	3647/3626/**3631**	3624/3606/**3614**	912/11/**128**
B&B	53827/13600/**24966.7**	545090/284028/**413599.5**	549/0/**98.2**
40/40	[0]	[0]	[11]
Gap (%)	58.5/21.4/**42.9**	79.1/17.5/**48.5**	17.9/0.0/**4.1**
Time	3622/3614/**3619**	3622/3606/**3614**	3611/34/**1864**
B&B	10601/3652/**7486.6**	474001/285001/**374823.7**	1596/0/**220**

Table 10.3 shows the results of solving the LP relaxations of the four RAP models. The format of the table is identical to that of Table 10.1.

Without any valid inequality, the LP gap of RAP-Das is similar to LP-MET-Das. The inclusion of valid inequalities reduces the gap somewhat, but the LP relaxation remains too weak to provide a good bound on integer optimum. For LP-RAP-MG+, the effect of valid inequalities is much more pronounced. Although there is a large variation in the gap values, the average is below 30% in most cases. This may still be too large, if the sole purpose of solving the LP is to obtain a lower bound to integer optimum. For B&B, however, we will soon see that the valid inequalities

Table 10.3 Results of solving the LP relaxations of the RAP models.

Network $N/\|D\|$	RAP-Das+ Max/min/**ave.**	RAP-MG+ Max/min/**ave.**	RAP-F2 Max/min/**ave.**	RAP-MT Max/min/**ave.**
10/5				
Gap (%)	82.7/47.9/**63.0**	50.6/0.0/**15.7**	2.3/0.0/**0.4**	0.0/0.0/**0.0**
Time	<0.1/<0.1/<**0.1**	<0.1/<0.1/<**0.1**	<0.1/<0.1/<**0.1**	<0.1/<0.1/<**0.1**
10/10				
Gap (%)	74.3/48.0/**62.7**	32.9/0.0/**11.1**	1.5/0.0/**0.2**	0.0/0.0/**0.0**
Time	<0.1/<0.1/<**0.1**	<0.1/<0.1/<**0.1**	<0.1/<0.1/<**0.1**	0.1/<0.1/<**0.1**
20/10				
Gap (%)	84.2/54.5/**71.0**	44.8/4.3/**26.8**	13.6/0.0/**2.6**	0.6/0.0/<**0.1**
Time	0.1/<0.1/<**0.1**	0.1/<0.1/<**0.1**	0.4/0.1/**0.2**	1.9/0.3/**1.3**
20/20				
Gap (%)	79.6/60.4/**70.6**	39.5/5.0/**18.0**	7.3/0.0/**3.0**	0.4/0.0/<**0.1**
Time	0.1/<0.1/<**0.1**	0.1/<0.1/<**0.1**	2.9/0.7/**1.6**	12.1/0.5/**4.4**
30/15				
Gap (%)	84.5/67.1/**76.1**	55.0/16.8/**31.2**	15.1/1.5/**7.4**	0.0/0.0/**0.0**
Time	0.3/0.1/**0.2**	0.6/0.2/**0.4**	7.5/2.4/**4.8**	98.2/35.9/**62.5**
30/30				
Gap (%)	81.3/68.6/**75.4**	40.8/11.3/**20.1**	14.4/0.4/**5.8**	0.3/0.0/<**0.1**
Time	0.3/0.2/**0.2**	0.7/0.2/**0.4**	20.2/8.8/**13.2**	227.3/37.8/**145.6**
40/20				
Gap (%)	85.4/74.9/**79.6**	48.9/21.3/**34.7**	16.9/2.0/**8.4**	0.0/0.0/**0.0**
Time	1.3/0.5/**0.9**	3.7/0.9/**2.0**	49.8/23.4/**37.3**	608.0/310.0/**449.8**
40/40				
Gap (%)	81.0/70.2/**76.8**	34.3/13.4/**22.4**	12.5/0.9/**6.3**	0.0/0.0/**0.0**
Time	1.3/0.9/**1.1**	4.9/0.5/**1.9**	288.0/61.0/**98.2**	2931.6/841.7/**1848.6**

and its effect to LP gap bring substantial benefit to obtaining integer optimum by RAP-MG+. As for computing time, LP-RAP-Das+ and LP-RAP-MG+ take only seconds to solve, and the time is not sensitive to $|D|$.

As expected, the bound from LP-RAP-F2 is much tighter than LP-RAP-Das+ and LP-RAP-MG+. The average gap is below 9% for all network groups. The tightness of the bounds comes at the price of much longer solution time, up to a factor of about 50 for the largest network group.

The multi-tree model RAP-MT is distinctive in its bounding capability. For five of the eight network groups, the LP gap is zero for all the instances. For the other groups, the maximum gap is 0.6%, and the average is only 0.1%. Thus solving LP-RAP-F2 yields an excellent starting point for getting integer optimum. The computing time, on the other hand, is the highest among all the models. Comparing LP-RAP-MG+ to LP-RAP-MT, for example, one can observe that the time difference is at a factor of 1,000. For some instances, the solution time of LP-RAP-MT gets close to the limit of one hour.

Experimental results of computing integer solutions of RAP are reported in Table 10.4. The table format follows that of Table 10.2.

Table 10.4 Results of computing integer solutions of RAP.

Network $N/\|D\|$	RAP-Das+ [Solved] Max/min/**ave.**	RAP-MG+ [Solved] Max/min/**ave.**	RAP-F2 [Solved] Max/min/**ave.**	RAP-MT [Solved] Max/min/**ave.**
10/5	[20]	[20]	[20]	[20]
Gap (%)	–	–	–	–
Time	<1/<1/**<1**	<1/<1/**<1**	<1/<1/**<1**	<1/<1/**<1**
B&B	98/0/**27.1**	20/0/**1.2**	0/0/**0**	0/0/**0**
10/10	[20]	[20]	[20]	[20]
Gap (%)	–	–	–	–
Time	<1/<1/**<1**	<1/<1/**<1**	<1/<1/**<1**	<1/<1/**<1**
B&B	448/26/**87.1**	2/0/**0.2**	0/0/**0**	0/0/**0**
20/10	[20]	[20]	[20]	[20]
Gap (%)	–	–	–	–
Time	202/4/**50**	11/<1/**5**	9/<1/**2**	3/2/**2**
B&B	19055/312/**4690.5**	552/0/**206.6**	144/0/**8.3**	0/0/**0**
20/20	[20]	[20]	[20]	[20]
Gap (%)	–	–	–	–
Time	470/12/**153**	14/<1/**4**	55/<1/**8**	8/2/**4**
B&B	53887/1637/**12780.5**	1010/12/**200.2**	460/0/**37.0**	0/0/**0**
30/15	[4]	[20]	[20]	[20]
Gap (%)	32.5/0.00/**14.7**	–	–	–
Time	3469/331/**3292**	327/30/**131**	325/8/**48**	78/18/**40**
B&B	52461/6408/**32358.0**	6547/299/**2133.5**	2167/0/**328.1**	0/0/**0**
30/30	[0]	[20]	[20]	[20]
Gap (%)	44.5/17.2/**27.6**	–	–	–
Time	3645/3625/**3632**	348/12/**91**	1194/8/**182**	244/43/**98**
B&B	87621/37900/**54232.1**	6774/226/**1633.1**	2913/0/**330.7**	2/0/**0.1**
40/20	[0]	[11]	[19]	[20]
Gap (%)	91.2/57.5/**77.0**	12.5/0.0/**3.2**	3.69/0.0/**0.2**	–
Time	3628/3619/**3622**	3617/586/**2694**	3619/31/**827**	653/141/**435**
B&B	12701/4249/**7683.5**	14819/1627/**8691.2**	5603/2/**1582.0**	0/0/**0**
40/40	[0]	[13]	[10]	[20]
Gap (%)	86.3/50.0/**70.1**	10.2/0.00/**1.4**	11.2/0.00/**2.6**	–
Time	3627/3619/**3623**	3616/76/**2177**	3629/289/**2395**	3606/1023/**2066**
B&B	19501/6808/**12852.2**	14337/398/**6328.3**	3464/3/**583.3**	0/0/**0**

Examining Table 10.4, we observe that the behavior of RAP-Das+ is similar to MET-Das. The model can deliver integer optimum for small networks. For $N \geq 30$, the final gap between the lower and upper bounds is very large, except for a few instances for which optimum is reached. The number of B&B nodes is highest for this model, as a result of the weak LP.

RAP-MG+ clearly outperforms RAP-Das+. For $N \leq 30$, RAP-MG+ is solved to optimality for all the instances. The model also admits integer optimum for more than half of the cases for $N = 40$. Moreover, for instances that are not solved to optimality, the average final gap is pretty small. There are two reasons for the good performance of RAP-MG+. First, the valid inequalities turned out to speed up B&B significantly, in particular in obtaining high-quality integer solutions. Second, the solver (CPLEX) is able to utilize the structure of RAP-MG+ and strengthen further the LP by its built-in cut generator.

Model RAP-F2 has a performance comparable to that of RAP-MG+, although the former admits slightly more instances to be solved to optimality. The B&B nodes used is clearly fewer for RAP-F2. This, per se, does not necessarily mean an advantage. One can indeed observe that the final gap values as well as the times for getting integer solutions are in the same range for these two models. Note that the time values in Table 10.4 do not include those for solving LPs. Thus the overall time requirement of RAP-F2 is higher.

RAP-MT is the only model that could be solved to integer optimality for all the instances within the time limit. The extremely sharp LP bound of RAP-MT virtually eliminates the need of B&B enumeration – Integer optimum is found and verified without B&B for all but one instance, which required two B&B nodes. As for solution time, the model performs well in comparison to the other models. However, recall again that the LP time of RAP-MT is highest among the models, and therefore the overall time of RAP-MT is sometimes longer than those of RAP-MG+ and RAP-F2.

10.7 Concluding Remarks

In this chapter, we have presented a number of compact integer programming models for two types of closely-related power-optimal trees in ad hoc wireless networks. We have provided theoretical analysis of the strengths of the models in LP bound. Experiments on networks of various sizes have been presented to shed light on the models' performance in bounding and obtaining integer optimum.

Several concluding remarks can be made from the analysis and experiments. First, the models vary significantly in their LP strengths. For small networks, this is not a crucial aspect at all for approaching integer optimum. When the network size grows, however, the model choice becomes a crucial aspect, and an LP-stronger model tends to be more efficient in enabling optimal or close-optimal solutions. Second, the experimental results illustrate the trade-off between the time required for bounding and the size of the enumeration tree. For the two types of optimal trees considered in the chapter, the LP tightness pays off in terms of scalability. Third, for some models, valid inequalities are very useful for enhancing the performance, and sometimes a weak model combined with valid inequalities may be a better choice than a larger model with a tight LP-bound.

There are several directions of complementary investigations. One is a more extensive study of the multi-tree model, including bounding procedures based on this model for very large instances. Another topic is variations of RAP-F2 and RAP-MT, obtained by adapting models for the Steiner tree problem [45], and their performance study. Structural properties and further developments of solution procedures for the non-compact models in [10, 40] compose another line of study. A further topic is the adaption of the models to other optimal-tree problems (e.g., [43]) in wireless networks.

Acknowledgements The work of the first author is supported by the Research Council of Norway under contract 174982/I30. The work of the second author is supported by the Swedish Research Council (Vetenskapsrådet) and CENIIT, Linköping University, Sweden.

References

1. Aggélou, G.N.: An integrated platform for ad hoc GSM cellular communications. In Ilyas. M. (ed.) The Handbook of Ad hoc Wireless Networks, pp. 189–203. CRC press (2002).
2. Ahuja, R.K., Magnanti, T.L., Orlin, J.B.: Network Flows: Theory, Algorithms, and Applications. Prentice Hall, Upper Saddle River, NJ (1993)
3. Akyildiz, I. F., Su, W., Sankarasubramaniam, Y., Cayirci, E.: Wireless sensor networks: a survey. Computer Networks **38**, 393–422 (2002).
4. Akyildiz, I. F., Wang, X., Wang W.: Wireless sensor networks: a survey. Computer Networks **47**, 445–487 (2005).
5. Althaus, E., Călinescu, G., Mandoiu, I., Prasad, S., Tchervenski, N., Zelikovsky, A.: Power efficient range assignment in ad-hoc wireless networks. In: Proceedings of IEEE WCNC '03, pp. 1889–1994 (2003).
6. Althaus, E., Călinescu, G., Mandoiu, I., Prasad, S., Tchervenski, N., Zelikovsky, A.: Power efficient range assignment in ad-hoc wireless networks. Wireless Networks **12**, 287–299 (2006).
7. Altinkemer, K., Salman, F.S., Bellur, P.: Solving the minimum energy broadcasting problem in ad hoc wireless networks by integer programming. In: Proceedings of the second workshop on modeling and optimization in mobile, ad hoc, and wireless Networks (WiOpt '04), pp. 48–54 (2004).
8. Ambühl, C.: An optimal bound for the MST algorithm to compute energy efficient broadcast trees in wireless networks. In: Lecture Notes in Computer Science 3580, pp. 1139–1150 (2005).
9. Ambühl, C., Clementi, A., Ianni, M.D., Rossi, G., Monti, A., Silvestri, R.: The range assignment problem in non-homogeneous static ad-hoc networks. In: Proceedings of the IEEE 18th International Parallel and Distributed Processing Symposium (2004).
10. Bauer, J., Haugland, D., Yuan, D.: Analysis and computational study of several integer programming formulations for minimum-energy multicasting in wireless ad hoc networks. Networks **52**, 57–68 (2008).
11. Bauer, J., Haugland, D., Yuan, D.: New results on the time complexity and approximation ratio of the broadcast incremental power algorithm. Information Processing Letters **109**, 615–619 (2009).
12. Bauer, J., Haugland, D., Yuan, D.: A fast local search methods for minimum energy broadcast in wireless ad hoc networks. Operations Research Letters **37**, 75–79 (2009).
13. Baumann, F., Buchheim, C. (2009). Compact and extended formulations for range assignment problems. Zentrum für Angewandte Informatik Köln, Lehrstuhl Jünger, Germany. http://www.zaik.uni-koeln.de/ paper/unzip.html?file=zaik2009-582.pdf Cited 23 July 2009.
14. Čagalj, M., Hubaux, J.-P., Enz, C.: Minimum-energy broadcast in all-wireless networks: NP-completeness and distribution issues. In: Proceedings of ACM MobiCom '02, pp. 172–182 (2002).
15. Čagalj, M., Hubaux, J.-P., Enc, C.: Energy-efficient broadcasting in all-wireless networks, Mobile Networks and Applications **11**, 177–188 (2005).
16. Caragiannis, I., Kaklamanis, C., Kanellopoulos, P.: Power consumption problems in ad-hoc wireless networks. In Lecture Notes in Computer Science 2909, pp. 252–255 (2004).

17. Calinescu, G., Kapoor, S., Sarwat, M.: Bounded-hops power assignment in ad hoc wireless networks. Discrete Applied Mathematics **154**, 1358–1371 (2006).
18. Chu, T., Nikoladis, I.: Energy efficient broadcast in mobile ad hoc networks. In: Proceedings of the 1st International Conference on Ad-Hoc Networks and Wireless (ADHOC-NOW), pp. 177–190 (2002).
19. Clementi, A., Penna, P., Silvestri, R.: Hardness results for the power range assignment problem in packet radio networks. In: Lecture Notes in Computer Science 1671, pp. 197–208. Springer (1999).
20. Clementi, A., Penna, P., Silvestri, R.: The power range assignment problem in radio networks on the plane. In: Lecture Notes in Computer Science 1770, pp. 651–660. Springer (2000).
21. Clementi, A., Ianni, M.D., Silvestri, R.: The minimum broadcast range assignment problem on linear multi-hop wireless networks. Theoretical Computer Science **299**, 751–761 (2003).
22. Clementi, A., Penna, P., Silvestri, R.: On the power assignment problem in radio networks. Mobile Networks and Applications **9** 125–140 (2004).
23. Conti, M., Giordano, S.: Multi-hop ad hoc Networking: the reality. IEEE Communications Magazine **45** 88–95 (2007).
24. Conti, M., Crowcroft, J., Passarella, A.: Multi-Hop Ad Hoc Networks from Theory to Reality. Nova Science Publishers (2007).
25. Cormen, T.H., Leiserson, C.E., Rivest, R.L., Stein, C. Introduction to Algorithms, 2nd Edition. MIT Press and McGraw-Hill (2001).
26. Das, G.K., Das, S., Nandy, S.C.: Range assignment for energy efficient broadcasting in linear radio networks. Theoretical Computer Science **352**, 332–341 (2006).
27. Das, G.K., Ghosh, S., Nandy, S.C.: An efficient heuristic algorithm for 2D h-hops range assignment problem. In: Proceedings of IEEE GLOBECOM '04, pp. 1051–1055 (2004).
28. Das, A.K., Marks, R.J., El-Sharkawi, E., Arabshahi, P., Gray, A.: Minimum power broadcast trees for wireless networks: integer programming formulations. In: Proceedings of IEEE INFOCOM '03, pp. 1001–1110 (2003).
29. Das, A.K., Marks, R.J., El-Sharkawi, E., Arabshahi, P., Gray, A.: Optimization methods for minimum power bidirectional topology construction in wireless networks with sectored antennas. In: Proceedings of IEEE GLOBECOM '05, pp. 3962–3968 (2005).
30. Das, A.K., Marks, R.J., El-Sharkawi, E., Arabshahi, P., Gray, A.: r-shrink: a heuristic for improving minimum power broadcast trees in wireless networks. In: Proceedings of IEEE GLOBECOM '03, pp. 523–527 (2003).
31. Fortnow, L.: The Status of the P Versus NP Problem. Communications of the ACM **52**(9), 78–86 (2009).
32. ILOG CPLEX 10.0, Users' Manual, 2006.
33. Kang, I., Poovendran, R.: Iterated local optimization for minimum energy broadcast. In: Proceedings of the 3rd IEEE International Symposium on Modeling and Optimization in Mobile, Ad Hoc, and Wireless Networks (WiOpt '05), pp. 332–341 (2005).
34. Khoury, B.N., Pardalos, P.M., Hearn, D.W.: Equivalent formulations for the Steiner problem in graphs. In Du, D.Z. et al. (eds.), Network Optimization Problems, pp. 111–123. World Scientific Publishing, Singapore, 1993.
35. Kirousis, L.M., Kranakis, E., Krizanc, D., Pelc, A.: Power consumption in packet radio networks. In: Lecture Note in Computer Science 1200, pp. 363–374 (1997).
36. Klasing, R., Navarra, A., Papadopoulos, A., Pérennes, S.: Adaptive broadcast consumption (ABC), a new heuristic and new bounds for the minimum energy broadcast routing problem. In: Lecture Notes in Computer Science 2042, pp. 866–877, Springer (2004).
37. Liang, W. Constructing minimum-energy broadcast trees in wireless ad hoc networks. In: Proceedings of the 3rd ACM International Symposium on Mobile Ad Hoc Networking & Computing (MOBIHOC '02), pp. 112–122, 2002.
38. Min, M., Prokopyev, O.A., Pardalos, P.M.: Optimal solutions to minimum total energy broadcasting problem in wireless ad hoc networks. Journal of Combinatorial Optimization **11**, 59–69 (2006).

39. Montemanni, R., Gambardella, L.M.: Exact algorithms for the minimum power symmetric connectivity problem in wireless networks. Computers & Operations Research **32**, 2891–2904 (2005).
40. Montemanni, R., Gambardella, L.M., Das, A.K.: Models and algorithms for the MPSCP: an overview. In Wu, J. (ed.) Handbook on Theoretical and Algorithmic Aspects of Sensor, Ad Hoc Wireless, and Peer-to-Peer Networks, pp. 133–146. Auerbach Publications (2006).
41. Navarra, A.: Tighter bounds for the minimum energy broadcasting problem. In: Proceedings of the 3rd International Symposium on Modeling and Optimization in Mobile, Ad Hoc, and Wireless (WiOpt 2005), pp. 313–322 (2005).
42. Nguyen, G.D.: General algorithms for construction of broadcast and multicast trees with applications to wireless networks. Journal of Communications and Networks **7**, 263–277 (2005).
43. Papadimitriou, I., Georgiadis, L.: Minimum-energy broadcasting in multi-hop wireless networks using a single broadcast tree. Mobile Networks and Applications **11**, 361–375 (2006).
44. Perkins, C.E.: Ad Hoc Networking. Addison-Wesley (2001).
45. Polzin, T., Daneshmand, S.V.: A comparison of Steiner tree relaxations. Discrete Applied Mathematics **112**, 241–261 (2001).
46. Ramanathan R., Rosales-Hain, R.: Topology control of multihop wireless networks using transmit power adjustment. In Proceedings of IEEE INFOCOM '00, pp. 404–413 (2000).
47. Wan, P.-J., Călinescu, G., Li, X.-Y., Frieder, O.: Minimum-energy broadcast routing in static ad hoc wireless networks. Wireless Networks **8** 607–617 (2002).
48. Wan, P.-J., Călinescu, G., Yi, C.-W.: Minimum-power multicast routing in static ad hoc wireless networks. IEEE/ACM Transactions on Networking **12**, 507–514 (2004).
49. Wieselthier, J.E., Nguyen, G.D., Ephremides, A.: On the construction of energy-efficient broadcast and multicast trees in wireless networks. In: Proceedings of IEEE INFOCOM '00, pp. 585–594 (2000).
50. Wieselthier, J.E., Nguyen, G.D., Ephremides, A.: Algorithms for energy-efficient multicasting in static ad hoc wireless networks. Mobile Networks and Applications **6**, 251–263 (2001).
51. Wieselthier, J.E., Nguyen, G.D., Ephremides, A.: Distributed algorithms for energy-efficient broadcasting in ad hoc networks. In: Proceedings of IEEE MILCOM '02, pp. 820–825 (2002).
52. Wieselthier, J.E., Nguyen, G.D., Ephremides, A.: Energy-efficient broadcast and multicast trees in wireless networks, Mobile Networks and Applications **7**, 481–492 (2002).
53. Yuan, D.: Computing optimal or near-optimal trees for minimum-energy broadcasting in wireless networks. In: Proceedings of the third international symposium on modeling and optimization in mobile, ad hoc, and wireless networks (WiOpt '05), pp. 323–331 (2005).
54. Yuan, D: An integer programming approach for the minimum-energy broadcast problem in wireless networks. In: Proceedings of the International Network Optimization Conference (INOC), pp. 643-650 (2005).
55. Yuan, D., Bauer, J., Haugland, D.: Minimum-energy broadcast and multicast in wireless networks: an integer programming approach and improved heuristic algorithms. Ad Hoc Networks **6**, 696–717 (2008).

Chapter 11
Improving Network Connectivity in Ad Hoc Networks Using Particle Swarm Optimization and Agents

Abdullah Konak, Orhan Dengiz, and Alice E. Smith

Abstract In a mobile ad hoc network (MANET) the nodes can serve as both routers and hosts and can forward packets on behalf of other nodes in the network. This functionality allows the MANET to form an instant, autonomous telecommunication network without an existing infrastructure or a central network control. This chapter introduces a dynamic MANET management system to improve network connectivity by using controlled network nodes called agents. Agents have predefined wireless communication capabilities similar to the other nodes in the MANET. However, the agents' movements, and thus their locations, are dynamically determined to optimize network connectivity. A particle swarm optimization (PSO) algorithm is used to choose optimal locations of the agents during each time step of network operation.

11.1 Introduction

Mobile wireless ad hoc networks (MANET) are instantaneous, autonomous telecommunication networks that provide service to users wherever and whenever the service is needed. The communication depends on wireless links that are formed between the mobile nodes. A link is established between two nodes if they are within each other's wireless communication range. A MANET topology is dynamic and ex-

Abdullah Konak
Department of Information Sciences and Technology, Penn State Berks, Tulpehocken Road P.O. Box 7009, Reading, PA 19610-6009, USA, e-mail: auk3@psu.edu

Orhan Dengiz
DnD Technical Solutions, Tunus Cad. 87-5, Kavaklidere Ankara 06680, Turkey, e-mail: orhan.dengiz@gmail.com

Alice E. Smith
Department of Industrial and Systems Engineering, 3301 Shelby Center, Auburn University, Auburn, AL 36849, USA, e-mail: smithae@auburn.edu

J. Kennington et al. (eds.), *Wireless Network Design: Optimization Models and Solution Procedures*, International Series in Operations Research and Management Science 158, DOI 10.1007/978-1-4419-6111-2_11,

pected to change often and unpredictably due to arbitrary movements of its nodes. Therefore, a MANET might be disconnected intermittently during its mission time. This chapter introduces an approach to maintain global connectivity in MANETs by using controlled MANET nodes, called agents. The function of an agent node is to augment network connectivity. During a mission time of a MANET, the user nodes are assumed to move freely. However, their current and past location data are available, and a mobility prediction model can be used to predict their future locations. Based on the predicted locations of user nodes, the agent nodes are directed to move to new locations to improve connectivity among the user nodes. Although networks operate in continuous time, to make the optimization tractable, the continuous time operation is sliced into T discrete time steps.

Let $G_t(N_t, E_t)$ denote a MANET with node set N_t and arc set E_t at time t. Node set N_t includes two types of nodes: user nodes (UN_t) and agent nodes (AN_t). User nodes are nodes that demand network services. Agent nodes are responsible for helping the user nodes experience the best network service possible. At the beginning of time t, it is assumed that the current and past locations (x_{ik}, y_{ik}) are known for all nodes $i \in UN_t$ and $k = 1,...,t$. The decision problem at time t is to determine the best new location $(x_{i(t+1)}, y_{i(t+1)})$ for each agent node $i \in AN_t$ to maximize a measure of network connectivity. Note that it will take some time for agent nodes to move from their current locations at the beginning of time t to their new locations, and during this time user nodes may also move to new positions. Between time t and $t+1$, each agent node i moves to its new location (i.e., $(x_{i(t+1)}, y_{i(t+1)})$). It is assumed that at the beginning of each time period, new location data becomes available, and the new best locations of agent nodes should be determined again. Therefore, the decision problem should be solved as quickly as possible when new location data becomes available at the beginning of each time period, and the decisions about agents' next locations should be propagated to them.

This chapter presents a particle swarm optimization (PSO) algorithm to direct the motion of the mobile agents in MANETs where the global state of the network can be tracked using a GPS tracking system. In GPS tracking systems, each node is equipped with a GPS receiver, and the location data is periodically transmitted to a central location using low frequency radio or a satellite modem embedded in the node [17]. The proposed approach in this chapter includes a mobility prediction model to predict the future locations of the nodes, and the PSO algorithm determines the new locations of the agent nodes based on this prediction. The proposed PSO algorithm is tested under dynamic scenarios using a simulation environment. Among application areas of the proposed approach are network-centric warfare, fleet tracking, and search and rescue operations.

11.2 Background

In the literature, several approaches have been proposed to address the challenges in MANETs due to unpredictable user node movements. One of the major problems is the accessibility of the centralized network services used by all network nodes when the network is disconnected. A solution approach to this problem involves replicating network services [28, 29] and critical data [18] at multiple nodes and dynamically deploying these nodes to disconnected partitions of the network. Another problem is delivery of data packets across disconnected network partitions. To address this problem, a few papers [9, 31] propose using special agent nodes that can buffer packets until their destination nodes are reachable. These agent nodes may move arbitrarily or systematically, and when they are connected to a network partition, they deliver their payloads. This approach, however, is mainly applicable to delay-tolerant data networks. Alternatively, several papers propose topology control by modifying node trajectories or power level nodes. In [21], the intermediate nodes between a source and a destination node modify their trajectories to ensure the delivery of packets, and an algorithm to minimize the trajectory modifications is proposed. Goyal and Caffery [14] propose first determining critical links whose failures cause partitioning of the network, and then supporting these links by either modifying the trajectory of the nodes involved in the critical links or bringing an outside node to reinforce them. Kim et al. [20] use transmission power management schemes to strengthen the network topology across the critical nodes of a network.

There has been very limited work in the literature to improve network connectivity in MANETs through mobile agents. Ou [23] uses forwarding nodes that can adjust their locations to assist network partitions in an ad hoc network. When deployed initially, forwarding nodes move randomly to discover network partitions. It is assumed that each node has GPS capability and the nodes can exchange location data. When a forwarding node receives a service request, it will move to the service area if it is available (i.e., it is not servicing other nodes). However, the paths of the forwarding nodes are not optimized in this approach. Chadrashekar et al. [6] define the problem of achieving connectivity of a disconnected ground MANET by dynamically placing unmanned air vehicles (UAVs) which function as relay nodes. Then, they develop a heuristic algorithm for the single-UAV problem. Zhu et al. [32] also propose using UAVs equipped with communication capabilities to provide services to ground-based MANETs. They define a gradient-based algorithm to determine the location of a single UAV to maximize two network connectivity measures, global message connectivity and worst-case connectivity. Both connectivity measures are defined on a spanning tree so the optimization problem is to determine the most reliable Steiner tree of the network, where the UAV is considered a Steiner point in the network. Recently, Hauert et al. [15] propose the deployment of a swarm of UAVs for search and rescue missions. During a mission, UAVs are expected to maintain direct or indirect connection to their base-station through the ad hoc network that they form. An ant-colony approach [10] is developed to control the UAV MANET.

11.3 Proposed MANET Management System and Problem Description

The user nodes in the MANET move at their own will and it is assumed that their future positions are unknown. However, the location data of users and agents are assumed to be available to the agent control system at all times. Tracking user and agent node locations is technically possible by the GPS tracking technology discussed in Section 1. Arc set E_t at time t depends on the locations of the nodes and their transmission ranges as follows: $E_t = \{(i,j) : \min(R_{it}, R_{jt}) \geq d_{ijt} \quad \forall i,j \in N_t\}$ where R_{it} is the wireless transmission range of node i and d_{ijt} is the Euclidean distance between nodes i and j at time t. Each arc (i,j) has a transmission capacity u_{ijt} (in bits per second-bps), which is also a function of the distance between nodes i and j at time t.

11.3.1 Objectives and Evaluating Network Connectivity

The overall objective is to increase the connectivity of a MANET. In the ideal case, there should be at least one path between every user node pair in the network. Therefore, the primary objective is to connect each user node to every other one. This overall connectivity objective can be expressed as:

$$O_{1t} = 2 \times \frac{\sum\limits_{i,j \in UN_t : i<j} z_{ijt}}{nu_t \times (nu_t - 1)} \tag{11.1}$$

where $z_{ijt} = 1$ if there is a path between user nodes i and j at time t, and $z_{ijt} = 0$, otherwise, and nu_t is the number of user nodes at time t. Intuitively, equation (11.1) is the percentage of user node pairs that are connected at time t.

Equation (11.1) is a meaningful objective function if the network is disconnected. For a connected network, however, equation (11.1) is not sensitive with respect to changes in agent node locations. In other words, equation (11.1) will have the value of one for a possibly infinite number of agent node location combinations as long as the network is connected. However, each of these combinations may result in different network topologies, link capacities and network performance. A good measure to quantify the connectivity of a connected network is the capacity of the cut set with the minimum capacity among all cut sets of the network. Let $MaxFlow(G_t, i, j)$ be the maximum data (in bps) that can be sent from node i to node j through all possible paths between these two nodes on network G_t. $MaxFlow(G_t, i, j)$ provides a good measure of the connectivity between nodes i and j, as well as the quality of the links between them. $MaxFlow(G_t, i, j)$ can be calculated by solving a maximum flow problem with node i as the source and node j as the sink on capacitated network G_t. The minimum capacity cut set of a network can be determined by solving $MaxFlow(G_t, i, j)$ for all node pairs i and j of the network. However, solving the maximum flow problem is a computationally expensive operation, particularly for large problems. Since the network operates dynamically in real time, a good solution must be quickly found during each time step so that agent nodes can be properly deployed to their new locations. To reduce the computational time, the maximum flows between clusters of nodes, instead of between individual nodes, are considered in the approach herein. Let $C_{1t},...,C_{Kt}$ denote K clusters of user nodes at time t. The second criteria to gauge network connectivity is given as:

$$O_{2t} = \min_{\substack{i=1,...,K-1,\\ j=i+1,...,K}} \{MaxFlow(G_t, C_{it}, C_{jt})\} \tag{11.2}$$

In the calculation of $MaxFlow(G_t, C_{it}, C_{jt})$, first a dummy node d_i is connected to all nodes of cluster C_{it} using directional arcs with unbounded capacities and a dummy node d_j is connected to all nodes of cluster C_{jt} in a similar fashion, and then the maximum flow between d_i and d_j is calculated. O_{2t} improves the weakest connection between the clusters of user nodes. In addition, it is a responsive objective function for connected networks. For disconnected networks, O_{2t} is not required to be evaluated since it is always zero. In summary, O_{1t} is used to compare two solutions with disconnected topologies, and O_{2t} is used to compare two solutions with connected topologies.

11.3.2 Future User Location Prediction

The objectives defined in the previous section are considered for each time step t. However, the proposed system is expected to maximize network connectivity during a mission time of T time periods. Therefore, predicting the future locations of user nodes and deploying agent nodes based on these predictions may improve the average connectivity during the mission time. In the literature, there are a few papers addressing the location prediction problem in MANETs. Wang and Chang [30], Su et al. [26] and Tang et al. [27] utilize a simple location and velocity based expected position to help the routing protocols in MANETs. Ashbrook and Starner [2] propose a Markov model to predict the next important station which a user will be located at. This approach is only valid over the specific area that is used for learning. Mitrovic [22] develops a neural network model to predict car motion. This neural network is trained using specific maneuvers on certain road conditions. However, the proposed neural network approach can be adapted for more generalized motion patterns. Creixell and Sezaki [8] report promising results to predict pedestrian trajectories using a time series method. Kim et al. [19] propose a double-exponential smoothing approach to predict velocity and direction of mobile users. In a more recent study, Huang and Zaruba [16] propose a method that enables non-GPS equipped ad hoc nodes to estimate their approximate locations by using information from GPS equipped nodes.

Any of the location prediction methods cited above can be used in the proposed MANET management system, but in this chapter the double-exponential smoothing approach is selected since it is computationally efficient and it can be used to predict node velocities. The predicted location of user node i at time $t+H$ is given as:

$$\hat{x}_{i(t+H)} = \bar{x}_{it} + Hv^x_{it}$$
$$\hat{y}_{i(t+H)} = \bar{y}_{it} + Hv^y_{it}$$

where $\bar{x}_{it}$ and $\bar{y}_{it}$ are the smoothed locations of user node i along the x-axis and y-axis, respectively, and v^x_{it} and v^y_{it} are the corresponding estimated velocities. $\bar{x}_{it}$ and v^x_{it} are updated based on the past locations as follows:

$$\bar{x}_{it} = \alpha_1 x_{it} + (1-\alpha_1)\bar{x}_{i(t-1)}$$
$$v^x_{it} = \alpha_2(\bar{x}_{it} - \bar{x}_{i(t-1)}) + (1-\alpha_2)v^x_{i(t-1)}$$

where α_1 and α_2 are smoothing parameters between 0 and 1 for the expected location and velocity, respectively. Note that $\bar{y}_{it}$ and v^y_{it} are also updated in a similar fashion.

11.3.3 Optimization Problem and Deployment Decision

At the beginning of any time t, the locations of user nodes (UN_t) and agent nodes (AN_t) are known. The overall decision problem is to relocate the agents from their current positions at time t to their new locations at time $t+1$. However, considering predicted user locations further in the future than the next time period and deploying the agents accordingly may improve the connectivity of the network over the long term. Therefore, the optimization problem is formulated to determine the future locations of the agents to maximize network connectivity at time $t+H$. In other words, the PSO algorithm searches for the best location of each agent node i at time $t+H$. Let $(x_{i(t+1)}, y_{i(t+1)})$ denote the deployment decision of agent node i at time t. The distance that an agent node can travel during a time period is limited by $D_{\max}$, and the deployment decision of an agent node at a time period determines its starting location at the following time period. Therefore, the optimization problem at time t is expressed as follows:

Problem $ALOC(t,H)$:

$$Max \quad z = O_{1(t+H)} + O_{2(t+H)}$$

$$\sqrt{\left(x_{i(t+H)} - x_{it}\right)^2 + \left(y_{i(t+H)} - y_{it}\right)^2} \leq H \times D_{\max} \quad \forall i \in AN_t$$

The outcome of problem $ALOC(t,H)$ is the best location $(x^*_{i(t+H)}, y^*_{i(t+H)})$ of each agent node i at time $t+H$. However, due to the dynamic nature of the problem, at time $t+1$, which is the start of a new optimization cycle, the new location information of user nodes becomes available and the optimization problem will be solved again with this new information. Therefore, the outcome of problem $ALOC(t,H)$

must be implemented during the single time period between t and $t+1$. Based on $(x^*_{i(t+H)}, y^*_{i(t+H)})$, the deployment decision at time t is as follows:

$$\begin{aligned} x_{i(t+1)} &= x_{it} + (x^*_{i(t+H)} - x_{it})/H \\ y_{i(t+1)} &= y_{it} + (y^*_{i(t+H)} - y_{it})/H \end{aligned} \tag{11.3}$$

Equation (11.3) means that each agent node i is deployed at time t with a commitment to be at location $(x^*_{i(t+H)}, y^*_{i(t+H)})$ at time $(t+H)$ if the direction and velocity of the agent are not changed. Note that the velocities and directions of agent nodes are assumed to be constant during their deployment between times t and $t+1$.

11.4 Particle Swarm Optimization and Implementation

11.4.1 Particle Swarm Optimization (PSO)

PSO is a population-based meta-heuristic which emulates the social behavior of species that live in the form of swarms in nature. These swarms are capable of exchanging valuable information such as food locations in the habitat. PSO was developed by Eberhart and Kennedy [11]. In PSO, the aim of particles is to search for the optimal point in a continuous search space as they constantly move from one point to another according to their velocities. For a problem with n decision variables, each particle j in swarm S is represented by three vectors $\mathbf{X}_j = \{x_{1j}, ..., x_{nj}\}$, $\mathbf{P}_j = \{p_{1j}, ..., p_{nj}\}$, and $\mathbf{V}_j = \{v_{1j}, ..., v_{nj}\}$. Vector $\mathbf{X}_j$ represents the current position of particle j, $\mathbf{P}_j$ represents the position of the particle's historic best fitness, and $\mathbf{V}_j$ is the velocity vector that defines the direction and magnitude of the particle's movement in the search space. $\mathbf{V}_j$ is used to update $\mathbf{X}_j$ each iteration as follows:

$$\mathbf{X}_j \leftarrow \mathbf{X}_j + \mathbf{V}_j \tag{11.4}$$

As particles swarm over the search space, they communicate with each other, broadcasting their best found positions and learning about the best found positions of others. Based on these interactions as well as their own past experiences, particles adjust their velocities in each iteration as follows:

$$\mathbf{V}_j \leftarrow \omega \mathbf{V}_j + \phi_1 \cdot (\mathbf{P}_j - \mathbf{X}_j) + \phi_2 \cdot (\mathbf{G}_{N(j)} - \mathbf{X}_j) \tag{11.5}$$

where $N(j)$ is the set of particles which particle j shares information with (i.e., the neighborhood of particle j) and $\mathbf{G}_{N(j)} = \{g_{1j}, ..., g_{nj}\}$ is the best position discovered so far by the particles in $N(j)$. Vectors ϕ_1 and ϕ_2 are vectors of n uniform random numbers between 0 and φ_1 and between 0 and φ_2, respectively. Parameters φ_1 and φ_2 are called cognition and social coefficients, respectively, and are very important for facilitating convergence in PSO. According to an empirical study [24] as well as theoretical results [7], it is recommended to set φ_1 and φ_2 such that $\varphi_1 + \varphi_2 \leq 4$ in order to keep the growth in velocity magnitudes under control. Note that the operator $(\cdot)$ in equation (11.5) denotes element-by-element vector multiplication.

Parameter ω is called the inertia constant and is used to balance local versus global search. In a standard PSO, ω is usually set to one, which encourages exploration of new areas in the search space. For small values of ω such as $\omega < 0.4$, the search shifts to the intensification mode as particles tend to oscillate between $\mathbf{P}_j$ and $\mathbf{G}_{N(j)}$. Dynamically updating ω, starting with a value close to one for global search at the initial stages of the search and then gradually reducing it for local search at the final stages, has been shown to improve the performance of PSO algorithms [25]. Selecting a proper neighborhood structure $N(j)$ depends on the problem type. A common approach is to use a global neighborhood structure where each particle shares information with every other particle in the swarm (i.e., $N(j) = S$). The global neighborhood structure leads to fast convergence, but it may increase the chance of being trapped at a local optimum.

The pseudo code of the standard PSO is presented in Fig. 11.1 for a maximization objective function $f()$. In the pseudo code, $U(a,b)$ represents a random sample from the uniform distribution on $[a,b]$ and $\mathbf{G}^* = \{g_1,...,g_n\}$ is the best position found so far by all particles. The initial positions of particles are randomly determined between the maximum and minimum values of each decision variable ($x_i^{\max}$ and $x_i^{\min}$, respectively). Alternatively, particles can also be systematically initialized in the search space. Particle velocities are also randomly initialized between the maximum and minimum velocity limits ($v_i^{\max}$ and $v_i^{\min}$, respectively). Velocity limits usually depend on the ranges of the decision variables.

$$
\begin{array}{l}
\textbf{for } \text{each particle } \mathbf{X}_j \in S \textbf{ do } \{ \\
\quad x_{ij} \leftarrow U(x_i^{\min}, x_i^{\max}) \quad \text{for } i = 1,...,n \\
\quad v_{ij} \leftarrow U(v_i^{\min}, v_i^{\max}) \quad \text{for } i = 1,...,n \\
\quad \mathbf{P}_j \leftarrow \mathbf{X}_j \\
\quad \textbf{if } f(\mathbf{X}_j) > f(\mathbf{G}_{N(j)}) \textbf{ then } \mathbf{G}_{N(j)} \leftarrow \mathbf{X}_j \\
\quad \textbf{if } f(\mathbf{X}_j) > f(\mathbf{G}^*) \textbf{ then } \mathbf{G}^* \leftarrow \mathbf{X}_j \\
\} \\
\textbf{while}(\text{Stop is false})\{ \\
\quad \textbf{for } \text{each particle } \mathbf{X}_j \in S \textbf{ do } \{ \\
\quad\quad \textbf{for } i = 1,...,n \textbf{ do } \{ \\
\quad\quad\quad v_{ij} \leftarrow v_{ij} + U(0,\varphi_1)(p_{ij} - x_{ij}) + U(0,\varphi_2)(g_{ij} - x_{ij}) \\
\quad\quad\quad x_{ij} \leftarrow x_{ij} + v_{ij} \quad \} \\
\quad\quad \textbf{if } f(\mathbf{X}_j) > f(\mathbf{G}_{N(j)}) \textbf{ then } \mathbf{G}_{N(j)} \leftarrow \mathbf{X}_j \\
\quad\quad \textbf{if } f(\mathbf{X}_j) > f(\mathbf{P}_j) \textbf{ then } \mathbf{P}_j \leftarrow \mathbf{X}_j \\
\quad\quad \textbf{if } f(\mathbf{X}_j) > f(\mathbf{G}^*) \textbf{ then } \mathbf{G}^* \leftarrow \mathbf{X}_j \\
\quad \}\} \\
\text{return } \mathbf{G}^*
\end{array}
$$

Fig. 11.1 Pseudo code of the standard PSO.

PSO has successfully been applied to problems in the continuous domain. It has few parameters that require adjustment, which makes the development process relatively easy and fast. Implementation is also easy due to its simple but robust mechanism. PSO has also been applied to dynamic problems. Eberhart and Shi [12] test PSO on nonlinear optimization problems where the optimal points change during the search. They show the ability of PSO to track changing objectives. Carlisle and Dozier [5] modify the memory property (**P** vector maintenance strategy) of the swarm for dynamically changing environments. When a change in the problem environment is detected, all memory positions (**P**) are reevaluated and set to either the old memory or to the current particle position, whichever is better. Their results show that PSO can successfully track a time dependent objective function.

11.4.2 The Optimization Procedure

In order to solve Problem ALOC(t,H) using PSO, each particle j in swarm S is represented by three vectors: $\mathbf{X}_{jt} = \{(x_{ijt}, y_{ijt}) : i \in AN_t\}$, $\Delta\mathbf{X}_{jt} = \{(\Delta x_{ijt}, \Delta y_{ijt}) : i \in AN_t\}$, and $\mathbf{X}^*_{jt} = \{(x^*_{ijt}, y^*_{ijt}) : i \in AN_t\}$. Vector $\mathbf{X}_{jt}$ represents the current position of particle j in the solution space at time t, vector $\mathbf{X}^*_{jt}$ is the position of particle j with the best fitness found, and $\Delta\mathbf{X}_{jt}$ is the velocity vector that defines the direction and magnitude that the particle will travel in the solution space if not disturbed. The particle velocities are updated in each iteration as follows:

$$\begin{aligned} \Delta x_{ijt} &\leftarrow \omega\Delta x_{ijt} + U(0,\varphi_1)(x^*_{ijt} - x_{ijt}) + U(0,\varphi_2)(x^*_{it} - x_{ijt}) \quad \forall i \in AN_t \\ \Delta y_{ijt} &\leftarrow \omega\Delta y_{ijt} + U(0,\varphi_1)(y^*_{ijt} - y_{ijt}) + U(0,\varphi_2)(y^*_{it} - y_{ijt}) \quad \forall i \in AN_t \end{aligned} \tag{11.6}$$

where (x^*_{it}, y^*_{it}) denotes the location of agent node i according to the best solution found so far ($\mathbf{X}^*_t$) during the search at time t. In other words, the PSO uses a global neighborhood to update the particle velocities. The PSO employs a dynamic inertia coefficient ω. In this chapter, ω is initially set to 1.5 and decreased geometrically by a coefficient of 0.98 at each iteration. Although the common practice is to let the inertia coefficient decrease monotonically, from the preliminary experimentation it was found beneficial to occasionally reset it to its initial value. This improves the ability to escape local optima by "exciting" the particles every now and then, helping them swarm to other regions. The inertia coefficient is reset to its original value of 1.5 with a probability of 0.02, for each particle at each iteration step. With a relatively high inertia coefficient, the current direction and magnitude of the particle's motion are weighted heavily. As the iterations advance and the inertia coefficient is decreased, changing the direction and magnitude of the particle velocities toward $\mathbf{X}^*_{jt}$ and $\mathbf{X}^*_t$ becomes easier. The swarm particle velocities are limited within $[-D_{\max}, D_{\max}]$. After updating particle velocity $\Delta\mathbf{X}_{jt}$, the new position of each particle $j \in S$ is calculated as follows:

$$\begin{aligned} x_{ijt} &\leftarrow x_{ijt} + \Delta x_{ijt} \quad \forall i \in AN_t \\ y_{ijt} &\leftarrow y_{ijt} + \Delta y_{ijt} \quad \forall i \in AN_t \end{aligned} \tag{11.7}$$

Once new position vector $\mathbf{X}_{jt}$ is determined for particle j, the objective functions are evaluated for the new position, and $\mathbf{X}^*_{jt}$ and $\mathbf{X}^*_t$ are updated accordingly. The first step in the evaluation of $\mathbf{X}_{jt}$ is to form capacitated network $G_t(N_t, E_t)$. Then, O_{1t} can be calculated. If $O_{1t}(\mathbf{X}_{tj}) = 1$, then all user nodes are connected and O_{2t} is evaluated. To do so, K clusters of user nodes are determined using a clustering algorithm based on the Kruskal Spanning Tree Algorithm (see pages 515-516 of [1]). Then, the maximum flows between each node cluster pair are computed and the minimum of them is used as O_{2t}. There can be cases where two different disconnected networks have the same value of O_{1t} (i.e., $O_{1t}(\mathbf{X}_{jt}) < 1$ and $O_{1t}(\mathbf{X}_{jt}) = O_{1t}(\mathbf{X}^*_{jt})$). In such cases, a tertiary criterion is used to break ties between two disconnected networks with the same value of O_{1t} as follows. First a depth-first search is used to identify all disconnected partitions of the network such that user nodes in a network partition are connected, but they are not connected to the nodes in other network partitions. Then, the midpoint between the closest nodes of two network partitions is calculated for each partition pair. These midpoints are called attraction points since deploying the agent nodes close to the attraction point between two disconnected partitions may connect them. Finally, the distances between agents and attraction points are calculated and the minimum of these distances is used as the tie breaker between two disconnected networks with the same O_{1t}. This tertiary criterion is expressed as follows:

$$O_{3t} = \min_{\substack{i \in AN_t \\ j \in A_t}} \{\sqrt{(x_{it} - x_{jt})^2 + (y_{it} - y_{jt})^2}\} \tag{11.8}$$

where A_t is the set of attraction points at time t. The particle evaluation and update procedure of the PSO is given in Fig. 11.2.

Procedure EvaluateUpdateParticle()
Evaluate $O_{1t}(\mathbf{X}_{jt})$
if $O_{1t}(\mathbf{X}_{jt}) < 1$ and $O_{1t}(\mathbf{X}_{jt}) > O_{1t}(\mathbf{X}^*_{jt})$ **then** $\mathbf{X}^*_{jt} \leftarrow \mathbf{X}_{jt}$
if $O_{1t}(\mathbf{X}_{jt}) < 1$ and $O_{1t}(\mathbf{X}_{jt}) = O_{1t}(\mathbf{X}^*_{jt})$ **then**{
    Evaluate $O_{3t}(\mathbf{X}_{jt})$
    if $O_{3t}(\mathbf{X}_{jt}) < O_{3t}(\mathbf{X}^*_{jt})$ **then** $\mathbf{X}^*_{jt} \leftarrow \mathbf{X}_{jt}$
    }
if $O_{1t}(\mathbf{X}_{jt}) = 1$ **then**{
    Evaluate $O_{2t}(\mathbf{X}_{jt})$
    if $O_{2t}(\mathbf{X}_{jt}) > O_{2t}(\mathbf{X}^*_{jt})$ **then** $\mathbf{X}^*_{jt} \leftarrow \mathbf{X}_{jt}$
    }
if $O_{1t}(\mathbf{X}^*_{jt}) + O_{2t}(\mathbf{X}^*_{jt}) > O_{1t}(\mathbf{X}^*_t) + O_{2t}(\mathbf{X}^*_t)$ **then** $\mathbf{X}^*_t \leftarrow \mathbf{X}^*_{jt}$

Fig. 11.2 Solution evaluation procedure of the PSO algorithm.

At the beginning of a time step t, the inputs to the PSO algorithm are: (i) the best deployment decision from the previous interval (i.e., $\mathbf{X}^*_{t-1} = \{(x^*_{i(t-1)}, y^*_{i(t-1)}) : \forall i \in AN_{t-1}\}$), and (ii) the predicted user node locations at time $t + H$. The deployment decisions of the agents at time $(t-1)$ become their starting locations at time t. Each particle j is initialized as described in procedure InitializeSwarm() given in Fig. 11.3.

Procedure InitializeSwarm()
for each particle $\mathbf{X}_j \in S$ **do** {
$x_{ijt} \leftarrow x^*_{i(t-1)} + U(-H \times D_{\max}, HD_{\max}) \quad \forall i \in AN_t$
$y_{ijt} \leftarrow y^*_{i(t-1)} + U(-H \times D_{\max}, HD_{\max}) \quad \forall i \in AN_t$
$\Delta x_{ijt} \leftarrow U(-D_{\max}, D_{\max}) \quad \forall i \in AN_t$
$\Delta y_{ijt} \leftarrow U(-D_{\max}, D_{\max}) \quad \forall i \in AN_t$
}

Fig. 11.3 Particle initialization procedure of the PSO algorithm.

11.4.3 Semi-Intelligent Agent Node Behavior

The agent location optimizer requires agent nodes to be an integral part of the MANET in order to improve the objective functions. Since the main objective is to improve the experience of user nodes, agent nodes are not included in O_{1t} and O_{2t}. This means that an isolated agent node, which is an agent node with a node degree of zero or one, will have no effect on the objective functions. Note that to maximize O_{1t} and O_{2t} agent nodes must have two or more arcs. In dynamic settings, however, agent nodes can become isolated due to sudden changes in the directions and velocities of user nodes. The PSO algorithm may not guide an isolated agent to better locations since the objective functions are insensitive to small changes in the positions of isolated agents. If an agent becomes disconnected from the network, it has to be able to move independently and rejoin the MANET again to be properly utilized. A semi-intelligent behavior of agent nodes is adopted as described below.

Assume that for position $\mathbf{X}_{jt}$, particle j has an agent node i which is isolated at time t. Velocity Δx_{ijt} of this agent node is reset as follows:

$$\begin{aligned} \Delta x_{ijt} &\leftarrow x_{A_{ijt}} - x_{ijt} \\ \Delta y_{ijt} &\leftarrow y_{A_{ijt}} - y_{ijt} \end{aligned} \tag{11.9}$$

$x_{A_{ijt}}$ and $y_{A_{ijt}}$ are the x and y coordinates of the attraction point with respect to the agent node i in $\mathbf{X}_{jt}$. The attraction point for an isolated agent i depends on whether the network is connected or not. If the network is connected, then the attraction point is the last geometric center of the network and if the network is disconnected then it is the closest A_t that is used in the calculation of O_{t3}. By resetting the velocity of isolated agents, they are stimulated to rejoin the rest of the network. Considering this semi-intelligent behavior the overall procedure of the PSO algorithm is given in Fig. 11.4. In the pseudo code, $c_{\max}$ represents the maximum number of the PSO iterations during a time period. Inertia constant ω is randomly reset to its original value by a particle with a probability of 0.02.

```
Procedure PSO()
InitializeSwarm()
c ← 1; ω ← 1.5
while (c < c_max){
   c ← c+1
   ω ← 0.98ω
   for each particle j ∈ S do {
      Update ΔX_jt using (11.6)
      Calculate new particle position X_jt using (11.7)
      EvaluateUpdateParticle()
      Reset the velocity of isolated agents using (11.9)
      if U(0,1) < 0.02 then ω ← 1.5
   }
}
Return X*_t
```

Fig. 11.4 Overall procedure of the PSO algorithm.

11.5 Computational Experiments

The proposed PSO algorithm is tested using both dynamic and static scenarios. For the dynamic scenarios, the movements of user nodes are simulated using a random mobility model over a mission of T time periods. The details of this mobility model are given in section 11.5.1. To evaluate the performance of the MANET management system and the PSO algorithm in dynamic scenarios, two performance metrics are used based on the objective functions of the problem as follows:

$$P_1 = \frac{\sum_{t=1}^{T}(O_{1t})}{T} \qquad P_2 = \frac{\sum_{t=1}^{T} \min_{i,j \in UN_t: j>i} \{MaxFlow(G_t,i,j) : MaxFlow(G_t,i,j)>0\}}{T} \tag{11.10}$$

Performance metrics P_1 and P_2 are indicators of the algorithm performance over the entire simulation time span T. P_1 is the average user connectivity ratio and P_2 is the average all-pair minimum bandwidth in bits per second during the simulation time span. In dynamic scenarios, the performance of the PSO algorithm and the proposed MANET management system are tested as described in the procedure given in Fig. 11.5. Note that Problem $ALOC(t,H)$ is solved for the predicted locations of user nodes, and performance metrics P_1 and P_2 are evaluated using the actual locations of user nodes. Similarly, the clusters of user nodes are also determined in each time period based on the predicted locations of user nodes. A minimum spanning tree-based clustering algorithm (see [1] pages 515-516 for details) is used to identify K clusters of user nodes. First, the minimum spanning tree of user nodes is determined in the full mesh network of user nodes, where the rectilinear distances between user nodes are used as the arc weights, and then K disconnected clusters of user nodes are obtained by deleting (K-1) arcs of the minimum spanning tree one by one in decreasing order of their distances. The maximum flows between the pairs of user nodes (or user clusters) are calculated by an implementation of the highest-label push-relabel algorithm [13] available from BOOST C++ libraries, which is a peer-reviewed, freely available software library collection [3]. The PSO code was implemented in the C programming language and all runs were performed on Linux PCs with Dual 3.0 GHz Intel Xeon E5450 Quad-Core Processors and 32 GB RAM.

for $t = 1,\ldots,3$ **do** {
  Using the mobiliy model, generate locations (x_{it}, y_{it}) $\forall i \in UN_t$
}
for $t = 3,\ldots,T$ **do** {
  Predict $(\hat{x}_{i(t+H)}, \hat{y}_{i(t+H)})$ $\forall i \in UN_t$
  Determine K clusters, $C_{1t},\ldots,C_{Kt}$, based on the predicted user locations
  Solve Problem $ALOC(t,H)$ to determine $(x^*_{i(t+H)}, y^*_{i(t+H)})$ $\forall i \in AN_t$
  Deploy mobile agent i to $(x_{i(t+1)}, y_{i(t+1)})$ $\forall i \in AN_t$
  Simulate actual locations $(x_{i(t+1)}, y_{i(t+1)})$ $\forall i \in UN_t$
  Create network $G_{(t+1)}$
  Evaluate $G_{(t+1)}$
}
Calculate P_1 and P_2

Fig. 11.5 Procedure for testing dynamic problems.

11.5.1 Mobility Simulation Environment

To test the performance of the PSO algorithm under dynamic scenarios, a mobility simulation model was created to randomly generate the movements of user nodes. In the simulation model, each user node i is assigned to a random starting point (x_{i0}, y_{i0}) and to a random destination point (x_{iT}, y_{iT}), and each follows a path with random perturbations to its destination. Let $\mathbf{v}_{it}$ be the unit vector representing the direction of the motion of user node i at time t. At each time period, each user node i travels a random distance $U(d_{\min}, d_{\max})$ in the direction of vector $\mathbf{v}_{it}$ which is calculated as follows:

$$\mathbf{v}_{it} \leftarrow \begin{cases} \left(\alpha_v \mathbf{v}_{i(t-1)} + (1-\alpha_v)\frac{(x_{iT},y_{iT})-(x_{it},y_{it})}{|(x_{iT},y_{iT})-(x_{it},y_{it})|}\right) \begin{bmatrix} \cos(\theta) & \sin(\theta) \\ -\sin(\theta) & \cos(\theta) \end{bmatrix} & \text{if } U(0,1) < p, \\ \alpha_v \mathbf{v}_{i(t-1)} + (1-\alpha_v)\frac{(x_{iT},y_{iT})-(x_{it},y_{it})}{|(x_{iT},y_{iT})-(x_{it},y_{it})|} & \text{otherwise.} \end{cases}$$

where α_v is a weight factor for the current motion direction, θ represents the random changes in the direction of the motion, and p determines how frequently the direction is perturbed. The initial directions of user nodes are also randomly determined. This motion simulation model is a version of the Random Waypoint Model [4], which is frequently used in the literature as a benchmark mobility model to evaluate MANET management protocols. By changing parameters $d_{\min}$, $d_{\max}$, α_v, θ, and p, very complex user motions can be achieved. The simulation parameters for the test problems are: $d_{\min} = 0.01$, $d_{\max} = 0.05$, $\alpha_v = 0.95$, $\theta = U(-\pi/4, \pi/4)$, and $p = 0.1$ This motion scenario models the case where users search for their destinations or make their way around forbidden areas or obstacles, which is a reasonable representation of a search/rescue or a military operation.

Although the proposed MANET management system and the PSO algorithm can handle cases where the number of user and agent nodes change during the simulation, it is assumed that the number of agent and user nodes is constant during the simulation. The simulation area is a two-dimensional rectangular area with lower-left corner coordinates of $(0,0)$ and upper-right corner coordinates of $(5,5)$. The simulation dimensions are set so that the wireless transmission ranges of all MANET nodes are one unit distance ($R_{it} = 1.0$). Without loss of generality, normalized transmission ranges and data rates are used in the simulation. Using the Euclidian distance d_{ijt} between two nodes i and j at time t, the link capacity u_{ijt} in terms of data rate in bits per second (bps) is approximated with a continuous function as follows:

$$u_{ijt} = 10^4 \left(1 + e^{10 \cdot (d_{ijt} - 0.5)}\right)^{-1} \tag{11.11}$$

Static test problem data specifies the locations of user nodes, which are randomly selected within the simulation area.

11.5.2 *The Effect of Number of Agents, Future Location Prediction, and Clustering*

To study the effect of the input parameters of the proposed MANET system, two test problem groups were used. Five 10-user and five 20-user dynamic test problems were randomly generated with $T = 100$ as described in the test problem simulation environment. Each random problem was solved five times using the PSO algorithm with different random number seeds for the MANET system input parameter values of the number of agents $na = \{1, 2, 3\}$, the number of prediction horizons $H = \{0, 1, 2, 3, 4, 5, 6\}$, and the number of clusters $K = \{4, 6\}$. For problems with 10 user nodes, node clustering was not used (i.e., $K = 10$). In other words, 25 replications were performed for each combination of na, H, and K (a total of 1050 runs for the 20-user group and 525 for the 10-user group). The swarm size was 40 in all runs and the PSO algorithm was run for $c_{\max} = 50$ iterations during each time step of the simulation. In real-world cases, parameter $c_{\max}$ depends on how frequently the problem is solved and deployment decisions are propagated to agent nodes. An ANOVA model was used for each problem group to study the effect of na, H, and K on the performance of the proposed MANET management system.

Figures 11.6 and 11.7 present the main effect plots for the 10-user and 20-user problem groups, respectively. As expected, both network connectivity (P_1) and minimum bandwidth (P_2) significantly improve with the increasing number of agents.

As seen in the main effect plots, location prediction has a significant effect on the performance of the network (ANOVA p-values of significance P_1: 0.0 for both problem groups; P_2: 0.065 for 10-node and 0.020 for 20-node problems). Based on these results, predicting future locations of user nodes is particularly important for improving overall network connectivity. Network connectivity performance metric P_1 significantly improves from $H = 0$ (no prediction) to $H = 1$ (prediction for the next time period). For the test problems studied, H=3 or 4 provides the best results. It should be noted that although several hundred runs were performed for the ANOVA, the results of this experiment should not be generalized. H is a parameter that should be selected by network managers based on empirical data and experience. Various mobility models or prediction techniques may yield different results. Nonetheless, the results of this experiment demonstrate that incorporating a user location prediction model into the proposed MANET management system may improve overall network performance in dynamic problem scenarios.

Another interesting result of the experiment with dynamic problems is that the number of user node clusters K has no statistically significant effect on either P_1 or P_2 (p-values of 0.3 and 0.14, respectively). The main justification for using clusters of user nodes instead of individual nodes in evaluating the weakest connection of a network is computational tractability, particularly for dynamically solving large problems. Using a cluster of user nodes makes the evaluation of O_{t2} independent of the number of user nodes. Therefore, very large networks might be efficiently evaluated in limited time periods.

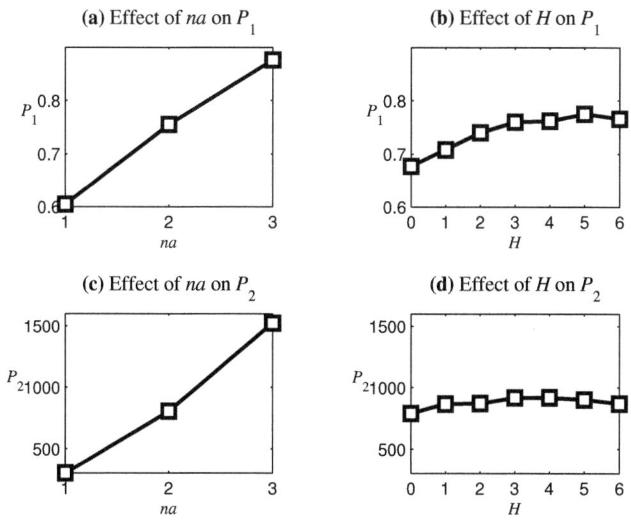

Fig. 11.6 Main effect plots for 10-user problems.

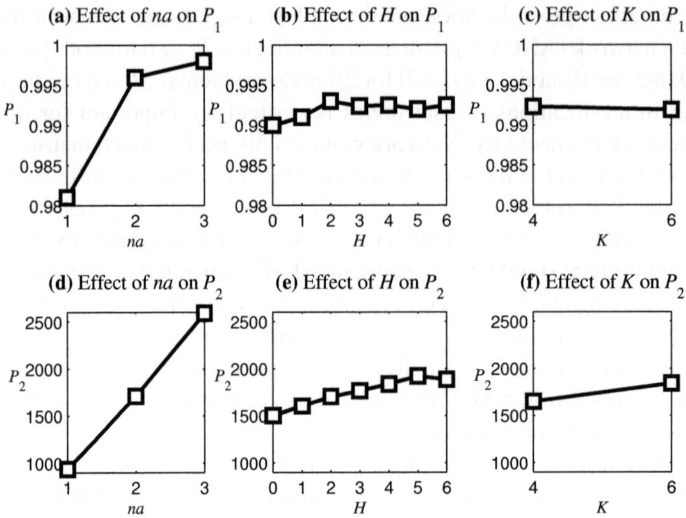

Fig. 11.7 Main effect plots for 20-user problems.

11.5.3 Comparative Performance of the PSO Algorithm

In this study, the performance of the PSO algorithm is compared with respect to a MIP formulation in several static test problems. The agent deployment problem, Problem $ALOC(t,H)$, is a very difficult non-linear MIP problem to solve to optimality. An approximate MIP formulation of the static problem was developed by only considering O_{2t} as given below (Problem *ALOC*). To do so, the nonlinear arc distance versus capacity relationship and the Euclidean arc distances were modeled using eight piecewise linear approximations. Although the MIP formulation is computationally intractable for real-world sized problems, the main objective of this comparison is to provide a benchmark for the performance of the PSO algorithm.

Because Problem *ALOC* is for static cases, time index t is omitted in the decision variables and problem parameters. In addition, the upper bound ($D_{\max}$) constraint on the distance that agent nodes can travel is irrelevant for static cases. Problem *ALOC* is a multi-commodity network flow problem where variable f_{ijst} represents the flow from the source node s to the sink node t on arc (i,j). Node set N includes both user nodes (UN) and agent nodes (AN). Arc set E includes two types of arcs, user node arcs (i.e., $\{(i,j) : \min(R_i,R_j) \geq d_{ij} \quad \forall i,j \in UN, i \neq j\}$) and agent node arcs (i.e., $\{(i,j) : \forall i \in N, \forall j \in AN, i \neq j\}$). Because the locations of user nodes are known (let $(\hat{x}_i,\hat{y}_i)$ represent the location of user node i for all $i \in UN$), the capacities of the user node arcs are constant in Problem *ALOC* due to constraints (11.24) and (11.25). On the other hand, the capacities of agent node arcs depend on the locations of agent nodes.

The objective function of Problem *ALOC* is to maximize the minimum of the maximum flows between the user node pairs. Given the locations of the user nodes $(\hat{x}_i, \hat{y}_i)\ \forall i \in UN$, the decision variables of Problem *ALOC* include the locations of agent nodes (i.e, $(x_i, y_i)\ \forall i \in AN$). Constraints (11.13)-(11.15) are the node-flow balance constraints to calculate the maximum flow U_{st} from the source node s and to the sink node t for $\forall s,t \in UN$ and $s < t$. Constraints (11.16)-(11.20) are used to calculate the squared-Euclidian distances sd_{ij} of agent arcs (i, j) for $i \in N, j \in AN, i < j$. In constraint (11.20), squared-distances $(d^x_{ij})^2$ and $(d^y_{ij})^2$ along the x and y axes, respectively, are approximated by piecewise linear functions while solving the problem in CPLEX. Constraints (11.21) and (11.22) determine the capacity u_{ij} of each agent arc (i, j) based on (11.11), which is also approximated by piece-wise linear functions. Constraint (11.23) ensures that the flow variables do not exceed the arc capacities.

Problem *ALOC*:

$$\text{Mazimize } z = U$$

$$st.$$

$$U \le U_{st} \quad \forall s < t \in UN \tag{11.12}$$

$$\sum_{\{j:(i,j)\in E\}} f_{ijst} - \sum_{\{j:(j,i)\in E\}} f_{jist} = 0 \quad \forall s < t \in UN, \forall i \in N \setminus \{s,t\} \tag{11.13}$$

$$\sum_{\{j:(s,j)\in E\}} f_{sjst} - \sum_{j:(j,s)\in E\}} f_{jsst} = U_{st} \quad \forall s < t \in UN \tag{11.14}$$

$$\sum_{\{j:(t,j)\in E\}} f_{tjst} - \sum_{\{j:(j,t)\in E\}} f_{jtst} = -U_{st} \quad \forall s < t \in UN \tag{11.15}$$

$$d^x_{ij} \ge x_i - x_j \quad \forall i \in N, j \in AN, i < j \tag{11.16}$$

$$d^x_{ij} \ge x_j - x_i \quad \forall i \in N, j \in AN, i < j \tag{11.17}$$

$$d^y_{ij} \ge y_i - y_j \quad \forall i \in N, j \in AN, i < j \tag{11.18}$$

$$d^y_{ij} \ge y_j - y_i \quad \forall i \in N, j \in AN, i < j \tag{11.19}$$

$$(d^x_{ij})^2 + (d^y_{ij})^2 \le sd_{ij} \quad \forall i \in N, j \in AN, i < j \tag{11.20}$$

$$u_{ij} - 10^4 \left(1 + e^{10(\sqrt{sd_{ij}} - 0.5)}\right)^{-1} \le 0 \quad \forall i \in N, j \in AN, i < j \tag{11.21}$$

$$u_{ji} \le u_{ij} \quad \forall i \in N, j \in AN, i < j \tag{11.22}$$

$$f_{ijst} \le u_{ij} \quad \forall (i, j) \in E \tag{11.23}$$

$$x_i = \hat{x}_i \quad \forall i \in UN \tag{11.24}$$

$$y_i = \hat{y}_i \quad \forall i \in UN \tag{11.25}$$

$$u_{ij}, x_i, y_j, f_{ijst} \ge 0$$

Problem *ALOC* was solved by CPLEX v11 with a limit of eight hours CPU time. CPLEX was not able to find feasible solutions for problems with more than ten user nodes and five agents within the allowed CPU time limit. Fig. 11.8 shows the comparisons on static cases of ten randomly generated 5-user and 10-node problems with three and five agents. In these problems, the locations of users were randomly generated in a way that connecting all users were possible. In all problems, both user and agent nodes had a communication range of one unit distance. For each problem instance, the PSO was run for ten random replications with a swarm size of 40 for c_{max}=1000 iterations. Therefore, the box-plots in Fig. 11.8 include 100 data points for the PSO and ten data points for CPLEX. The majority of 5- and 10- node problems with three agents were solved to optimality within the CPU time limit. These static problems with three agents demonstrate that the PSO performed as well as the MIP formulation. In some replications of these problems, the PSO finds slightly better solutions than the MIP formulation. These differences can be attributed to the linear approximation in the MIP formulation. As seen in Fig. 11.8, the PSO algorithm perform significantly better than the MIP formulation in the problems with five agents. Increasing the number of agents from three to five significantly increases the search space of the problem and these problems could not be solved to optimality within the CPU time limit. The solutions found by the PSO are superior to the best solutions found by CPLEX over eight hours of CPU time. The CPU time requirement of the PSO depended on the problem. In random 5-node problems with three agents, for example, 1000 PSO iterations require CPU times ranging from 1.3 to 80.0 seconds with an average of 31.9 seconds. For random 10-node problems with five agents, the minimum, maximum, and average CPU times are 21.9, 855.2, and 316.5 seconds, respectively. In these problems, a significant percent of the computational effort is due to calculation of maximum flows. A more efficient approach to calculating all-pairs maximum flow (e.g., the all-pairs minimum value cut algorithm [1]) may provide significant computational savings.

11.6 Conclusions

In this chapter, a new model is proposed to conceptualize an autonomous topology optimization for mobile ad hoc networks. The proposed approach relocates mobile agents to help maintain a suitable level of communication service in the network. The representation of wireless ad hoc network communications as network flows and optimization using a maximum flow model is a novel approach. This representation is very responsive to small changes in topology when evaluating network connectivity and performance. Also, it can be used with any signal attenuation model when calculating the data flow rates.

The dynamic nature of the problem is a challenge, but it also enables the optimizer to gain additional information by leveraging the information obtained during the optimization at previous time steps. The usefulness of predicting user locations to improve network connectivity is demonstrated. A particle swarm algorithm is developed to solve the modeled dynamic optimization problem. Computational experiments show that the particle swarm algorithm is very promising and well suited to this problem.

The proposed approach, while developed for dynamic topology optimization, easily adapts to a static scenario by increasing the agent velocity constraints. The static scenario is useful when users want to improve an existing system of sensors or communication hubs already positioned in the field, or when designing a new static system. The proposed approach could also be used for "what-if" purposes before launching an actual network in the field. The simulation is useful to plan for the most efficient number of mobile agents to deploy under a certain scenario, and to consider cost / benefit tradeoffs.

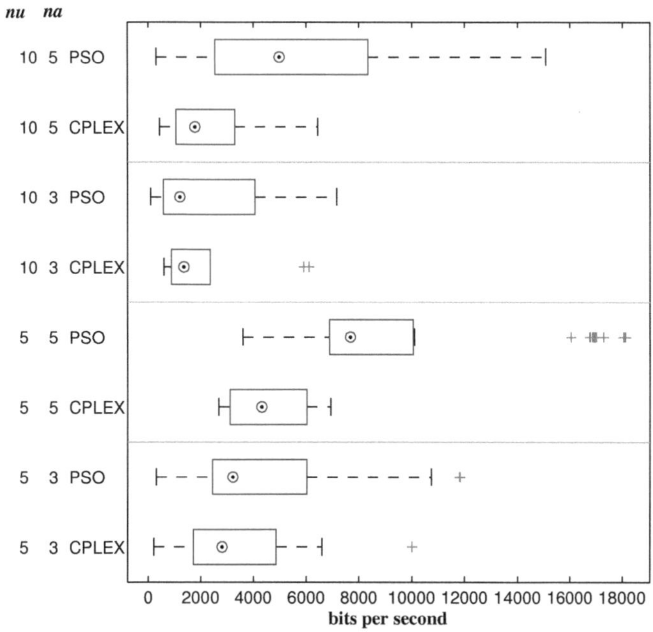

Fig. 11.8 Comparison of the PSO and MIP on ten random problems of each size. (nu = the number of users, na = the number of agents.)

References

1. Ahuja, R.K., Magnanti, T.L., Orlin, J.B.: Network Flows, Theory, Algorithms, and Applications. Prentice-Hall, Inc., Upper Saddle River, New Jersey (1993)
2. Ashbrook, D., Starner, T.: Learning significant locations and predicting user movement with gps. In: Proceedings of the 6th International Symposium on Wearable Computers (ISWC 2002), pp. 101–108 (2002)
3. BOOST.org: Boost c++ libraries (2006)

4. Broch, J., Maltz, D.A., Johnson, D.B., Hu, Y.C., Jetcheva, J.: A performance comparison of multi-hop wireless ad hoc network routing protocols. MobiCom'98. Proceedings of Fourth Annual ACM/IEEE International Conference on Mobile Computing and Networking, pp. 85–97. ACM, New York, NY, USA (1998)
5. Carlisle, A., Dozier, G.: Tracking changing extrema with adaptive particle swarm optimizer. In: Proceedings of the 5th Biannual World Automation Congress, vol. 13, pp. 265–270 (2002)
6. Chadrashekar, K., Dekhordi, M., Baras, J.: Providing full connectivity in large ad-hoc networks by dynamic placement of aerial platforms. Tech. Rep. TR 2004-21, University of Maryland (2004)
7. Clerc, M., Kennedy, J.: The particle swarm - explosion, stability, and convergence in a multidimensional complex space. IEEE Transactions on Evolutionary Computation **6**(1), 58–73 (2002)
8. Creixell, W., Sezaki, K.: Mobility prediction algorithm for mobile ad hoc network using pedestrian trajectory data. In: Proceedings of the IEEE Region 10 Conference, (TENCON 2004), vol. B, pp. 668–671 (2004)
9. Davis, J.A., Fagg, A.H., Levine, B.N.: Wearable computers as packet transport mechanisms in highly-partitioned ad-hoc networks. In: Proceedings of Fifth International Symposium on Wearable Computers, pp. 141–148. IEEE Comput. Soc, Los Alamitos, CA, USA (2001)
10. Dorigo, M., Maniezzo, V., Colorni, A.: Ant system: optimization by a colony of cooperating agents. IEEE Transactions on Systems, Man and Cybernetics, Part B (Cybernetics) **26**(1), 29–41 (1996)
11. Eberhart, R.C., Kennedy, J.: A new optimizer using particle swarm theory. In: Proceedings of Sixth International Symposium on Micro Machine and Human Science, p. 39–43. Nagoya (1995)
12. Eberhart, R.C., Shi, Y.: Tracking and optimizing dynamic systems with particle swarms. In: Proceedings of the 2001 Congress on Evolutionary Computation, vol. 1, pp. 94–100 (2001)
13. Goldberg, A.V., Tarjan, R.E.: A new approach to the maximum-flow problem. Journal of the ACM (JACM) **35**(4), 921–940 (1988)
14. Goyal, D., Caffery Jr, J.: Partitioning avoidance in mobile ad hoc networks using network survivability concepts. In: Proceeedings of Seventh IEEE Symposium on Computers and Communications (ISCC'02), pp. 553–558. Taormina-Giardini Naxos, Italy (2002)
15. Hauert, S., Winkler, L., Zufferey, J.C., Floreano, D.: Ant-based swarming with positionless micro air vehicles for communication relay. Swarm Intelligence **2**(2-4), 167–188 (2008)
16. Huang, R., Zaruba, G.V.: Location tracking in mobile ad hoc networks using particle filters. Journal of Discrete Algorithms **5**(3), 455–470 (2007)
17. Kaplan, E., Hegarty, C.: Understanding GPS: Principles and Applications Second Edition. Artech House, Norwood, MA, USA (2006)
18. Karumanchi, G., Muralidharan, S., Prakash, R.: Information dissemination in partitionable mobile ad hoc networks. In: Proceedings of the 18th IEEE Symposium on Reliable Distributed Systems, pp. 4–13. IEEE Comput. Soc, Los Alamitos, CA, USA (1999)
19. Kim, I., Jang, S., Lee, J.: Qlp-lbs: quantization and location prediction-based lbs for reduction of location update costs. Frontiers of High Performance Computing and Networking - ISPA 2007 Workshops. Proceedings ISPA 2007 International Workshops SSDSN, UPWIN, WISH, SGC, ParDMCom, HiPCoMB, and IST-AWSN. (Lecture Notes in Computer Science vol. 4743), pp. 101–108. Springer-Verlag, Berlin, Germany (2007)
20. Kim, T.H., Tipper, D., Krishnamurthy, P., Swindlhurst, A.: Improving the topological resilience of mobile ad hoc networks. In: Proceedings of the 7th International Workshop on the Design of Reliable Communication Networks, pp. 191–197. Alexandria, Virginia, USA (2009)
21. Li, Q., Rus, D.: Communication in disconnected ad hoc networks using message relay. Journal of Parallel and Distributed Computing **63**(1), 75–86 (2003)
22. Mitrovic, D.: Short term prediction of vehicle movements by neural networks. In: Proceedings of the 3rd International Conference on Knowledge-Based Intelligent Information Engineering Systems, pp. 187–190 (1999)

23. Ou, C.H., Ssu, K.F., Jiau, H.C.: Connecting network partitions with location-assisted forwarding nodes in mobile ad hoc environments. In: 10th Pacific Rim International Symposium on Dependable Computing (PRDC'04), p. 239–247. Papeete, Tahiti, French Polynesia (2004)
24. Ozcan, E., Mohan, C.K.: Particle swarm optimization: surfing the waves. In: Proceedings of the 1999 Congress on Evolutionary Computation-CEC99, *Proceedings of the 1999 Congress on Evolutionary Computation-CEC99 (Cat. No. 99TH8406)*, vol. 3, pp. 1939–44. IEEE, Piscataway, NJ, USA (1999)
25. Shi, Y., Eberhart, R.C.: Fuzzy adaptive particle swarm optimization. In: Proceedings of the 2001 Congress on Evolutionary Computation (IEEE Cat. No.01TH8546), vol. 1, pp. 101–106. IEEE, Piscataway, NJ, USA (2001)
26. Su, W., Lee, S.J., Gerla, M.: Mobility prediction and routing in ad hoc wireless network. International Journal of Network Management **11**(1), 3–30 (2001)
27. Tang, J., Xue, G., Zhang, W.: Reliable routing in mobile ad hoc networks based on mobility prediction. In: Proceedings of the IEEE International Conference on Mobile Ad-hoc and Sensor Systems, pp. 466–474 (2004)
28. Wang, K.H., Baochun, L.: Efficient and guaranteed service coverage in partitionable mobile ad-hoc networks. In: Proceedings of IEEE INFOCOM 2002 Conference on Computer Communications. Twenty-First Annual Joint Conference of the IEEE Computer and Communications Societies (Cat. No.37364), vol. 2, pp. 1089–1098. IEEE, Piscataway, NJ, USA (2002)
29. Wang, K.H., Baochun, L.: Group mobility and partition prediction in wireless ad-hoc networks. In: Proceedings of 2002 IEEE International Conference on Communications, vol. 2, pp. 1017–1021. IEEE, Piscataway, NJ, USA (2002)
30. Wang, N.C., Chang, S.W.: A reliable on-demand routing protocol for mobile ad hoc networks with mobility prediction. Computer Communications **29**(1), 123–135 (2005)
31. Zhao, W., Ammar, M., Zegura, E.: A message ferrying approach for data delivery in sparse mobile ad hoc networks. In: Proceedings of the 5th ACM international symposium on Mobile ad hoc networking and computing, pp. 187–198. ACM, Roppongi Hills, Tokyo, Japan (2004)
32. Zhu, H., Swindlehurst, A.L., Liu, K.J.R.: Smart deployment/movement of unmanned air vehicle to improve connectivity in manet. In: 2006 IEEE Wireless Communications and Networking Conference, pp. 252–257. Vegas, NV, USA (2006)

Part IV
Optimization Problems in the Operation of Wireless Networks

Chapter 12
Simulation-Based Methods for Network Design

Marina Aguado, Jasone Astorga, Nerea Toledo and Jon Matias

Abstract In the last few years, the evaluation of large scale applications across mobile wireless networks has raised a crucial challenge to the research community. Simulation modeling methods can be quite powerful in gaining insight into the behavior of complex systems. The main advantages of simulation modeling include the ability to evaluate scenarios not easily achievable through empirical methods (i.e. scalability testing of networks) and the ability to modify models to test system sensitivity and tune performance. Another important application of simulation modeling techniques is to perform meaningful comparative studies of different technologies, protocols, or new implementations to determine which communication technology performs better from the QoS point of view in a specific scenario. Simulation modeling is a crucial step in the design, development, test and evaluation of any network implementation or strategy, providing the necessary support to the usual demand for "what if" analysis and studies in the network environment. This chapter describes a taxonomy of the different simulation techniques available in the market, introduces statistical concepts for validating the results of a simulation study and describes case studies illustrating the role of simulation in wireless network design.

Marina Aguado
ETSI Bilbao, Depto. Electrónica y Comunicaciones, Al. Urquijo S/N, CP. 48013 Bilbao, Spain, e-mail: marina.aguado@ehu.es

Jasone Astorga
ETSI Bilbao, Depto. Electrónica y Comunicaciones, Al. Urquijo S/N, CP. 48013 Bilbao, Spain, e-mail: jasone.astorga@ehu.es

Nerea Toledo
ETSI Bilbao, Depto. Electrónica y Comunicaciones, Al. Urquijo S/N, CP. 48013 Bilbao, Spain, e-mail: nerea.toledo@ehu.es

Jon Matias
ETSI Bilbao, Depto. Electrónica y Comunicaciones, Al. Urquijo S/N, CP. 48013 Bilbao, Spain, e-mail: jon.matias@ehu.es

J. Kennington et al. (eds.), *Wireless Network Design: Optimization Models and Solution Procedures*, International Series in Operations Research and Management Science 158, DOI 10.1007/978-1-4419-6111-2_12,

12.1 Introduction

In the last few years, the evaluation of large scale applications across mobile wireless networks has raised a crucial challenge to the research community. Actually, when facing the necessity to design or evaluate a specific network implementation different options appear, namely measurement, emulation, simulation, or analytical modeling. Therefore, one of the most important steps in the start-up evaluation process is to identify the most suitable technique for each scenario. In order to do so, and before setting up a test bed or measurement experiment, writing complex equations for an analytical model, or building up elaborate simulation scenarios, the problem to be solved must be clearly stated. As the old saying goes "a problem well stated is half solved".

For this purpose, the first issue to be considered is the global aim being sought, which is, in fact, the analysis of the system's performance rather than a full set of numerical statements describing its operation. Bearing this point in mind, it must be noted that the specific tool or methodology used to reach that aim depends on the nature of the problem under study and the time scale of interest. In this sense, test beds provide realism but are coupled to the physical circumstances where the experiment is carried out and frequently they constitute a complex, expensive, and time-consuming approach. Additionally, when a new concept is proposed (e.g., a new protocol, technology or algorithm) in the research field, the possible approaches are restricted to analytical models, simulation, and emulation. More specifically, when a mobile wireless network is involved in a complex multi-factor and multimode scenario, then an analytical model may not be feasible due to the high level of complexity. In fact, analytical modeling requires in general many simplifications and assumptions compared to a simulation process, which can incorporate more details and require fewer assumptions.

Simulation modeling methods can be quite powerful in gaining insight into the behaviour of complex systems. The main advantages of simulation modeling include the ability to evaluate scenarios not easily achievable through empirical methods (i.e. scalability testing of networks) and the ability to modify models to test system sensitivity and tune performance [14]. When using simulation modeling, many beginner analysts tend to model the behaviour of the system under evaluation as close as possible to the real world without paying attention to the problem they are expected to solve. Assuming they are even able to complete such a complex simulation, this approach may result in a costly, time consuming, and fruitless simulation product. Therefore, when using a simulation tool for evaluation or design purposes, the idea is not to model the full detailed "scenario" but to keep clearly in mind the specific question we seek to answer. On the other hand, we cannot forget that when implementing a system level simulator — a simulator that considers many users, many possible cells, and different layers of the protocol stack — the signal propagation must be characterized correctly and accurately enough for the purposes of the study.

Another important application of simulation modeling techniques is to perform meaningful comparative studies of different technologies, protocols, or new implementations to determine which communication technology performs better from the QoS point of view in a specific scenario. Therefore, we can conclude that simulation modeling is a crucial step in the design, development, test and evaluation of any network implementation or strategy, providing the necessary support to the usual demand for "what if" analysis and studies in the network environment. Figure 12.1, based on [10], represents the inputs and outputs expected from a simulation framework (block diagram).

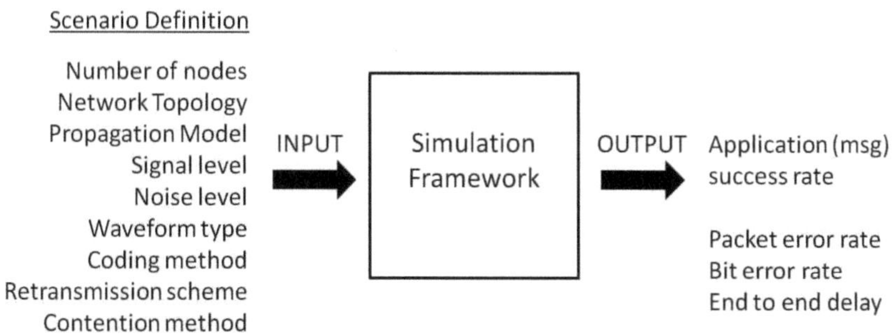

Fig. 12.1 Block Diagram for a Simulation Framework

It is commonly understood that simulation analysis is a necessary stage prior to actual production and network deployment. However, it is also fair to point out that we should not consider simulation modeling as a substitute for empirical testing. Analytical models, field measurements, prototyping, etc. are all critical components in the life cycle of a successful network design. And in fact they are part of the validation process since, in accordance to [9], the three rules of validation are:

1. Do not trust the results of a simulation model until they have been validated by analytical modeling or measurements.
2. Do not trust the results of an analytical model until they have been validated by a simulation model or measurements.
3. Do not trust the results of a measurement until they have been validated by simulation or analytical modeling.

The rest of this chapter is structured as follows. Section 12.2 describes a taxonomy of the different simulation techniques available in the market. Section 12.3 introduces the concepts of verifying statistical validity of discrete event simulation, random seed values and confidence intervals. Section 12.4 surveys related work on simulation for wireless network design, and, to conclude, Section 12.5 presents case studies that aim to cover the full set of different areas in wireless networking where a simulation tool plays an important role.

12.2 Simulation Methodologies: a Taxonomy

Different strategies for simulation modeling may be identified; each of them presenting different characteristics, advantages and disadvantages:

- *Emulation*: an emulator is a piece of software that allows running a program in a different platform (hardware architecture or operating system) than the one it was originally developed for. Unlike a simulator, which only tries to reproduce the behavior of a program, an emulator tries to model precisely the device or system being emulated.
- *Monte-Carlo simulation*: this is a static simulation methodology typically used for the evaluation of numerical expressions. Monte Carlo methods are based on the systematic sampling of random variables, and they include a set of techniques that allow for the resolution of mathematical and physical problems by the execution of random repetitive tests. However, in practice the random tests are usually replaced by the result of certain calculations performed with random numbers.
- *Discrete event simulation*: this type of simulation allows for modeling systems that evolve over time and whose state variables are discrete and change instantaneously in separate points along time. A discrete system can be characterized by the fact that the system activities cause discrete changes to the value of the state variables. Typically, these changes take place at particular time instants known as events. The description of the state of such a system includes, apart from the values of the variables, the possible transitions between events. Therefore, a model of a discrete event system basically consists of three elements: the state variables, the events that change the value of those variables and the relationships between events.
- *Trace driven simulation*: this type of simulation is used to model dynamic systems and it is characterized by using a real trace as entry data. A trace is in fact a register of events, ordered in time, which is obtained from a real system. This approach has the advantage of providing high credibility as there is a great similarity between the model and the real system. However, it is also true that such models are more complex and not as representative of the real system since the trace has a finite length.

Another classification makes reference to the support elements involved in the specific simulation. In accordance to this, three types of simulations are identified. These include *software*, *hardware*, and *hybrid approaches*. Hybrid simulation approaches combine software and hardware simulation methods. In particular, hardware-in-the-loop (HITL) simulations can be quite powerful for testing existing hardware platforms and gaining insight into their performance characteristics while modeling other aspects of the system in software. One of the case studies described in Section 12.5 makes use of this technique.

Another global classification is the distinction between *system level simulation* versus *link level simulation*. System-level simulation concentrates on evaluation of the performance when multiple base stations (BSs) and mobile stations (MSs) operate together. Link-level simulation generally focuses on single transmitter-to-receiver link performance. The results from the link-level simulation are usually adopted as inputs to the system-level simulation. System-level simulation has been widely used to evaluate system performance with incorporation of link-level simulation curves. In order to reduce the simulation complexity, the effects of link-level technologies are abstracted into several curves that are used in a system-level simulation. An example of this mapping is covered in the WiMAX case study in Section 12.5.2.

After identifying which type of simulation is most suitable or available to solve the specific problem, the next step is to choose the specific modeling and simulation tool. It is always possible to build one's own simulation tool but normally in-house development is more time consuming than using commercially available tools. When using a commercial tool, it is necessary to first make a comprehensive reading of the main limitations and assumptions of the simulation framework and the specific model to be used in the simulation. After analyzing specific model features, it is good practice to test model limitations and verify major assumptions. Often a specific tool is chosen because it has a mature simulation model for the specific communication technology that is the subject of the simulation study. This is the case in the Host Identity Protocol (HIP) case study introduced in Section 12.5.4 in this chapter.

To summarize, Figure 12.2 groups the different available simulation tools in accordance with the OSI layer they cover [10].

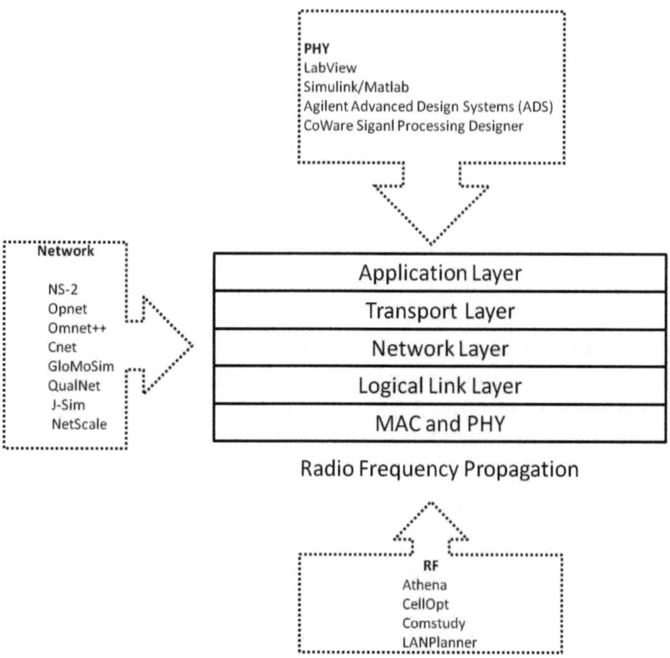

Fig. 12.2 Simulation Tools

12.3 Validating Discrete Event Simulations: the Random Number Seed and the Confidence Interval

A simulation process is a statistical experiment. In order to represent a behavior that is not known for a specific element, the simulation environment makes use of stochastic modeling with random number streams to introduce changes in the element's operating environment (i.e., its input). However, it is impossible for a computer program, by its very nature, to generate truly random numbers. Thus, the simulation framework relies on starting a *pseudo-random number generator* in different states to exhibit unpredictable behavior. The initial state of the generator is known as the *random number seed.* Each particular simulation run with a specific random number seed value will produce a different behavior and yield new and different results. To account for all statistical behavior that the system might exhibit, the simulation model must be run multiple times with different seed values. As more seeds are chosen, standard behavior should be observed more frequently than anomalous behavior. Intuitively, the more simulation runs performed, the more confidence we have in the results obtained from the study. In other words, the goal is to ensure that a simulation system produces valid statistical behavior by making a sufficient number of runs, each with a different random seed value. One of the important tasks a simulation designer must carry out is that of deciding how many separately seeded trials to run in a simulation experiment so that there will be enough data to make a strong statement about the system's typical behavior.

Suppose that a number of simulations of a system have been run with different random number seeds to obtain n samples of the statistic X, $(x_1, x_2, x_3, \ldots, x_i, \ldots, x_n)$. Statistic X may take on many values and its precise distribution is unknown. In order to calculate, μ, the true mean of the random variable X, which represents the typical behavior of the modeled system with regard to the statistic X, we should run a large number of simulations (theoretically, an infinite number would be required). However, since it is not usually possible to run a very large number of simulations to determine μ, it is interesting to determine the degree of precision with which the mean value, $\bar{x}$, of an n-sample set approximates μ. This determines whether the value can be used with confidence to make statements about the typical behavior of the modeled system.

Consider an experiment consisting of a collection of samples of X, each of size n, where $\bar{x}$ is the average value of the n samples corresponding to one trial of the experiment. If the experiment is performed many times with different seed values, then the resulting average value of $\bar{x}$, which we denote by $\overline{X}$, has its own distribution. In accordance with the central limit theorem, regardless of X's actual distribution, as the number of samples grows large, the random variable $\overline{X}$ has a distribution that approaches that of a normal random variable with mean μ, the same mean as the random variable X itself. The theorem further states that if the true variance of X is σ^2, the variance for statistic $\overline{X}$ is $\frac{\sigma^2}{n}$. To conclude, if sufficient sample size n has been chosen, normally $n > 30$, it is possible to work with a normal distribution which has known properties instead of working with an unknown distribution.

As a rule of thumb, two to three dozen runs of a simulation with different random seed values provide enough data to achieve a sufficient degree of confidence in the obtained results. This means that there will be enough information to build a useful *confidence interval* around the simulation results. If the standard deviation of the random variable, σ, is known, then it is possible to estimate the probability that the random sample $\overline{x}$ falls within a particular distance of μ. A confidence interval for μ represents the interval of real numbers such that the probability that μ lies within this interval has a particular value (90%, 95%, and so on). A 95% confidence interval for μ means that there is a 95% certainty level that the true mean of X lies within the intervals bounds. It is clear then that the confidence interval widens as the degree of confidence increases. The calculation of the confidence interval relies on the fact that the distribution of the statistic $\overline{X}$ is normal. Once the desired confidence interval is specified it is necessary to identify the corresponding number of standard deviations from the average, $Z = \frac{(x-\mu)}{\sigma}$, as depicted in Figure 12.3. The desired Z value is determined by referring to the *standard normal curve* (i.e., the probability density function for a normal random variable with a mean of zero and a standard deviation of one). For example, as shown in Figure 12.4, the interval from -1.645 to 1.645 contains 0.90% of the total area under the standard normal curve. Thus, a 0.90% confidence interval for μ ranges from μ minus 1.645 standard deviations to μ plus 1.645 standard deviations. Table 12.1 gives the confidence intervals for μ when σ is known. In this example, $n = 36$, $\sigma = 3$, and $\overline{x} = 24.2$.

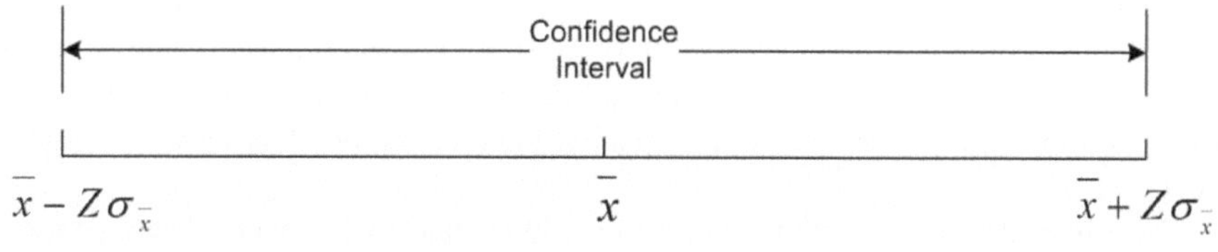

Fig. 12.3 Confidence Interval Representation

Table 12.1 Example confidence interval calculations

Desired	Z	Formula	Calculations	Interval
90%	1.65	$\overline{x} \pm 1.65\frac{\sigma}{\sqrt{n}}$	$24.2 \pm 1.65\frac{3}{\sqrt{36}}$	23.375 to 25.025
95%	1.96	$\overline{x} \pm 1.96\frac{\sigma}{\sqrt{n}}$	$24.2 \pm 1.96\frac{3}{\sqrt{36}}$	23.220 to 25.180
99%	2.58	$\overline{x} \pm 2.58\frac{\sigma}{\sqrt{n}}$	$24.2 \pm 2.58\frac{3}{\sqrt{36}}$	22.910 to 25.490

It is important to realize that some scenarios might need many more than 30 runs to ensure valid results. A specific decision regarding the number of runs must be made by observing the variation in the results. To compute a confidence interval it is necessary to run the simulation experiment with the same configuration parame-

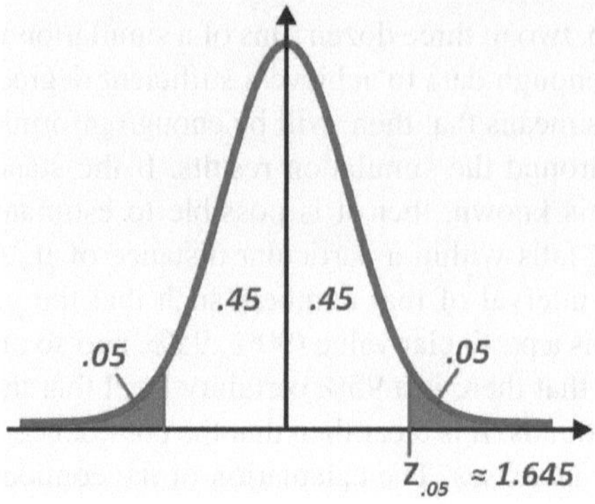

Fig. 12.4 Standard Normal Curve

ters each time, but with varying random number seed values. The total number of seed values chosen (i.e., the number runs) is critically important in determining the confidence interval, but the actual seed values themselves are not as important.[1] As an example, Figure 5(a) represents the obtained results with different seed values in the OPNET Modeler simulation framework [21] for a specific statistic (WiMAX delay) in a simulation scenario when the mobile node modifies its speed from 30 to 120 Km/h. Once all executions have taken place, multiple values for the same abscissa value are observed in Figure 5(b), each value corresponding to an experiment with the same abscissa parameter value (speed) but a different seed value as input [22]. It is then possible to select a desired confidence interval using the previous formula. For each abscissa value, the specific interval is plotted as shown in Figure 5(c).

As a final general recommendation, it is important to be aware that a general common goal to be achieved by any simulation study is that similar results can be obtained in different emplacements under similar constraints. Following the recommendations below helps to ensure that the experiment may be repeated somewhere else achieving similar results:

a. Identify and clearly characterize the type of data traffic involved in the simulation.
b. Specify the simulation platform and version.
c. Specify the number of simulation runs.
d. Specify confidence intervals.
e. Specify correctly the network topology and the values for the different configuration parameters for each node.

[1] Although the sequence of numbers produced by a pseudo-random number generator may appear to be random, the procedure used to generate the sequence from a given seed value is actually deterministic. Thus, the sequence will eventually repeat, and the number of pseudo-random numbers generated before the sequence repeats can be affected by the random number seed value.

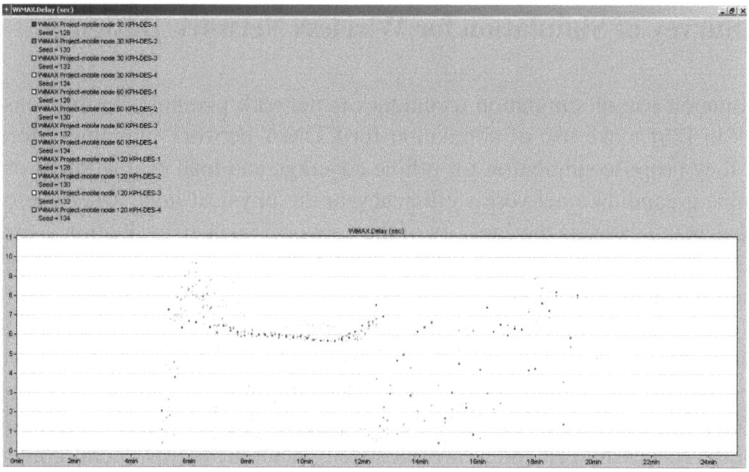

(a) OPNET graph representing multiple scalar values (multiple seeds) for each abscissa value

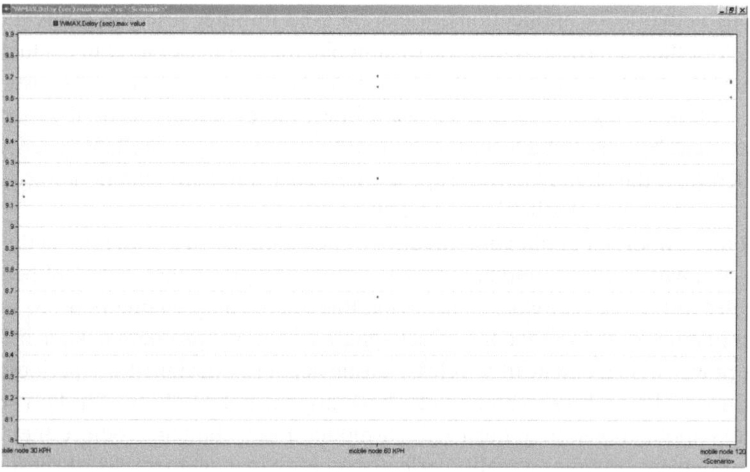

(b) Graphical confidence interval representation in the OPNET Modeler Simulation framework.

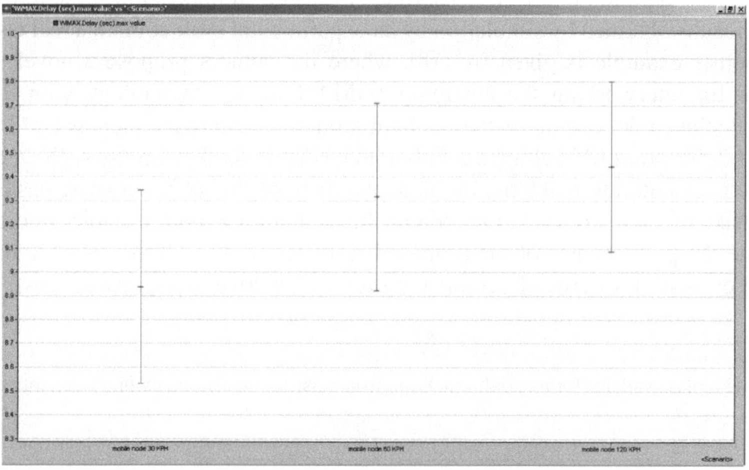

(c) WiMaX Delay 80% Confidence Seeds.

Fig. 12.5 Confidence interval representation in OPNET.

12.4 Survey of Simulation for Wireless Network Design

One common use of simulation techniques is network planning. In this sense, the authors in [26] make use of simulation for CDMA network planning. More precisely, they propose simulation for offline coverage and load evaluation in order to determine a candidate network's efficiency in the physical layer when the overall service demand exceeds the capacity of the network. Such an evaluation is essential to determine prior to deployment if a potential network configuration can provide adequate physical performance for different numbers of users, located in different points over time. Similarly, in [11] the OPNET simulation tool is used to perform a simulation of the network-traffic characteristics that a mobile WiMAX network to be deployed in a near future will have to face. Additionally, a demand estimation and a VoIP performance assessment is also carried out. The obtained results are crucial for the successful cell and network-capacity planning of any operator interested in deploying a WiMAX mobile network.

Regarding network planning, the use of simulation is not restricted to communications networks or applications; it can be used in other research fields. For example, for efficient power system expansion planning, it is essential to evaluate and minimize the total cost of the grid expansion including the construction costs associated with investing in new transmission lines, the operation costs of generators, and the stand-by costs which are the auxiliary service costs for reliable energy supply. Therefore, the authors in [6] propose a new cost simulation methodology which allows for efficient and reliable cost evaluation.

Another important application field for which simulation is essential is the evaluation of novel and emerging approaches or technologies which have not been implemented yet in the real world. Following this approach, the authors in [3] use the ns2 simulation tool[2] to prove the effectiveness of a novel protocol designed for efficient power management in wireless sensor networks designed for periodic data acquisition. Although sensor networks have a great potential and can be used for countless applications, they must still surmount a key challenge before their massive deployment becomes a reality: effective energy management. Most sensors operate on non-rechargeable batteries, and thus the lifetime of the sensor network is bounded by the duration of the sensor batteries. The authors confirm the reliability of the results obtained by simulation by an experimental evaluation in a real testbed.

Another example is given by [18], where the authors propose a novel architecture for interworking the emerging WiMAX mobile technology with the already deployed 3G cellular networks built around technologies such as UMTS and CDMA2000 using IP Multimedia Subsystem (IMS). As they argue, such an architecture is essential to make the initial deployment of WiMAX networks easier and thus make the real world exploitation of this technology more feasible. In order to validate the performance of the proposed architecture, the authors present an analytical evaluation of crucial parameters such as handover delay, packet loss, jitter,

[2] The ns2 tool is available for free online at http://nsnam.isi.edu/nsnam/index.php/User_Information.

and signaling cost during vertical handover. Finally, they corroborate the reliability of the obtained results by using the OPNET discrete event simulation tool.

Similarly, simulation is also widely used to carry out effective performance evaluations of wireless systems in a fast and inexpensive way. In this sense, the authors in [7] propose a new system-level simulation tool that allows for the performance evaluation of OFDMA networks considering the impact of different key factors such as data traffic and resource allocation strategies. In the same way, the authors in [8] study the impact of the network size on the performance of Wireless Networked Control Systems (WNCSs) over mobile ad hoc networks (MANETs). In fact, the concept of WNCSs over MANETs is a novel research area with many potential applications. However, from the technical point of view such systems still present some challenges that must be solved before their real-world deployment is feasible: unpredictable network packet delay and dropouts, random node movements, etc. As the research and evaluation of such environments using real-world test beds is expensive and time consuming, an effective alternative is to use simulation tools. Following this approach, the authors in [27] present a co-simulation study using SIMULINK and OPNET in order to simulate both the system dynamics and the behavior of the network simultaneously.

Apart from wireless networks, performance evaluations can also be carried out in other environments. For example, the authors in [19] have used simulation to evaluate and compare three different types of optical multicast protection approaches. In optical networks it is difficult to predict when and where a failure may occur. Such failures are catastrophic for the network operation. That is why the implementation of fast and efficient recovery mechanisms is crucial in such environments. Thanks to the obtained simulation results, the authors concluded that the best recovery approach depends on the static or dynamic nature of the affected traffic.

As an alternative approach to the use of simulation tools, the authors in [17] [22] propose the use of agent-based modeling tools for modeling large and complex modern networks such as self-organized distributed networks or peer-to-peer (P2P) environments. Simulation is an essential tool for the design and development of new emerging network approaches, especially when they are based on distributed and self-organizing schemas for their operation.

12.5 Case Studies

As discussed in the previous section, a simulation framework may cover a different set of areas related to wireless networking design. This section covers the most relevant ones and introduces some examples for each of them. The simulation frameworks chosen are OPNET Modeler [21], OMNeT++ [20] and an emulation tool named Click Router [12]. Initially, OPNET was designed as a software package for discrete event network simulation. However, this simulator now integrates with other types of simulators (e.g., waveform and RF propagation simulators). In the specific case of the WiMAX model implementation in Section 12.5.2, the physical layer behavior takes into account pre-calculated results from a MATLAB [25] implementation.

12.5.1 Traffic Modeling Application

Accurately modeling traffic in the network helps produce an optimal application and network design. Different data traffic simulation technology strategies may be followed. A simulator framework may generate traffic in the network explicitly, simulating all packets and queues in detail. As a consequence high-fidelity results are obtained. On the other hand, potentially longer simulation run times are obtained. Other times, the simulation tool may make use of an analytical strategy, simulating the data traffic at a more abstract level making use of mathematical models. This implies faster simulation run times. There are also some simulators that exploit hybrid solutions making use of a combination of both explicit and analytical modeling techniques. As an example, in [24], mathematical traffic models are introduced for the most common application services. One of the traffic models studied is the web browsing (HTTP) traffic model. Although most measurements of HTTP traffic show that the large majority of page responses consist of relatively small objects, the distribution of page sizes is heavy-tailed. That is, infrequent but very large page objects constitute a significant proportion of overall transmitted bytes.

Most of the simulation frameworks allow the final user to select between a set of standard application models (e-mail, ftp, http, database, voice, video conferencing, etc.) and to configure the source (user or group of users), destination (user or servers), traffic patterns, and traffic representation for each service. Each standard application also allows specific configuration settings. Some simulation tools make use of specific development tools that allow the end user to create customized applications.

In [1] the authors describe a train-control application built with OPNET's ACE Whiteboard tool. The system is a multi-tier application between the train, several radio block control centers, and the control center. The different train control routines (movement authority request, position report, etc.) are modeled in the simulation tool. When creating this application, it was necessary to define the tiers that would participate in message exchanges and that later on would be identified as nodes within the simulation scenario. Figure 12.6 depicts the messages exchanged between the different tiers. The size and HEX format of these messages as well as the existent dependency between them, and the logical sequence that they follow are properly modeled in the simulation. The message processing time in each tier was obtained from a real trace log. By creating these routines and building this new application in the simulation tool, it was possible to identify the data traffic profile for this service. The first graph in Figure 12.7 shows the traffic sent by the train and generated by the train control application.

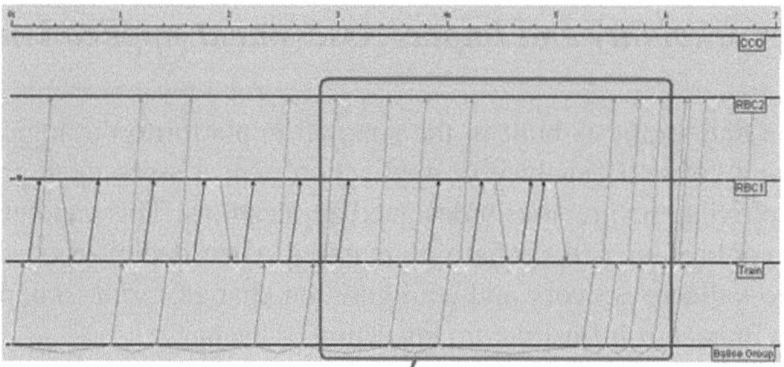

ID	Source	Destination	Bytes	Tag	Description	Subtask	Connection	Depends On	Processing Time	User Time
18	Train	RBC2	32		Position Report	<None>	3	ID: 17	0.114819	0.000000
19	Train	RBC1	32		Position Report	<None>	2	ID: 18	0.110000	0.000000
20	RBC2	Train	100		Movement Authority	<None>	3	ID: 18	0.154819	0.000000
22	Train	RBC2	22		ACK	<None>	3	ID: 20	0.124819	0.000000
23	Balise Group	Train	32		Position Info	<None>	1	ID: 17	0.000000	0.800000
24	Train	RBC2	32		Position Report	<None>	3	ID: 23	0.134819	0.000000
25	Train	RBC1	32		Position Report	<None>	2	ID: 24	0.120000	0.000000
26	Balise Group	Train	32		Position Info	<None>	1	ID: 23	0.000000	0.450000
27	Train	RBC2	32		Position Report	<None>	3	ID: 26	0.104819	0.000000
28	Train	RBC1	32		Position Report	<None>	2	ID: 27	0.090000	0.000000
29	RBC1	Train	32		General Message	<None>	2	ID: 28	0.124819	0.000000
30	Train	RBC1	22		ACK	<None>	2	ID: 29	0.124819	0.000000
31	Train	RBC1	18		Termination of a communication session	<None>	2	ID: 30	0.120000	0.000000
32	RBC1	Train	18		ACK of a communication session	<None>	2	ID: 31	0.104819	0.000000
33	Balise Group	Train	32		Position Info	<None>	1	ID: 26	0.000000	1.320000
34	Train	RBC2	32		Position Report	<None>	3	ID: 33	0.084819	0.000000
35	Balise Group	Train	32		Position Info	<None>	1	ID: 33	0.000000	0.210000
36	Train	RBC2	32		Position Report	<None>	3	ID: 35	0.104819	0.000000
37	RBC2	Train	100		Movement Authority	<None>	3	ID: 36	0.144819	0.000000

Fig. 12.6 Details of the Train Control Service Modeling Process

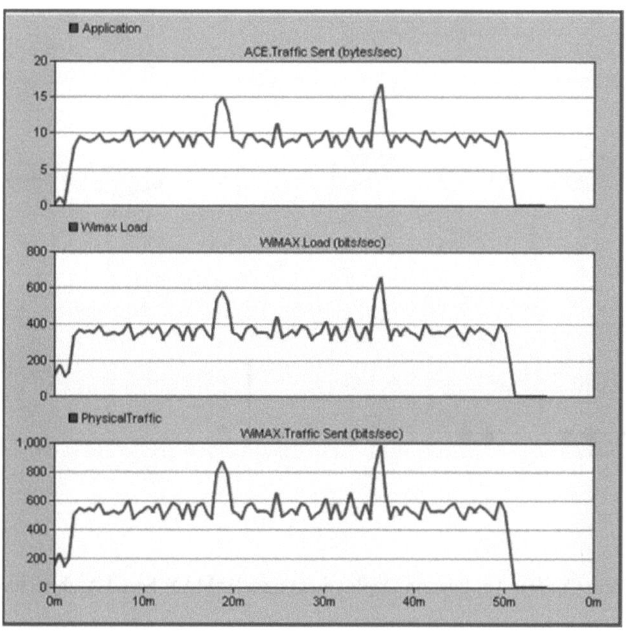

Fig. 12.7 Train Control Application Profile

12.5.2 Networkability and Impact Assessment Application

Once that the data traffic is built in the simulation platform, the logical next step is to carry out a "what if" analysis or application impact assessment predicting the impact of deploying the previously designed applications. This can help determine whether the application will perform as required when deployed on the network. The idea is to validate network and configuration changes prior to deployment, or to predict the impact of infrastructure migration or change.

For example, the train control customized data application described above is normally born by a GSM-based infrastructure. A set of quality requirements, namely EIRENE (European Integrated Railway Radio Enhanced Network) requirements, is currently specified and standardized for the train control application. In order to know if the same quality of service requirements will be met when deployed in a different infrastructure, i.e., a WiMAX-based architecture as shown in Figure 12.8, a new simulation scenario was built using this technology and the quality of service requirements were compared. Further details of this architecture may be found in [2].

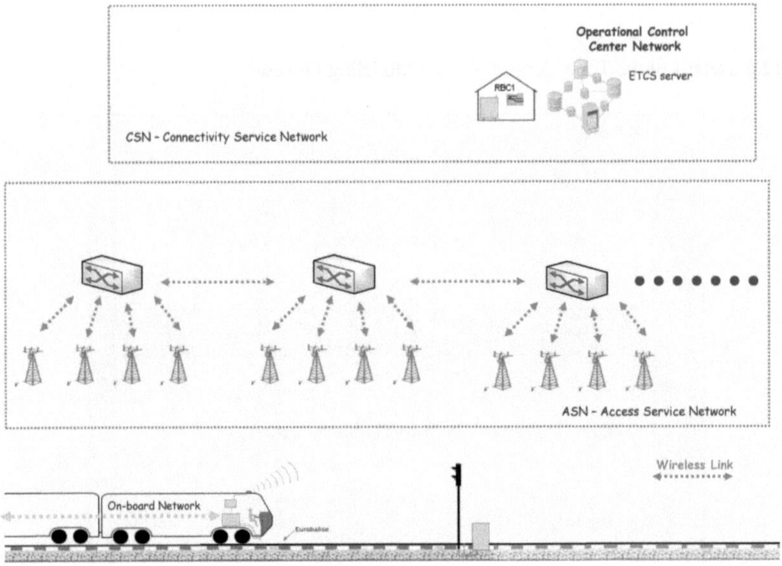

Fig. 12.8 A Train Control Application Deployment on a WiMAX-based Architecture

The results obtained in the simulation platform are summarized in Table 12.2 and clearly validate that the new proposed architecture meets the specific QoS requirements for train control services.

Table 12.2 Results Obtained in Simulation Platform Compared to QoS Requirements.

QoS Parameter	EIRENE Value	WiMAX-based Arch. Value
Connection establishment delay of mobile originated calls (CED)	< 8.5s (95%) ≤ 10s (100%)	< 1.5s (100%)
Maximum end-to-end transfer delay (30 byte data block) (TED)	≤ 0.5s (99%)	< 0.35s (100%)
Average end-to-end transfer delay	≤ 400 to 500 ms (100%)	< 130 ms
Network registration delay (NRD)	≤ 30s (95%), ≤ 35s (99%), ≤ 40s (100%)	
Handover effective time (between BSs)	> 300 ms	25 ms
Transmission data rate	≥ 24 Kbps	Upto 10 Mbps

12.5.3 Network Operations and Capacity Planning

There is a constant trend for networks to increase in size and in the number of services they provide. These services have very clearly defined performance requirements. A simulation study provides the possibility to know if these service level agreements and requirements are met in this new situation or in a specific critical situation. For example, we are currently studying the performance of the previously described train-control application together with a Voice Over IP (VoIP) application deployed on a WiMAX-based architecture in the situation where two trains are crossing. The goal is to validate the application service quality in a scenario where there are two trains with train control data traffic and voice over IP. In this scenario, the trains are travelling in a double line stretch and their routes cross at a given point. Similar studies may be undertaken to identify the total number of trains in a congested yard from the telecom point of view. The critical point under study relates to the moment where the trains cross and share the same base station. The WiMAX MAC traffic sent by each train is represented in Figure 12.9. The second graph is a detailed zoom from the previous graph where the train control service profile is added to the constant VoIP data traffic profile. The received data traffic is shown in Figure 12.10. Traffic data drops may be observed when handovers take place. Observing the obtained results, it can be concluded that in the new infrastructure a BS already serving a mobile node will not see its performance affected by the acceptance of a new mobile node. It will provide both demanded services with similar performance.

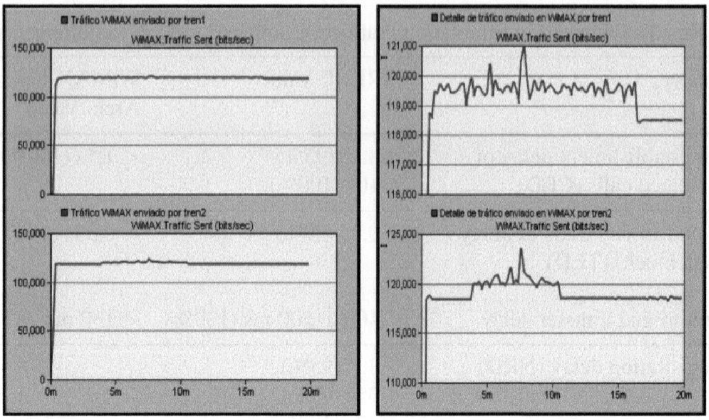

Fig. 12.9 WiMAX MAC traffic sent and detail on Train Control Traffic over VoIP traffic

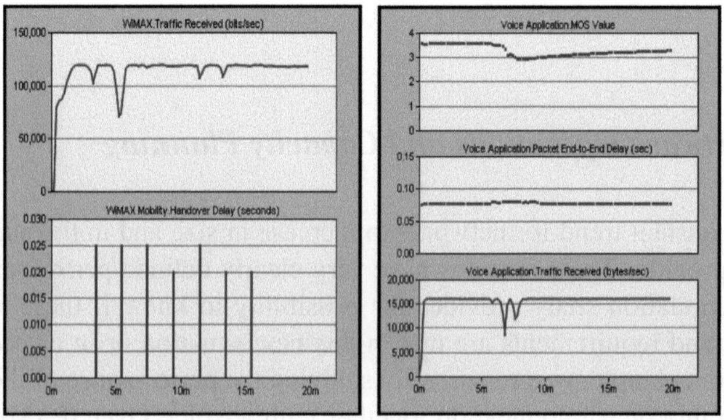

Fig. 12.10 Traffic drops during handover, handover delay, MOS, end-to-end delay and incoming traffic

12.5.4 *Network Research and Development*

Another important area of application for a network simulation tool is the design and evaluation of new network protocols, handover mechanisms, etc. As an example, in this section we introduce the evaluation of handovers using IEEE802.11 and

IEEE802.16 wireless access technologies when the Host Identity Protocol (HIP) [17] is used for managing mobility. HIP is an emerging end-to-end (e2e) security association establishment and maintenance protocol that gathers security, multihoming, and mobility management through the introduction of a new namespace in the TCP/IP stack. Therefore, this case study focuses on analyzing the feasibility of HIP when handovers take place. It provides insight into the mobility management protocol selection not only for the research community, but also for network operators.

The OMNeT++ [20] simulation tool was selected for this study because of its mature simulation model for HIP. OMNeT++ is an open source simulation platform that provides a modular and extensible architecture for modeling any communication context regardless its complexity. OMNeT++ does not provide specific components to be included in the simulation scenario, but modules that can be hierarchically combined to build specific network components. These modules are defined based on a set of frameworks. A framework is a package that contains functions and atomic modules to create different network functionalities and components. The development of frameworks is independent to the core of the OMNeT++ simulation tool. Consequently, customized modeling environments, that meet specific user needs, can be developed through the integration of available frameworks.

In this case study, in order to study HIP protocol in a multi-technology mobile environment, we built a new framework that integrates functionalities from two different frameworks: Inetmanet [20] and HIPSim++[3]. Inetmanet offers modules for modeling IEEE802.11 and IEEE802.16 technologies as well as modules for defining the mobility patterns of the mobile nodes, channel models, etc. On the other hand, the HIPSim++ framework provides a basic model of the HIP protocol. HIPSim++ has been specifically designed for modeling the HIP protocol based on the xMIPV6 framework where fundamental functionalities for modeling mobility management protocols are defined. The resultant integrated framework provides required modules and functionalities to model IEEE802.11 and IEEE802.16 wireless access technologies and handover processes managed with the HIP protocol. Figure 12.11 shows the main stages involved in the framework development to obtain the customized framework.

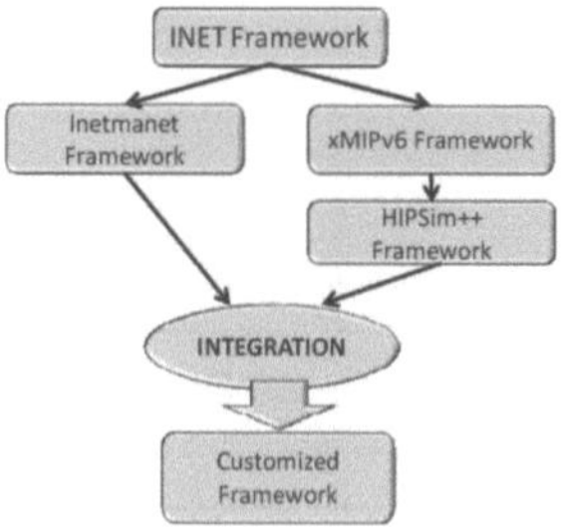

Fig. 12.11 Framework Integration

[3] The source code for HIPSim++ is available online at http://www.ict-optimix.eu/index.php/HIPSim.

After building this customized framework with the new modules and integrated functionalities, it was necessary to create new network components for the specific use cases. As an example, a new node with IEEE802.11 and IEEE802.16 interfaces supporting HIP protocol was introduced in the object palette. Additionally, we included other necessary elements (routers, a HIP DNS, a HIP aware node, and a video server) to model the handover scenario. Figures 12(a) and 12(b), and Tables 12.3 and 12.4, show the simulation scenarios and main configuration parameters used for analyzing WiFi and WiMAX-HIP-based handovers respectively.

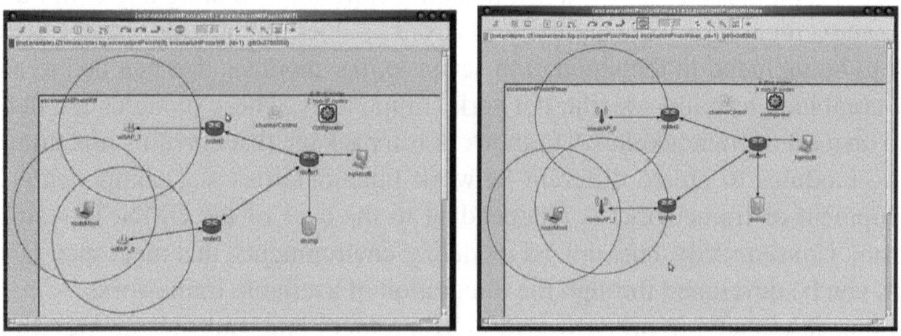

(a) WiFi (b) WiMAX

Fig. 12.12 Scenario configurations for horizontal handovers.

Table 12.3 Parameters for Simulation Depicted in Figure 12(a)

Parameter	Value
Frequency (GHz)	2,4
Bit rate (Mbps)	2
Processing time in AP switch-ethernet (μs)	2
Router advertisement interval (ms)	70
Video streaming traffic rate (packet/s)	100
Data packet size (B)	512

Table 12.4 Parameters for Simulation Depicted in Figure 12(b)

Parameter	Value
Frequency (GHz)	3,5
Bit rate (Mbps)	4
Service request timeout (s)	0.5
Scheduling algorithm	Weight round robin
Router advertisement interval (ms)	70
Video streaming traffic rate (packet/s)	100
Data packet size (B)	512

Using these scenarios we obtained the e2e complete handover and HIP handover latency measurements at different mobile node speeds from 5 Km/h to 300 Km/h with a linear mobility pattern. When referring to the complete handover process, we considered the time between the arrival of the last packet that arrived to the old IP address and the arrival of the first packet received at the new IP address. On the other hand, the latency of the HIP handover only considers the time needed for updating the end-to-end HIP association. Figures 12.13 and 12.14 show latency metrics for complete horizontal handover and HIP update processes for IEEE802.11 and IEEE802.16 technologies respectively at different mobile node speeds. The results obtained show that the time required for updating the HIP association represents a minimum part of the time required for handing over. Consequently the simulation study demonstrates that HIP hand over is a functionally successful solution to manage mobility; however research efforts should be focused on optimizing lower layer handover.

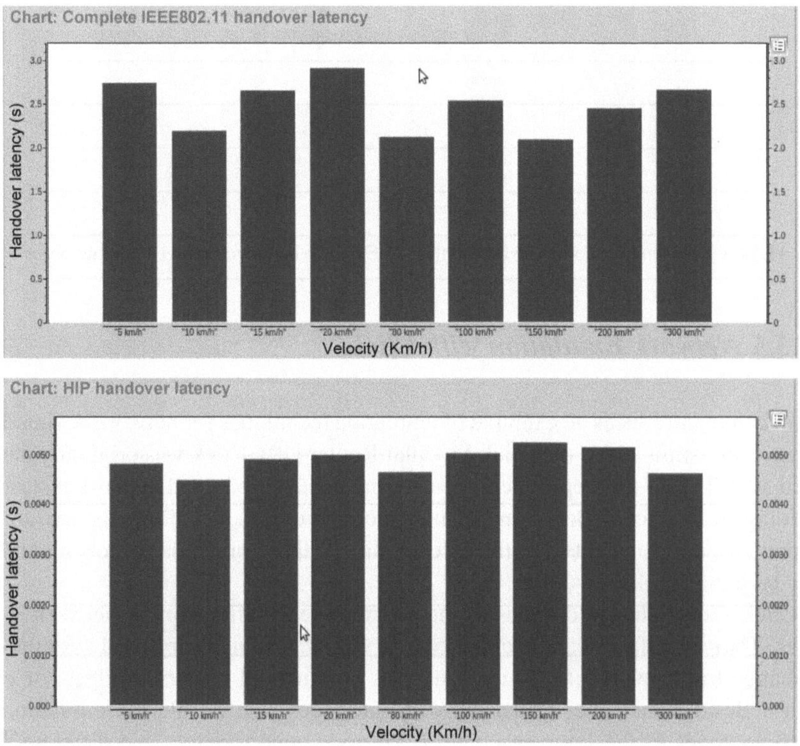

Fig. 12.13 Delay time for complete horizontal IEEE802.11 handover and HIP update process

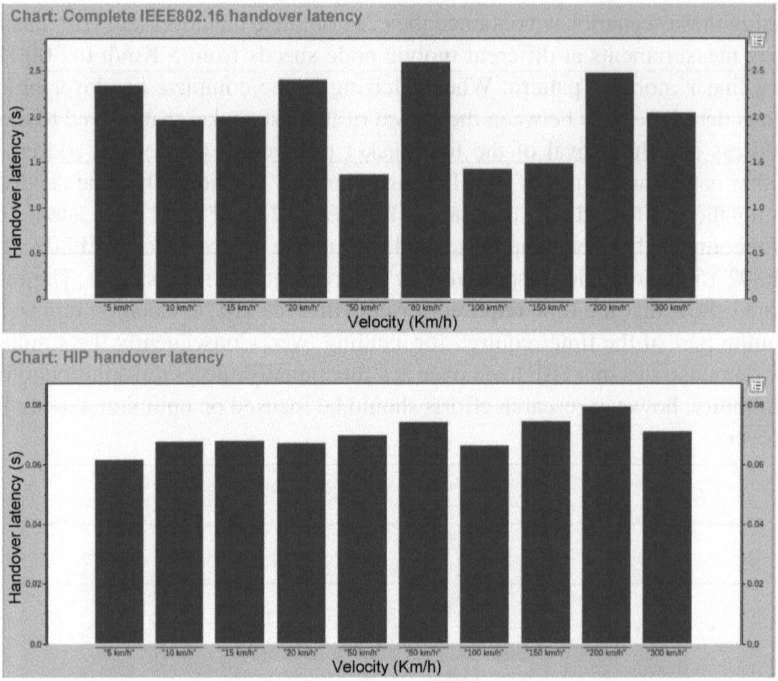

Fig. 12.14 Delay time for complete horizontal IEEE802.16 handover and HIP update process

12.5.5 Network Emulation with Click

This section introduces an example of emulation techniques for network design. The chosen emulation tool is the Click Modular Router [12]. Click was originally developed at MIT. This tool reproduces the external behavior of a real network instead of building an abstract model of one. It has a modular design that allows the reusability of previous developments and the extensibility of these previous models by adding new building blocks.

Click allows emulated nodes to interact with real traffic from/to network interfaces. These interfaces can be either real or virtual. The new emulated systems can exchange traffic with real devices, such as commercial routers, bridges, or end-nodes. Besides that, the virtual network interfaces also enable the emulation of multiple nodes in the same machine. Each node is considered as a different process by Click. By combining real and virtual interfaces, it is possible to build more complex scenarios with multiple emulation nodes running in several machines (in a distributed scenario) interconnected by real networks and interacting with commercial devices. One of the main advantages of this tool is the ability to modify the internal behavior of network devices. This is not an easy task when commercial products are involved. This tool is also a very useful and attractive one when testing new protocols in real scenarios interacting with real nodes. It enables the functional validation of new research proposals.

All these capacities turn Click into a valid candidate to emulate the behavior of Carrier-Grade Ethernet (CGE) technologies. In [26][27] [16, 15], we used Click Modular Router (1.8.0 release) to implement IEEE 802.1q (VLAN), 802.1ad (Q-in-Q) and 802.1ah (MAC-in-MAC) based on the EtherSwitch element (core package). Once these components were integrated, a modified version of the forwarding module was implemented to compare our proposed system's behavior versus the behavior of the standard system. The CGE elements allow the emulation of complex scenarios with a large number of emulated nodes, different interconnection schemes, and integration with real platforms as illustrated in Fig. 12.15. When using network emulation, multiple network schemes can be tested at a very low cost.

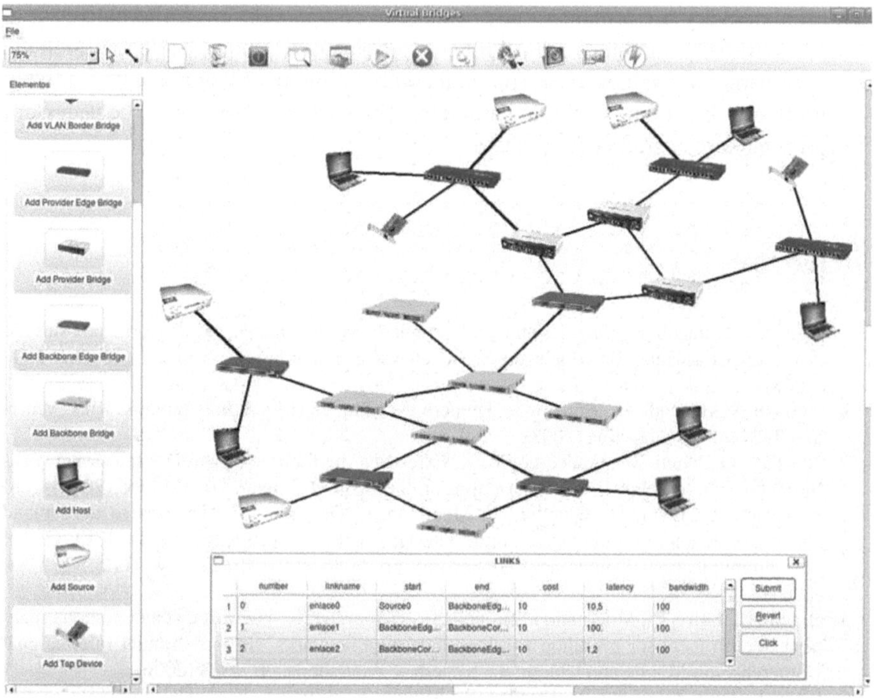

Fig. 12.15 Network Emulation with Click

Click has also been used for wireless network design research. For example, a wireless core package for WiFi (802.11 wireless networking) is available on-line at http://read.cs.ucla.edu/click/elements#wifi, and repository of papers using Click in variety of application areas such as multi-hop wireless body area networks (WBAN) [5, 13], multi-interface wireless mesh networks [4], and wireless grids [23] is available at http://read.cs.ucla.edu/click/publications.

12.6 Conclusions

To conclude, some useful tips to take into account when using a simulation tool for performance evaluation are listed below. The main idea is to follow a systematic approach detailed in the following steps:

- Define the system (identifying system boundaries)
- Establish goals
- List services and outcomes
- Select performance metrics related to the service being provided.
- List all the parameters that may affect the system performance. These parameters can be divided into two groups, those that will vary during the simulation (e.g., factors in a parametric study) and those that will not.

The goal of every performance study is either to compare different alternatives or to find the optimal parameter value. With simulations, it may be possible to search the space of parameter values for the optimal combination, but often it is not clear what the trade-offs are among different parameters [9]. In these cases, it will be necessary to carry out some parametric studies.

References

1. Aguado, M., Jacob, E., Saiz, P., Matias, J., Higuero, M., Berbineau, M.: Performance simulation study of a wimax based telecom architecture for train control systems. In: WCRR 2008 (2008)
2. Aguado, M., Onandi, O., Agustin, P., Higuero, M., Taquet, E.J.: WiMax on rails. IEEE Vehicular Technology Magazine (2008)
3. Anastasi, G., Conti, M., Francesco, M.D.: Extending the lifetime of wireless sensor networks through adaptive sleep. IEEE Transactions on Industrial Informatics **5**, 351–365 (2009)
4. Bouckaert, S., Letor, N., Blondia, C., Moerman, I., Demeester, P.: Distributed on demand channel selection in multi channel, multi interface wireless mesh networks. In: IEEE Global Telecommunications Conference (Globecom), pp. 5086–5091. IEEE, New York, NY, USA (2007)
5. Braem, B., Latré, B., Moerman, I., Blondia, C., Demeester, P.: The wireless autonomous spanning tree protocol for multihop wireless body area networks. In: 3rd Annual International Conference on Mobile and Ubiquitous Systems - Workshops, pp. 1 – 8 (2006)
6. Choi, J., Tran, T., Kwon, J., Thomas, R., Mount, T., Billinton, R.: Nodal probabilistic production cost simulation considering transmission system unavailability. IET Gener. Transm. Distrib. **2**, 32–42 (2008)
7. Fernekeß, A., Klein, A., Wegmann, B., Dietrich, K.: Modular system level simulator concept for OFDMA systems. IEEE Communications Magazine **47** (2009)
8. Hasan, M.S., Yu, H., Carrington, A., Yang, T.C.: Co-simulation of wireless networked control systems over mobile ad hoc network using SIMULINK and OPNET. Journal of IET Communications **3**, 1297–1310 (2009)
9. Jain, R.: The Art of computer systems performance analysis : techniques for experimental design, measurement, simulation and modeling. Wiley, New York (1991)
10. Kasch, W.T., Ward, J.R., Andrusenko, J.: Wireless network modeling and simulation tools for designers and developers. Comm. Mag. **47**(3), 120–127 (2009). DOI http://dx.doi.org/10.1109/MCOM.2009.4804397

11. Kim, B.H., Yun, J., Hur, Y., So-In, C., Jain, R., Al Tamimi, A.K.: Capacity estimation and TCP performance enhancement over mobile WiMAX networks. IEEE Comm. Magazine **47**, 132 – 141 (2009)
12. Kohler, E., Morris, R., Chen, B., Jannotti, J., Kaashoek, M.F.: The click modular router. ACM Transactions on Computer Systems **18**, 263–297 (2000). Software available online at http://read.cs.ucla.edu/click/
13. Latré, B., Braem, B., Moerman, I., Blondia, C., Reusens, E., Josephand, W., Demeester, P.: The wireless autonomous spanning tree protocol for multihop wireless body area networks. In: Fourth Annual International Conference on Mobile and Ubiquitous Systems: Networking and Services (MobiQuitous), pp. 1 – 8 (2007)
14. Maria, A.: Introduction to modeling and simulation. In: WSC '97: Proceedings of the 29th conference on Winter simulation, pp. 7–13. IEEE Computer Society, Washington, DC, USA (1997). DOI http://doi.acm.org/10.1145/268437.268440
15. Matias, J., Jacob, E., Aguado, M., Astorga, J.: Enhancing NGN's versatility for multi-service support: The bridging virtualization approach. In: 10th International Conference on Telecommuincations (ConTEL), pp. 367 – 374 (2009)
16. Matias, J., Jacob, E., Aguado, M., Astorga, J.: The bridging virtualization approach to next generation broadband access networks. Infocommunications Journal (To appear)
17. Moskowitz, R., Nikander, P., Jokela, P., Henderson, T.: Host identity protocol. http://tools.ietf.org/html/rfc5201. Accessed on August 7, 2010
18. Munasinghe, K., Jamalipour, A.: Interworked WiMAX-3G cellular data networks: An architecture for mobility management and performance evaluation. IEEE Transactions on Wireless Communications **8**, 1847–1853 (2009)
19. Niazi, M., Hussain, A.: Agent-based tools for modeling and simulation of self-organization in peer-to-peer, ad-hoc and other complex networks. IEEE Communications Magazine **47**, 163 – 173 (2009)
20. OMNeT++ Community: http://www.omnetpp.org/. Accessed on August 7, 2010
21. OPNET Technologies: http://www.opnet.com. Accessed on August 6, 2010
22. OPNET Technologies: Verifying Statistical Validity of Discrete Event Simulations. http://www.opnet.com/whitepapers/index_vsvodes.html. Accessed on August 6, 2010
23. Ramachandran, K., Kaul, S., Mathur, S., Gruteser, M., Seskar, I.: Towards large-scale mobile network emulation through spatial switching on a wireless grid. In: E-WIND '05: Proceedings of the 2005 ACM SIGCOMM workshop on Experimental approaches to wireless network design and analysis, pp. 46–51. ACM, New York, NY, USA (2005). DOI http://doi.acm.org/10.1145/1080148.1080158
24. Ramadas, K., Jain, R.: WiMAX system evaluation methodology. http://www.cse.wustl.edu/ jain/wimax/ftp/wimax_system_evaluation_methodology_v2_1.pdf (2008). Accessed on August 14, 2010
25. The MathWorks, Inc.: http://www.mathworks.com/products/matlab/. Accessed on August 7, 2010
26. Whitaker, R.M., Allen, S.M., Hurley, S.: Efficient offline coverage and load evaluation for CDMA network modeling. IEEE Vehicular Technology Society **58**, 3704 – 3712 (2009)
27. Zhang, F., Zhong, W.D.: Performance evaluation of optimal multicast protection approaches for combined node and link failure recovery. IEEE/OSA Journal of Lightwave Technology **27**, 4017–4025 (2008)

Chapter 13
Optimization of Wireless Broadband (WiMAX) Systems

Hang Jin and Li Guo

Abstract A WiMax system integrates a number of new technologies, such as multi-antenna technologies and adaptive modulation and coding schemes. Those new technologies potentially deliver high data rates that can satisfy customer requirements. However, proper optimization is necessary for system design to achieve the desired superior performance. This chapter discusses the optimization procedure of a WiMax system from various aspects and particularly the radio resource allocation optimization and scheduling optimization. The radio resource allocation includes the optimization of modulation and coding scheme, MIMO modes, power control and allocation and link adaptation. In scheduling optimization, several service and multi-user scheduler mechanisms used in a WiMax system are discussed. WiMax QoS policy is also presented for the scheduling optimization. Finally, several system-level experiments are presented. The experiment results show that proper optimization can boost the system throughput and satisfy various performance requirements.

13.1 Introduction

The mobile WiMAX technology, based on the IEEE 802.16e Air Interface Standard, is a broadband wireless solution that enables convergence of mobile and fixed broadband networks through a common wide area broadband radio access technology and flexible network architecture. The Mobile WiMAX Air Interface adopts Orthogonal Frequency Division Multiple Access (OFDMA) [5, 11] for improved multi-path per-

---

Hang Jin
Cisco Systems, 2200 East President George Bush Highway, Richardson, Texas 75082-3550, USA, e-mail: hangjin@cisco.com

Li Guo
Cisco Systems, 2200 East President George Bush Highway, Richardson, Texas 75082-3550, USA, e-mail: liguo2@cisco.com

J. Kennington et al. (eds.), *Wireless Network Design: Optimization Models and Solution Procedures*, International Series in Operations Research and Management Science 158,
DOI 10.1007/978-1-4419-6111-2_13,

formance in non-line-of-sight environments, and multiple-input-multiple-output to enhance the spectrum efficiency. Scalable OFDMA (SOFDMA) is introduced in the IEEE 802.16e Amendment to support scalable channel bandwidths from 1.25MHz to 20 MHz. Mobile WiMAX systems offer scalability in both radio access technology and network architecture, thus providing a great deal of flexibility in network deployment options and service offerings. Mobile WiMAX supports up to 63 Mbps downlink and 28 Mbps uplink data rate; end-to-end IP based QoS; ultimate scalability from 1.25MHz to 20MHz, robust operation with mobility and fading environments [1, 2, 11].

As a new broadband wireless access system, OFDMA-based WiMax system has a number of distinct features, including flexible subchannelization, adaptive modulation and coding, space-time coding and spatial multiplexing, dynamic packet-switch based air interface, and flexible network deployment such as fractional frequency re-use. All those features impose a requirement of careful optimization of the critical parameters and schemes in the system for obtaining superior performance. The optimization of mobile WiMAX systems is broadly categorized into the following three areas:

1. RF optimization
   The RF optimization is the key step in WiMAX networks deployment planning. The purpose of RF optimization is to provide reliable and ubiquitous RF coverage for the whole market. RF optimization consists of frequency re-use planning, site location selections, BTS height selection, and BTS antenna orientation selection (azimuth and down tilt selection). The optimization is an iterative procedure, which needs the interactions between optimization and design of each individual factor listed above.
2. Radio resource allocation optimization
   Radio resource allocation plays an important role in determining the overall system performances. Given a network configuration, the radio resources consist of:

   a. BTS transmitter power
   b. Total frequency bandwidth allocated to a BTS
   c. The total time duration ratio between downlink subframe and uplink subframe in TDD system, and the time duration of slot allocated to each user or service connection.
      The later two can be visualized as an optimization in a two-dimensional frequency-time space. In the case of mobile WiMAX, the minimum resource allocation unit is called a slot, which is a block of about 15 kHz in frequency and 200us in time in the frequency-time space. WiMAX support 5ms frame, and with 2:1 down link to up link ratio; there are about 420 slots for down link, and 175 slots for uplink. These down link and uplink slots need to distributed among hundreds users in an optimal way that the overall system performance metrics (capacity and latency) are optimized.
      BTS transmitter power varies from a few watts (micro/pico cells) to 10–20 watts (Macro cells). Mobile WiMAX employs OFDMA and there are multiple

data bursts for different users occurring at the same time. Users on the cell edge need more power to achieve the same performance of users in the cell center. There is a need for a fair scheme of power distribution among users to have optimal overall system performance and also ensure that calls of cell edge users are not dropped. It is worthwhile to mention that the radio resource allocation should also consider the channel fading. As the channels vary with time, the resource allocations need to be adjusted accordingly. This leads to a key topic: link adaptation. With link adaptation, one could dynamically adjust the resource allocations, including user modulation and coding scheme, to ensure near optimal system performance.

3. Scheduling optimization
   Another area where optimization plays a key role is in data scheduling. For each WiMAX BTS, there are multiple MS that request services with different QoS requirements. QoS is one of the key features of mobile WiMAX. BTS need to have a scheduling policy that can account for the various QoS requirements (priority, data rate and latency). The packets could result from various applications, for example, real time video, voice, web Browser, and file transfer (FTP). The QoS for these applications are quite different. The scheduler needs to schedule these packets differently so their QoS requirements can be met. Moreover, the scheduling is the solution to an optimization problem with a constraint that the physical channel can only handle certain throughput.
   RF optimization is a generic procedure that needs be done for all wireless networks, and it has been covered in chapter 5 of this book. This chapter will focus on the optimizations associated with the radio resource allocation and scheduling. The remainder of this chapter is organized as follows. Section 2 gives a brief description of WiMAX physical layer and lower MAC layer that are directly associated with radio resource allocation and scheduling. Section 3 describes the resource allocation and the optimization procedure. Section 4 describes the scheduling algorithm and some optimization consideration. The chapter is concluded in Section 5 with an actual simulation example.

13.2 Brief Description of WiMAX

13.2.1 WiMAX Physical Layer

Mobile WiMAX employs OFDMA as the multiple access technology for its air-interface. OFDM divides the bandwidth into multiple frequency sub-carriers. In an OFDMA system, the input data stream is divided into several parallel sub-streams of reduced data rate (thus increased symbol duration) and each sub-stream is modulated and transmitted on a separate orthogonal sub-carrier. The increased symbol duration improves the robustness of OFDM to delay spread. Furthermore, the introduction of the cyclic prefix (CP) can completely eliminate Inter-Symbol Interference

(ISI) as long as the CP duration is longer than the channel delay spread. The CP is a repetition of the last samples of data portion of the block that is appended to the beginning of the data payload. Introduction of CP renders very robust operation of OFDMA in the fading environments.

Mobile WiMAX uses OFDMA with 10.94kHz sub-carrier spacing. This sub-carrier spacing is used for all bandwidth (1.25MHz to 20MHz) with proper empty sub-carrier on the band edges. In the case of 10MHz bandwidth, there are total 1024 sub-carriers. Of these, the middle one (sub-carrier 512 is the DC carrier) and left 92 and right 91 sub-carriers are null and used for guard bands So there are total of 840 usable sub-carriers. Mobile WiMAX uses 5 ms frame, and supports full TDD duplex (FDD will be supported by standard IEEE802.16m, the future evolution standard for mobile WiMAX). Each 5ms frame consists of both the downlink sub-frame and uplink sub-frame. Proper guard times (RTG and TTG) occur between downlink and uplink sub-frames. Various downlink to uplink data ratios are supported by mobile WiMAX to have great flexibility in deployment and make it application specific.

The symbol duration of WiMAX is 91.4 us, then add 1/8 symbol as the CP, the total symbol duration is 102.9us. With 5 ms frame, there will be a total 47 symbols for both down and up links sub-frame, with the guard times (RTG and TTG).

The down link frame begins with a preamble (one symbol), followed by a zone that contains all the control messages (FCH and MAPS), and finally the actual data traffic. The uplink frame consists of ranging signals, feedback signals, and actual data traffic (Fig. 13.1).

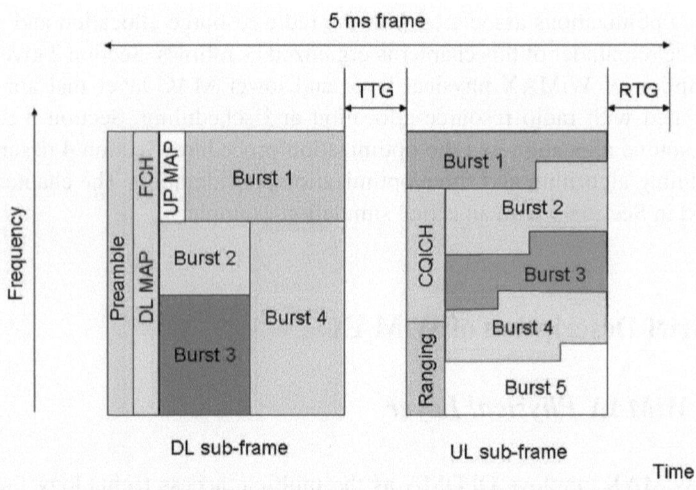

Fig. 13.1 WiMAX Frame Structure

The resource allocation unit in mobile WiMAX is called a slot. With PUSC permutation (Partial Usage of Subcarriers), for down link, one slot is a block spanning 28 sub-carriers in frequency and 2 symbols in time in the frequency-time space; for uplink, one slot is a block of 24 sub-carriers by 3 symbols. Mobile WiMAX employs a carrier permutation scheme that spreads the carriers of each slot over the whole bandwidth to have better frequency diversity. Thus the sub-carriers of each slot are spread over the whole band width.

For 10 MHz bandwidth systems, assuming 2:1 down link to uplink ratio, there are about 390 slots for down link traffic after removing the overhead associate with FCH and MAPs. The total slots for uplink is about 140, after removing the overheads associated with ranging and feedback channels.

The down link slot is further split into two clusters, and each cluster contains 14 sub-carriers in frequency and 2 symbols in time, forming 28 points in the two dimensional frequency-time space (Fig. 13.2). Among them, four points are the pilots for channel estimation. The uplink slot is split into 6 tiles, and each tile occupies 4 sub-carriers in frequency and 3 symbols in time, forming 12 points in the two-dimensional frequency-time space, among them, 4 points are the pilots for channel estimation (Fig. 13.3).

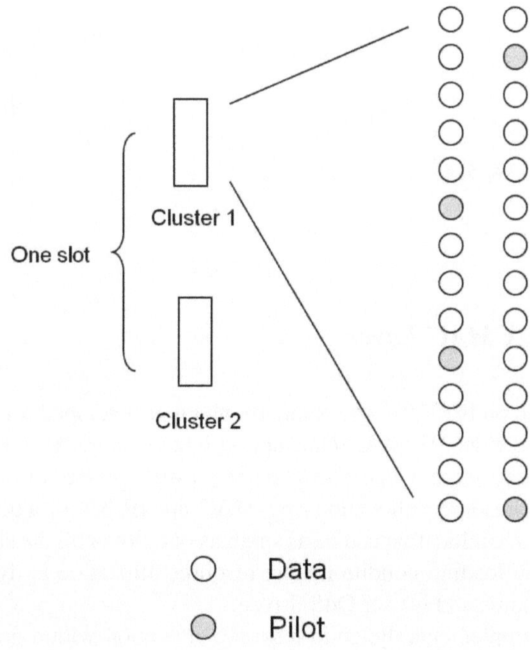

Fig. 13.2 Downlink PUSC Slot

WiMAX supports the following modulation coding schemes: 4QAM1/2, 4QAM3/4 16QAM1/2, 16QAM3/4, 64QAM1/2, 64QAM2/3, 64QAM3/4, and 64QAM5/6. The first number indicates the modulation order and the last number indicates the coding rate. The modulation coding scheme (MCS) can be dynamically adjusted based on the signal strength, channel condition and the interference level. Hybrid

ARQ (HARQ) and fast feedback channels are also supported to enhance the coverage, capacity and robustness of the systems.

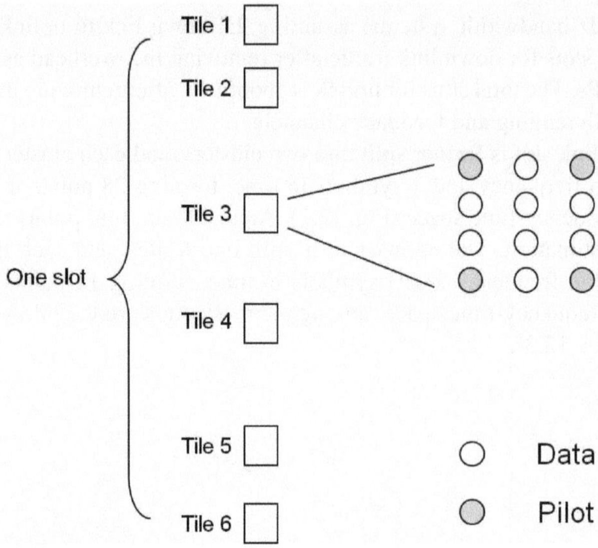

Fig. 13.3 Uplink PUSC Slot

13.2.2 WiMAX MAC Layer

WIMAX is based on IEEE802.16e standard which is developed for the applications of voice, data and video. WIMAX can support burst data traffic with high peak data rate as well as streaming video and latency-sensitive voice traffic over the same channel. All the resource allocations by MAC are IP based, and done frame by frame. The resource allocation can be as small as one slot or all the slots in the frame, depending on the loading condition. The resource allocation is dynamic, changes from frame to frame, and is user QoS driven.

As WiMAX implements dynamic resource allocation with a granularity of one slot (6 bytes with 4QAM MCS) and frame to frame link adaptation, WIMAX resource scheduler can meet various QoS requirements for a wide range of data services and applications.

The QoS requirements are accommodated via the following procedure:

1. Set up service flows with various QoS configurations
2. Classify the incoming data and map them onto proper service flows
3. Define a scheduling/resource allocation scheme for the data in the active service flows that honor the QoS requirements of the service flows.

The service flows play the key role in ensuring the QoS requirements. Each server flow is connected to its peer at the other end so there is a one-to-one mapping between service flows with the proper QoS setting. The QoS setting then defines how the scheduler/resource allocation allocates the resources for the data in the service flow. The QoS based service flows are applied for both downlink and uplink traffic.

The scheduling and resource allocation are done frame by frame, and the resource allocation messages come in at the beginning of each down link frame (MAP zone) so that the MS can check the allocation message, based on which, to look for the slots where its data are allocated. Moreover, there is a physical channel defined for the purpose of feedback, CQICH, so MS can feed back the signal quality very quickly. The BS uses this feed back signal quality to adjust the resource allocations: up or down grade the MCS, increase or decrease the throughput, MIMO mode selection, and HARQ setting. These all allow the system to work robustly while also optimizing the system performance.

The data is carried in WiMAX in the form of MPDU (MAC packet data unit). The WiMAX MAC uses a variable-length MPDU and offers a lot of flexibility to allow for their efficient transmission. The physical allocation is in the form of data burst. For the down link, the data bursts are rectangular blocks in the frequency-time space. For uplink, the bursts are the strips cascaded together (Fig. 13.1). The mapping between the MPDU and burst is flexible; multiple MPDU of various lengths can be aggregated into a single burst to save the allocation overhead. Also, the MPDU can be fragmented into multiple bursts if the size of MPDU is too big for a single burst. The flexibility in term of resource allocation reduces the overhead of the MAC. An MPDU consists of three parts: a generic MAC header, the payload and the optional CRC (Fig. 13.4). Moreover, there are sub-headers following the generic header for various MAC and report functions

Generic MAC header (6 bytes)	Payload (variable length)	CRC (optional)

Fig. 13.4 MAC PDU Format

The MAC generic header has a fixed length of 6 bytes, it contains a connection identifier (CID), the length of frame, and bits to qualify the presence of CRC, subheaders, and whether the payload is encrypted and if so, with which key. The MAC payload is either a transport or a management message. Besides MSDUs (MAC service data unit), the transport payload may contain bandwidth requests or retransmission requests. The type of transport payload is identified by the subheader that immediately precedes it. Examples of subheaders are packing subheaders and fragmentation subheaders. WiMAX MAC also supports ARQ, which can be used to request the retransmission of un-fragmented MSDUs and fragments of MSDUs.

13.3 Radio Resource Allocation Optimization

The radio resource allocation can be formulated as a classic optimization problem. The objective function is a function of the system performance metrics including capacity, latency and user experiences. The optimization constraints are the limited BS transmitter power and limited number of channels (slots) and user SLA (service layer agreement). The user SLA consists of average throughput, the peak throughput, the max latency, and response time (the duration between the times when user powers up MS and when the service is fully activated). These parameters are part of the service agreement at the point of the sale. The goal of radio resource allocation is to optimally allocate the available resource so the performance metrics are met, or the gap between the target metrics and those achieved is minimized. Generally, the optimization problem can be formulated as maximizing the sum data rate:

$$\arg\max_{c_{k,n}\ p_{k,n}} \sum_{k=1}^{K} \sum_{n=1}^{N} c_{k,n} \rho(p_{k,n}, \gamma_{k,n})$$

subject to

$$\sum_{k=1}^{K} \sum_{n=1}^{N} c_{k,n} \leq N$$

$$\sum_{k=1}^{K} \sum_{n=1}^{N} p_{k,n} \leq P_{Tx}$$

$$p_{k,n} \geq 0, \quad c_{k,n} = 0 \text{ or } 1 \ \forall k \ \forall n$$

where ρ_k^{min} is the minimum required data rate of user k or one service flow; P_{Tx} is the total transmit power of BTS; $\gamma_{k,n}$ is the channel quality of user k on the subcarrier n; $\rho(.)$ is the data rate which can be achieved with the transmit power $p_{k,n}$ and channel quality $\gamma_{k,n}$. It is a joint optimization problem on power allocation $p_{k,n}$ and subcarrier allocation $c_{k,n}$.

Based on the above problem formulation, we can see that the complexity of the radio resource allocation in WiMAX, (and more generally, in all wireless broadband systems) is due to the following:

1. Multiple users in the system
   Typically, there are over a hundred of active users at any given time, and each user has different physical condition (path loss and channel fading condition) and performances requirements. The resource allocation needs to be flexible and yet disciplined so the requirements of each user can be met.
2. Fading channel
   An inherent characteristic of the wireless channel is that the communication link between BS and MSs experience channel fading. Channel fading is the terminology describing the signal received at BS or MS receiver fluctuates over time and frequency. The channel fading causes the received signal level as well the quality

of the signal to fluctuate over time. In the case of mobility, the fluctuation can be very severe. For 5 ms frame, the signal level can swing as much as 10dB from frame to frame. So, the resource allocation needs be dynamical with time, and reacts very fast to account for the channel fading.

3. A dynamical mix of various traffic types
   The applications and services that are running in wireless broadband networks are very diverse, and have wide range of the requirements. Common internet traffic models and requirements are given in Table 13.1.

Table 13.1 The Common Internet Traffic Models and Their Requirements

Class	Application	Protocol	Bandwidth	Latency	Jitter
1	Multiplayer Interactive Gaming	rtPS Real time polling service	≈ 50 kbps	Low < 25 msec	N/A
2	VoIP and Video Conference	ertPS Extended real time polling service	32-64 kbps	< 160 msec	< 50 msec
3	Streaming Media	rtPS Real time polling service	5 kbps to 2 Mbps	N/A	< 100 msec
4	Web Browsing and Instant Messaging	nrtPS rtPS Non-real time polling service	10 kbps to 2 Mbps	N/A	N/A
5	Media Content Downloads	BE Best effort	> 2 Mbps	N/A	N/A

For typical WiMAX networks, the traffic is mixed with the various types described above. The resource allocation needs to account for the requirements of these individual requirements for various services and applications. The radio resource allocation consists of the following steps:

1. MCS selection
2. MIMO mode selection
3. Power allocation (power control)
4. Link adaptation

Each of these topics will be discussed next.

13.3.1 Modulation Coding Scheme Selection

There are a total of 8 different MCS supported in WiMAX (Table 13.2). Downlink supports MCS 0-7, and for uplink it is mandatory to support MCS 0-3 and optional to supports MCS 4-7. Different MCS requires different signal quality, i.e. CINR (carrier to interference-noise ratio).

For uplink, BS receives the uplink signals form all MSs, and from the uplink signals, BS can compute the CINR for each MS. WiMAX specifies that the CINR shall be computed only from the pilots embedded in the traffic.

Table 13.2 The MCS Supported in WiMAX and Their CINR Requirements

MCS Type	Modulation/Code Rate	CINR Required (dB)
0	4QAM1/2	6
1	4QAM3/4	9
2	16QAM1/2	12
3	16QAM3/4	15
4	64QAM1/2	18
5	64QAM2/3	20
6	64QAM3/4	21
7	64QAM5/6	23

Thanks to the uplink power control, MS only transmits sufficient power for certain allocation configuration (MCs and number of sub-channels). To determine the MCS for any future uplink burst allocation, BS not only needs to know the CINR of the current uplink burst, but also the MS transmitted power and the number of the uplink carriers of the current uplink data burst. For example, let the CINR for the current burst be 6dB, and the number of the sub-carrier equals 24 (one uplink channel). The MS transmitted power is 15dBm although its nominal output power is 24dBm. BS then Fig.s out that with 15dBm output and one sub-channel, MS can achieve 6dB CINR. So, potentially, MS can achieve 15dB with one channel or 12dB with two channels if MS transmits with its full power. So, when assigning the MCS for the future allocation, BS will take these factors into the computation. For example, if the next burst needs 4 channels, and BS could select MCS 1 (4QAM3/4) as MCS 1 requires 9dB CINR, and with the full output power (24dBm), MS can achieve 9dB CINR with 4 channels (when transmitting with 15dBm output for one channel, 6dB CINR can be achieved. With 24dBm output, the power for each channel will be 18dBm, leading to 9dB CINR). The CINR is computed from the current uplink traffic, and BS knows the number of the channels assigned for the current uplink burst as it was assigned by BS in the previous frames. However, the actual transmitted power of the MS needs be reported back from MS through the one of the sub-headers in the MAC PDU.

The downlink MCS selection follows almost the same procedure as the uplink. The difference is that the down link CINR values need be reported back from MS through the fast feedback channel (CQICH). However, the power allocation is more involved as the BS transmitted power is shared among multiple users. For example, assume there are two active users, and one which is closer to BS, denoted as MS-A and the other which is farther away from BS, denoted as MS-B. Since it is far from the BS, MS-B experiences 100 times (20dB) more path loss than MS-A, who is closer to BS. Hence, to support the same down link MCS, BS needs to allocate 100 time power per sub-channel to MS-B than MS-A so both MS-A and MS-B can achieve the same downlink CINR required. This near-far issue is a big factor in downlink MCS selection. Generally, MSs that are far away from BS are assigned lower MCS and fewer number of sub-channels than MS that are closer to BS to optimize the BTS power utilization.

13.3.2 MIMO Mode Selection

One of the key features of a WiMAX system is the use of multiple antennas at the transmitter and receiver. WiMAX supports two MIMO (multi-input, multi-output) modes: MIMO-A and MIMO-B. MIMO-A is a rank-1 diversity scheme, and MIMO-B a rank-2 spatial multiplexing scheme [4, 8, 9] The architecture of MIMO-A is illustrated in Fig. 13.5. As shown in the Fig., the symbols from the encoder and modulator are encoded into two paths through STC encoding. Then the symbols of each path are mapped onto the sub-channels of OFDMA downlink space-time coding (STC) [2, 4]. The symbols from two paths are passed through the IFFT and transmitted on two separate antennas.

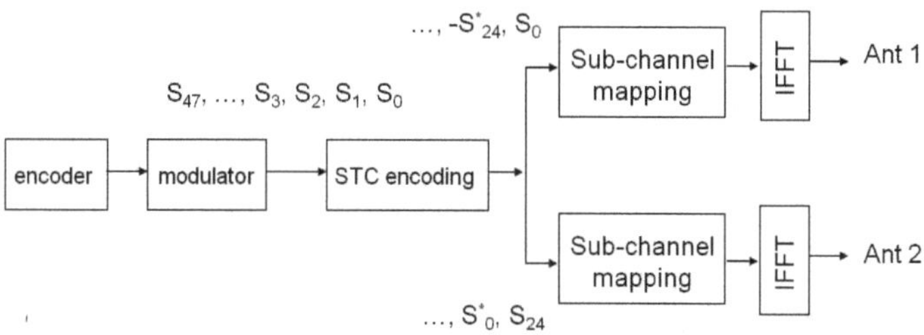

Fig. 13.5 System Architecture for MIMO-A (STC) in WiMAX

The symbol flow of the MIMO-B is illustrated in Fig. 13.6. MIMO-B uses spatial multiplexing with vertical encoding. As shown in Fig. 13.6, all the symbols in vertical encoding are from ONE FEC coded block. The modulated symbols are divided sequentially into two paths. Then the symbols of each path are mapped onto the sub-channels of the OFDMA down link. The symbols from the two paths are passed through the IFFT and transmitted using two separate antennas.

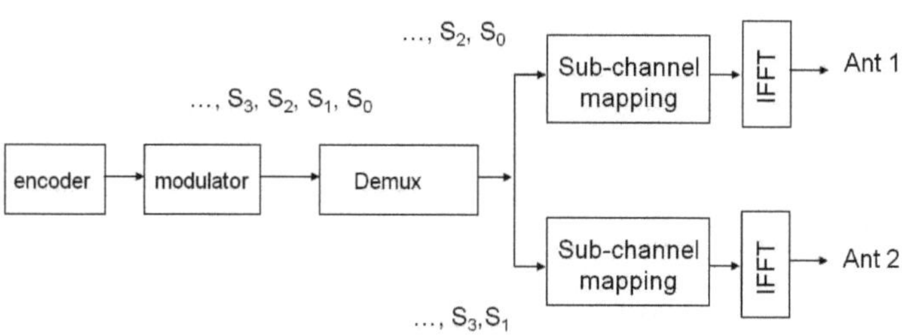

Fig. 13.6 System Architecture for MIMO-B (Spatial multiplexing) in WiMAX

Given the same path loss and BTS output power, the signal quality that can be achieved with MIMO-A is at least 3dB higher than that of MIMO-B. The reason for this difference is because MIMO-A is a rank-1 scheme in which the same symbol is transmitted twice while MIMO-B is a rank-2 scheme. So, the power per symbol is half in MIMO-B compared to MIMO-A. Furthermore, MIMO-A is a linear STC scheme (Alamouti scheme [4]), and there are no MIMO co-channel interferences in the MIMO decoding process. On the other hand, MIMO-B is a spatial multiplexing scheme, and depending on the spatial correlation of two co-channel data streams, there are co-channel interferences. Simulation shows that there are typically 6-8dB CINR drops when MIMO-A is switched to MIMO-B given the same channel conditions (Fig. 13.7).

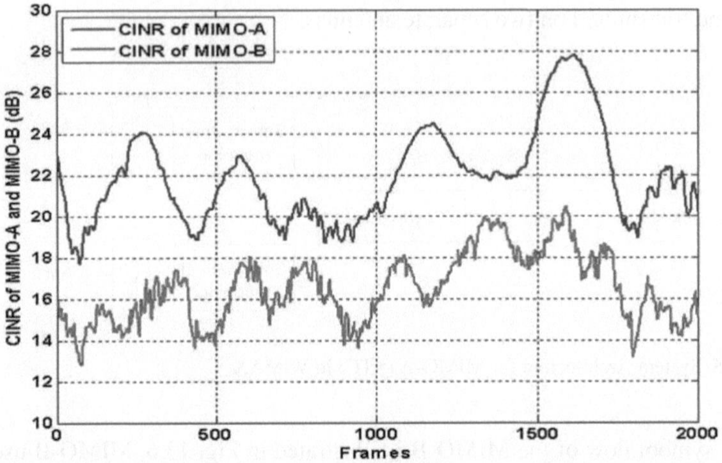

Fig. 13.7 CINR Differences Between MIMO-A and MIMO-B in a Fading Channel

Enabling MIMO-B improves the spectrum efficiency since two different data streams are transmitted on the same channel. On the other hand, switching to MIMO-B will cause the SNR to drop, and lead to downgrade of MCS for each data streams, which result in drops of the spectrum efficiency of each data stream. So, there are two contradicting effects resulting from activation of MIMO-B, and one needs to weigh these two effects to determine if there is a capacity gain. For example, if SNR=23dB is achieved with MIMO-A, so 64QAM5/6 is used, which has 5bits/Hz spectrum efficiency. If the MIMO mode is switched to MIMO-B, and assume the SNR is now dropped to 17dB. With SNR=17dB, 16QAM3/4 is used and 3bits/Hz/stream can be achieved. So, the total spectrum efficiency equals 6bits/Hz, a 20The key factor is the SNR degradation when the MIMO mode is switched from MIMO-A to MIMO-B. There is 3dB drop due to the power reduction with MIMO-B (half the power per symbol compared to MIMO-A). In additional, there are SNR drop due to the co-channel interferences between MIMO-B two data streams. The actual CINR drop depends on the spatial correlation between the two data streams. As a rule of thumb, MIMO-B is considered only when SNR is higher than 23dB, that is, the highest MCS (64QAM5/6) is achieved with MIMO-A.

13.3.3 Power Control

The power control in WiMAX, or any wireless broadband systems in general, is much more involved than cellular systems. The general purpose of power control in WiMax system is to deliver sufficient power for the MCS assigned to each MS and to reduce the interference caused to other users [3, 6]. The complexity results from the fact that multiple MCS are supported, and each MCS requires different power levels (different CINR). Also, MS can support multiple simultaneous data bursts with different MCS. So, the power control is not just user based, instead, it is burst based and MCS based. One user can have multiple power control settings, depending on the number of bursts and MCS setting. For down link, there is only closed loop power control, while for uplink, there are closed loop and open loop power controls.

13.3.3.1 Downlink Power Control

The purpose of the downlink power control is to ensure that MS gets just sufficient power for the MCS assigned, and ideally no more and no less. BS transmitter power is shared among all the MS, and in most cases, MS downlink MCS is limited by the power that BS can allocate to it. So, in a sense, it is the power allocation that actually determines the MCS assignment, not the other way around. Downlink power control can also help avoid unnecessary power transmission to reduce the multi-cell interferences.

The downlink power control follows the following procedure:

1. Set the max power density Pm as follows Pm = P -10*log10(N) + Pd Where P is the total BS transmitter power in dB, N is the total number of channels that the system supports, Pd is an additional factor, usually, set to 3dB
2. Collect the CINR values for all active MS
3. Given the previous power allocation and the associated CINR, BS computes the path losses of each MS
4. Allocate enough power and channels to those MS that need to meet their minimum SLA (throughput, latency and jitter).
5. Sort the path losses of all MS
6. Select the MS with the least path loss, and assign enough power to have the highest MCS. The power assignment can not exceed Pm.
7. Select the next MS, following the round robin scheme, until the BTS power is completely allocated.

13.3.3.2 UL Power Control

Uplink power control consists of two parts: the open loop power control and the closed loop power control. The goal of the uplink power control is to ensure that the uplink power densities received at the BTS from different MS are the same if they use the same MCS. For example, for MCS 0 (4QAM), the uplink power density is targeted at -125dBm measured in 10.938 kHz (The bandwidth of one carrier), and for MCS 3 (16AM), the power density is targeted at -116dBm measured in 10.938 kHz. The desired power density levels for different MCS are different to ensure that the received signal can have sufficient CINR for each MCS (Table 13.2).

According to WiMAX protocol, the uplink power control is done on a per MS basis. That is, there is only a single power control for each MS. If MS has multiple burst with different MCS, it is MS's reasonability to maintain different power densities among various bursts with different desired MCS. For example, let MS A have two bursts, where burst 1 uses 4QAM , and burst 2 uses 16QAM . Then, MS A needs to allocate its transmitter power in such a way that the power density for the second burst is 6dB higher than that of burst 1.

13.3.3.3 Uplink Open Loop Power Control

Due to the multi-path or/and MS mobility, the path loss between BS and MS varies with time. To maintain a desired constant received signal level at BS, the MS needs to vary its output power to compensate the path loss changes. The uplink open loop power control is done as follows:

- MS receives the systems parameters including the receive sensitivities, antenna gain, BS transmitted power, at the network entry (the systems parameters will be broadcast periodically by BS).
- MS detects the down link signal level (preamble power)
- From the detected downlink power, BS antenna gain, and BS transmitted power, MS computes the path loss as: $PL = PBS + BSantgain - Ppreamble$ where PBS is the BS transmitted power, $BSantgain$ is BS antenna gain, and $Ppreamble$ is the detected preamble power.
- MS then computes the required transmitted power $Ptx = Prx + PMCS + PL - BSantgain$ where Prx is the required BS sensitivity, $PMCS$ is the power offset for the desired MCS, (it is the additional CINR required for this MCS beyond that lowest MCS (4QAM)). Ptx is the current MS output power.

The open loop power control is done at MS side, and is transparent to BS.

Uplink Close Loop Power Control

Open loop power control can eliminate most of the signal variation due to the channel fading and MS movement. However, the open loop power control varies the MS transmitted power based on the down link signal variation, it does not factor in any uplink signal degradation due to the uplink interferences or other impairments. To rectify the issue, closed loop power control is added to help tune the uplink transmitted power. As the name implies, closed loop power control requires commands from BS to MS, and MS varies its output power accordingly. The closed loop power control is implemented as follows:

1. BS computes the CINR of the uplink traffic
2. If the CINR is higher than that required for the MCS assigned, BS sends a power control command to MS, asking MS to reduce its output power
3. If the CINR is lower than that required for the MCS assigned, BS sends a power control command to MS, asking MS to increase its output power.
4. MS receives the power control command, and reduces or increases its output power accordingly.

Compared to the open loop power control, closed loop power control is slower and requires additional overhead (the power control commands are sent down with power control IE (information element) in the down link broadcast channel, and has a total of 3 bytes data). While WiMAX supports both TDD and FDD duplex, the current system profile (profile 1.0) supports TDD only. In the case of TDD, open loop power control based on the down link signal variation can eliminate most of the uplink channel variation by utilizing the reciprocal property of the channel. Closed loop power control is used infrequently (once every 20 50 frames), and for the purpose of 'calibrating' the open loop power control.

13.3.4 Link Adaptation

The complexity of the radio resource allocation results mainly from the channel fading. In the case of mobility, the movement of MS not only causes the variation of the path loss, but also degrades the channel condition, a so-called Doppler effect. Channel fading and Doppler effects cause selective fading along both frequency and time. These effects are addressed in detail in Chapter 3. The communication link between BS and MS needs to be adapted to the variation of the channel.

Moreover, in the case of MIMO, the mode which is selected, MIMO-A or MIMO-B, depends on the signal quality and the MIMO channel condition. For example, MIMO-B operation requires certain de-correlation between two data streams, and if there is a strong correlation between the channels seen by different antennas, even when the signal qualities of each data stream is very high (high CINR), MIMO-B can not be supported due to the high MIMO co-channel interferences.

To optimize the systems throughput, the link adaptation technique adjusts the MCS and MIMO mode based on the channel conditions through

1. MCS adjustment
2. Down and up link powers adaptation
3. MIMO mode selection
4. Modifying the number of channels assigned

The link adaptation involves many variables, and is usually conducted in conjunction with the scheduling, which is addressed in the following section.

13.4 Scheduling Optimization

Scheduling is an optimization process where a fixed resource is distributed, or allocated, among multiple users with different QoS requirements according to certain rules. As the aggregated scheduled data needs to be transmitted over a wireless channel with a limited bandwidth, the system, as far as the scheduling is concerned, is often modeled as a 'leaky bucket' [7]. In the 'leaky bucket' model, the wireless system with limited bandwidth is considered as pipe with limited constant output rate and variable input rate. If the input rate is larger than output rate, some incoming packet would be discarded. The factors that need to be considered in the scheduling process includes the number of users, the QoS, requirements, and the channel conditions experienced by users. The common scheduling schemes are:

1. round robin
2. proportional fair
3. max systems throughput (max CINR)

13.4.1 Round Robin

A round robin scheduler allocates resource to all active users equally. The scheduler will look at the data queues of each user and allocate it with the same amount of resource if the user has data to send. Round robin scheme is the simplest scheduler, and can be implemented with minimum computation resources. Its drawback is that the scheduling does not make differentiations on different QoS requirements for different users and channel condition experienced by different users. By assuming all the users obtain the same share of BTS resource, the round robin optimization problem can be formulated as maximizing the sum data rate:

$$\arg\max_{c_{k,n}} \sum_{k=1}^{K} \sum_{n=1}^{N} c_{k,n} \rho(p_{k,n}, \gamma_{k,n})$$

subject to

$$\sum_{k=1}^{K} \sum_{n=1}^{N} p_{k,n} \leq P_{Tx}$$

$$\sum_{n=1}^{N} p_{k,n} = \sum_{n=1}^{N} p_{k_2,n} \quad \forall k_1 \in \{1,K\} \;\; k_2 \in \{1,K\}$$

$$\sum_{k=1}^{K}\sum_{n=1}^{N} c_{k,n} \le N$$

$$\sum_{n=1}^{N} c_{k,n} = \sum_{n=1}^{N} c_{k_2,n} \;\; \forall k_1 \in \{1,K\} \;\; k_2 \in \{1,K\}$$

$$p_{k,n} \ge 0, \;\; c_{k,n} = 0 \text{ or } 1 \;\; \forall k \; \forall n$$

where $p_{k,n}$ is the transmit power allocated to user k on subcarrier n and $c_{k,n}$ is the allocation of subcarrier n to user k. $c_{k,n} = 1$ means subcarrier n is allocated to user k. P_{Tx} is the total transmit power of BTS. $\gamma_{k,n}$ is the channel quality of user k on the subcarrier n. $\rho(.)$ is the data rate which can be achieved with the transmit power $p_{k,n}$ and channel quality $\gamma_{k,n}$. The equation $\sum_{n=1}^{N} c_{k,n} = \sum_{n=1}^{N} c_{k_2,n}$ imposes the constraint that the subcarriers are equally allocated to the users.

13.4.2 Proportional Fair

Proportional fair scheduler considers the channel conditions that each user experiences and schedules the data in such a way that each user has a fair share of the BS resources (power and channel). For example, there are two users, user A is far away from BS, and user B is very close to BS. To support the same throughputs, BS needs to allocate much more power to user A than user B to overcome the excessive path loss that user A is experiencing. Although user A and user B have the same throughput, from the resource allocation point of view, the scheduling is not fair as it allocates much more power to user A than user B. With proportional fair scheduling, BS will assign the same power to user A and B although user A is far away from BS. Proportional fair means the scheduling is fair on the side of BS resource allocation, every user has the same share of the BS resource.

The proportional fair optimization problem can be formulated as maximizing the sum data rate:

$$\arg\max_{c_{k,n}} \sum_{k=1}^{K}\sum_{n=1}^{N} c_{k,n}\rho(p_{k,n}, \gamma_{k,n})$$

subject to

$$\sum_{k=1}^{K}\sum_{n=1}^{N} p_{k,n} \le P_{Tx}$$

$$\sum_{n=1}^{N} p_{k,n} = \sum_{n=1}^{N} p_{k_2,n} \;\; \forall k_1 \in \{1,K\} \;\; k_2 \in \{1,K\}$$

$$\sum_{k=1}^{K}\sum_{n=1}^{N} c_{k,n} \leq N$$

$$\sum_{n=1}^{N} c_{k,n}\rho(p_{k,n},\gamma_{k,n}) \geq \rho_k^{min} \quad \forall k_1 \in \{1,K\}$$

$$p_{k,n} \geq 0, \;\; c_{k,n} = 0 \text{ or } 1 \;\; \forall k \; \forall n$$

where ρ_k^{min} is the minimum required data rate of user k or one service flow. With the constraint $\sum_{n=1}^{N} p_{k,n} = \sum_{n=1}^{N} p_{k_2,n}$, BS allocates same share of BS power to the users. On the other hand, with the constraint $\sum_{n=1}^{N} c_{k,n}\rho(p_{k,n},\gamma_{k,n}) \geq \rho_k^{min}$, BS meets the minimum data rate requirement for each user. Proportional fair scheduler takes the channel condition into consideration, and generates a sense of fairness in terms of resource allocation. However, it may starve the user on the cell edge and cause poor user experience to some users.

13.4.3 Max System Throughput

As implied by its name, the max system throughput is to schedule the resources to maximize the system throughput. To certain degree, max system throughput scheduler is the opposite of the round robin scheduler. Given the path loss distribution of all users, max system throughput scheduler allocates more resources to users closest to BS than those on cell edge as the power and spectrum efficiencies of the close-in users are much higher than users on the cell edge. From the overall systems point of view, max systems throughput scheduler makes best uses of the systems resources. However, it has the same issue as proportional fair scheduler, i.e., some users may be starved if they are on the cell edge and have poor channel conditions.

The max system throughput optimization problem can be formulated as maximizing the sum data rate:

$$\underset{c_{k,n}\; p_{k,n}}{\arg\max} \sum_{k=1}^{K}\sum_{n=1}^{N} c_{k,n}\rho(p_{k,n},\gamma_{k,n})$$

subject to

$$\sum_{k=1}^{K}\sum_{n=1}^{N} p_{k,n} \leq P_{Tx}$$

$$\sum_{k=1}^{K}\sum_{n=1}^{N} c_{k,n} \leq N$$

$$\sum_{n=1}^{N} c_{k,n}\rho(p_{k,n},\gamma_{k,n}) \geq \rho_k^{min} \quad \forall k_1 \in \{1,K\}$$

$$p_{k,n} \geq 0, \quad c_{k,n} = 0 \text{ or } 1 \quad \forall k \; \forall n$$

where p_k^{min} is the minimum required data rate of user k or one service flow. With the above constraints, BS first meets the minimum data rate requirement of each user and also tends to maximize the system throughput through power and subcarrier allocation.

13.4.4 WiMAX QoS Policy

The actual scheduler needs to consider the QoS, user channel condition and the overall system throughput at the same time. In WiMAX, QoS is administrated through the service flows. A service flow is a MAC transport service that provides unidirectional transport of packets either to UL packets transmitted by the MS or to DL packets transmitted by the BS. A service flow is characterized by a set of QoS parameters such as latency, jitter, and throughput assurances. When a packet arrives, it will be added into the proper service queues according to its QoS classifier. WiMAX supports 5 types of services: Unsolicited grant service (UGS), Real -time polling services (RtPS), Extended Real -time polling services (Ertps), Non Real-time polling services (nrtPS), and Best Effort service (BE).

13.4.4.1 Unsolicited Grant Service (UGS)

The UGS is designed to support real-time service flows that transport fixed-size data packets on a periodic basis, such as T1/E1 and Voice over IP without silence suppression. BTS should offer fixed-size grants on a real-time periodic basis, which eliminate the overhead and latency of MS requests and assure that grants are available to meet the flow's real-time needs.

13.4.4.2 Real-time Polling Services (rtPS)

The rtPS is designed to support real-time service flows that generate variable size data packets on a periodic basis, such as streaming video. BS should offer real-time, periodic, unicast request opportunities, which meet the flow's real-time needs and allow the MS to specify the size of the desired grant.

13.4.4.3 Extended Real-time Polling Services (ErtPS)

Extended rtPS is a scheduling mechanism, which builds on the efficiency of both UGS and rtPS. The BS shall provide unicast grants in an unsolicited manner like in UGS, thus saving the latency of a bandwidth request. However, as UGS allocations are fixed in size, ertPS allocations are dynamic. The BS may provide periodic allocations that may be used for requesting the bandwidth as well as for data transfer.

13.4.4.4 Non Real-time Polling Services (nrtPS)

The nrtPS offers unicast polls on a regular basis, which assures that the service flow receives request opportunities even during network congestion. The BS typically polls nrtPS MS on an interval on the order of one second or less. The difference between the rtPS and nrtPS is that in nrtPS the contention-based bandwidth request is allowed.

13.4.4.5 Best Effort Service (BE)

The intent of BE service grant scheduling is to provide efficient service for best effort traffic. In order for this service to work correctly, MS is allowed to use contention request opportunities. This results in the SS using contention request opportunities as well as unicast request opportunities. Since a MS can support multiple service at the same times, each MS could have multiple queues for different service flows. Also downlink and uplinks have separate queues (Fig. 13.8).

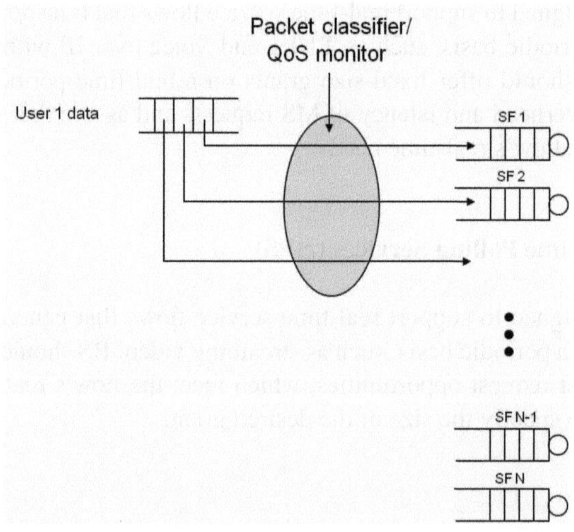

Fig. 13.8 Packet Classifications and Service Flows

At each frame, the scheduler looks at all the service flow queues, and follows the scheduler policy for resource allocation. The following is a typical example of the scheduler policy:

1. At each frame, BS scheduler examines all the service flow queues and prioritizes them as follows
   a. UGS/rtPS/ErtPS
   b. nrtPS
   c. BE
2. Schedule the proper bandwidth for all UGS/rtPS/ErtPS
3. For all nrtPS and BE, schedule enough bandwidth to meet the minimum SLA, using the proportional fair scheme.
4. For all the remaining data in the queues, schedule bandwidth using the maximum throughput scheme

For each catalogue, the scheduling priority goes to the MS which has tighter latency and jitter requirements.

13.5 System Simulations

A complete optimization procedure consists of two relatively separate stages of scheduling and radio resource allocation as depicted in Fig. 13.9. Ideally the optimization would be done from the end-to-end point of view, taking into account the interactions between scheduling and radio resource allocation. However, due to the practical constraints (the limited hardware resource, too complicated to have converge of the optimization, etc), the whole procedure is often partitioned into many sequential steps. For example, one could look at the data queues of all the active service flows, and schedule the data according to the QoS of each service flow without referring the user channel conditions. The scheduled data is then mapped into different MCS, based on the reported CINR and channel condition. The MIMO mode selection is done at the last stage (Fig. 13.9). The benefits of this partitioned procedure are that one can guarantee the convergence and can do it with limited hardware capability. The following is an example of the systems simulation on scheduling and resource allocation.

The network set up is given in Table 13.3. A total of 200 active users are assumed in the network, with the distributions of channel conditions given in Table 13.4. The traffic types and QoS categories are as shown in Table 13.5. The scheduling and resource allocation follow the flow chart of Fig. 13.9 and are broken into many sequential small steps. The simulations are done with time duration of a few hours so the users can go through all the channel status. The simulation results are summarized in Table 13.6.

Obviously, the simulation results meet or exceed WiMAX QoS requirements (Table 13.2). Above experiment is conducted to check the performance of WiMax sys-

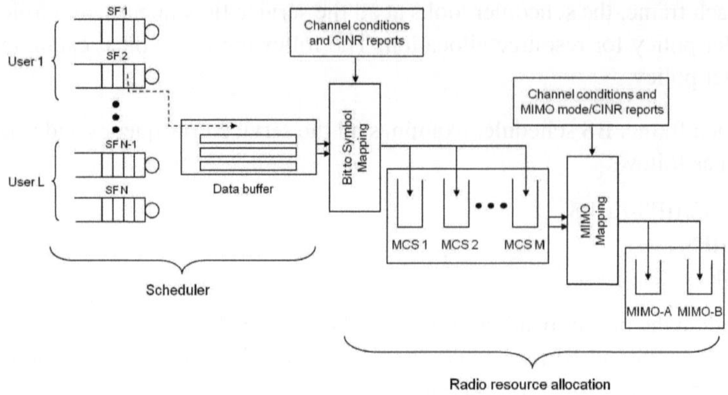

Fig. 13.9 Partition of Scheduling and Radio Resource Allocation

Table 13.3 Simulation Network Setup

Item	Value
Center frequency	2.5GHz
Cell division	3 sectors per site
Frequency reuse	3x3x3
Channel bandwidth	10MHz
DL:UL ratio	32:15:00
BTS antenna structure	8x8 (8 transmitters and 8 receivers) with 1/2 wavelength antenna spacing
MS antenna structure	1x2 (one transmitter, 2 receivers) with a half wavelength separation
BTS transmitted power per antenna	30dBm (total 39dBm with 8 transmitters)
BTS receiver noise Fig.	5dB
BTS antenna gain	16dBi
BTS height	40 meter
MS transmitted power	25dBm
MS receiver noise Fig.	6dB
MS antenna gain	0dBi
MS height	1.5 meter
Propagation model	COST231, suburban
Log normal shadow std	8dB
Indoor penetration loss	10dB

tem in a scenario with mixture of multiple services. The service performance metrics included packet delay and throughputs. From an operator view, fairness is also important since users expect to have the same experience regardless whether they are close to a BTS or at the cell edge. Another experiment is conducted to check the fairness performance in the scenario with multiple types of services and multiple types of service levels.

Table 13.4 Distribution of Users Channel Conditions

Mobility Case	Fading Channel Model	Percentage
Low mobility	ITU Ped A, 3km/H	40%
Low mobility	ITU Ped B, 3km/H	40%
High mobility	ITU Veh A, 90km/H	20%

Table 13.5 User Traffic QoS Distributions

Application	Traffic Category	Percentage
FTP	Best effort	30%
Web browser/HTTP	Interactive/real time	50%
VOIP	Real time	10%
Gaming	Interactive real time	10%

Table 13.6 Systems Simulation Results

DL systems throughput		17.7 Mbps
UL systems throughput		3.92Mbps
DL spectrum efficiency		2.66
UL spectrum efficiency		1.18
DL packet Delay (90% percentile)	VOIP	7ms
	Gaming	17ms
	FTP	72s (average download time)
	Web	2.5s
UL packet Delay (90% percentile)	VOIP	5ms
	Gaming	10ms
	FTP	52s (average upload time)

In this experiment, the distribution of user channel condition and the distribution of traffic QoS still follow the distribution listed in table 4 and 5, respectively. Additionally, multiple service levels are defined as shown in Table 13.7.

Table 13.7 Distribution of Service Levels

Service Level	Max. Service Rate(Mbps)	Sustain Rate (Kbps)	Percentage
Diamond Level	DL: 1/UL: 0.5	DL: 20/UL: 10	30%
Gold Level	DL: 0.5 /UL: 0.25	DL: 20/UL: 10	40%
Silver Level	DL: 0.25/UL: 0.128	DL: 20/UL: 10	30%

Each service level is defined by one max. service rate and one sustain rate. Each user is allocated one service level. The user throughput should never exceed the max. service rate, and the sustain rate is required for each CPE to maintain the wireless connection. The service level is realized through the radio resource allocation scheme. In this experiment, the policy for radio resource allocation is to:

1. Guarantee the packet delay for users with real-time service: VoIP and gaming
2. Guarantee the throughput of FTP and Web users to be within the max. service rate and sustain rate defined by the service level of each individual user.

So the basic procedure of the radio resource allocation is:

1. First allocate radio resource to Gaming and VoIP users to ensure the packet delay requirement;
2. Allocated radio resource to FTP and Web users, whose user throughput is less than the sustain rate defined in his service level;
3. Finally, allocate the remaining resource to those users whose throughput is lesser than the max. service rate but larger than the sustain rate.

In this experiment, 200 active users are simulated. The fairness metric defined in IEEE 802.16j [10] is employed in the experiment. In IEEE 802.16j, the fairness is defined as the reciprocal of the standard deviation of the normalized throughput

$$F = \frac{K}{\sqrt{\sum_{k=1}^{K} (\frac{x_k}{\bar{x}} - 1)^2}}$$

with the mean throughput $\bar{x} = \frac{1}{K} \sum_{k=1}^{K} x_k$. This metric grows with larger fairness, and is not bounded. The output of the experiment shows that the fairness of 200 users is 1.53, which is slightly better than Fair-Time scheduler.

From the above experiments, we can see that through proper design of radio resource allocation scheme in WiMax system, we can satisfy the various performance requirements.

13.6 Conclusion

In this chapter, we have discussed the optimization problem for radio resource management in OFDMA-based WiMax systems. Radio resource allocation in WiMax is a classic optimization problem with multiple factors. Various optimization solutions exist with different metrics and tradeoff between complexity and performance exist among those solutions. The dynamic features in most of the aspects of OFDMA-based WiMax systems impose difficulty on the optimization solution of a radio resource allocation problem. A Particular performance metric and requirement needs to be considered in the design of each radio resource allocation algorithm.

References

1. Air Interface for Fixed Broadband Wireless Access Systems. The Institute of Electrical and Electronics Engineers, Inc. (2004)
2. Air Interface for Fixed Broadband Wireless Access Systems. The Institute of Electrical and Electronics Engineers, Inc. (2009)
3. Ahmadi, S.: An overview of next-generation mobile WiMAXTechnology. IEEE Communications Magazine (2009)
4. Alamouti, S.: A simple transmit diversity technique for wireless communications. IEEE Journal on Selected Areas in Communications (JSAC) (1998)
5. Andrews, J.G.: OFDMA. Prentice Hall (2006)
6. Eklund, C., Marks, R., Stanwood, K., Wang, S.: IEEE Standard 802.16: A technical overview of the WirelessMANTM air interface for broadband wireless access. IEEE Communications Magazine (2002)
7. Ferguson, P., Huston, G.: Quality of Service: Delivering QoS on the Internet and in Corporate Network. John Wiley & Sons, Inc. (1998)
8. Gesbert, D., Bolcskei, H., and, G.D., Paulraj, A.: MIMO wireless channels: Capacity and performance prediction. IEEE Global Telecommunication Conf. (GLOBECOM00) **2** (2000)
9. L., Z., Tse, D.N.C.: Diversity and multiplexing: A fundamental trade-off in multiple antenna channels. IEEE Trans. Info. Theory **49** (2003)
10. Senarath, G., Tong, W.: Multihop system evaluation methodology performance metrics. IEEE C 802.16j-06/025 (2006)
11. Yagoobi, H.: Scalable OFDMA physical layer in IEEE 802.16 WirelessMAN. Intel Technology Journal (2004)

Chapter 14
Cross Layer Scheduling in Wireless Networks

Nitin Salodkar and Abhay Karandikar

Abstract A wireless channel poses significant challenges to resource management due to multipath fading. In this chapter, we review the impact of wireless channels on the design of scheduling algorithms at the link and network layers. It has been well recognized now that significant performance gains can be obtained by designing channel aware scheduling algorithms. Various scheduling schemes catering to different Quality of Service (QoS) objectives such as maximizing throughput, minimizing delay, or minimizing energy have been proposed in the literature. Most of these algorithms can be formulated as control problems within the framework of a Constrained Markov Decision Processes. We examine some of the representative work in this area with a specific focus on centralized scheduling in a single cell scenario with a single antenna system.

14.1 Introduction

Over the past decade, there has been a large scale proliferation of wireless communications technologies. While the first and second generation (1G and 2G) cellular systems were driven primarily by voice based cellular telephony, the recent years have witnessed a phenomenal increase in packet data applications like email, web browsing, peer to peer, and multimedia applications. These applications require Quality of Service (QoS) guarantees in terms of rate, delay, packet loss, etc. The time varying nature of the wireless channel due to multipath fading poses significant challenges

Nitin Salodkar
Department of Computer Science and Engineering, Indian Institute of Technology Bombay, Mumbai, 400076, India, e-mail: nitinsalodkar@gmail.com

Abhay Karandikar
Department of Electrical Engineering, Indian Institute of Technology Bombay, Mumbai, 400076, India, e-mail: karandi@ee.iitb.ac.in. This work was supported by TTSL-IIT Bombay Center for Excellence in Telecom.

J. Kennington et al. (eds.), *Wireless Network Design: Optimization Models and Solution Procedures*, International Series in Operations Research and Management Science 158, DOI 10.1007/978-1-4419-6111-2_14, © Springer Science+Business Media, LLC 2011 to radio resource management for providing QoS. In this chapter, we focus on the packet scheduling aspect of resource management.

Recent results suggest that significant performance gains can be obtained by exploiting wireless channel characteristics for the design of scheduling algorithms at the link and network layers. These algorithms are referred to as cross layer scheduling. Cross layer scheduling algorithms can be formulated as optimization problems where the objective is to optimize a given utility function such as energy subject to QoS constraints such as delay. The scheduler can be viewed as a controller where the control action comprises of determining the user to be scheduled, its transmission rate, and power depending upon the system state. These problems can often be cast within the framework of a Constrained Markov Decision Process (CMDP). The computational approaches to determine the optimal policy comprises of dynamic programming, linear programming, function approximation, reinforcement learning, stochastic approximation, etc. In this chapter, we examine some of the representative formulations to elucidate the nature of the problems being considered in the literature. The focus of this chapter is centralized scheduling in a cellular system with a single antenna.

We first begin by reviewing the characteristics of a wireless channel and its capacity. We then present a framework for cross layer scheduling algorithms, Finally, we review some techniques in fair opportunistic, power optimal, throughput optimal, and delay optimal scheduling algorithms.

14.2 Wireless Channel Capacity

Wireless channel is characterized by the decay of signal strength due to distance (path loss), obstructions due to objects such as buildings and hills (shadowing) and constructive and destructive interference caused by copies of the same signal received over multiple paths (multipath fading). In this chapter we primarily focus on multipath fading and its impact on the design of scheduling algorithms.

In multipath fading, multiple copies of the signal reach the receiver at various instants of time depending on the length of the path over which the signal traverses. The relative motion between the transmitter and receiver and/or movement of the reflecting objects results in random path length changes; consequently different multipath components have random amplitudes and phases. This results in time varying amplitude and phase of the received signal.

The time varying characteristics of the wireless channel can be modeled as a tapped delay line filter with each tap corresponding to a delay window during which different multipath components arrive at the receiver. In this chapter, we limit our discussion to *flat* fading channels where all the multipath components corresponding to a symbol arrive within the symbol duration. In such a case, the channel can be modeled using a filter with a single tap. Moreover, we also assume that this tap (or channel) gain remains constant for a block of symbols and changes only over block (termed as *slot* in this chapter) boundaries. This model is termed as *Block Fading Model.*

Under this model, if a user transmits a signal χ_n in slot n, then the received signal Y_n is given by:

$$Y_n = H_n \chi_n + Z_n, \tag{14.1}$$

where H_n corresponds to the time-varying channel (tap) gain due to fading and Z_n is the complex Additive White Gaussian Noise (AWGN) (with zero mean and variance N_0). Usually H_n is modeled as a zero mean complex Gaussian random variable. Let σ^2 denote the variance of H_n. Then $|H_n|$ is a Rayleigh random variable and $X_n = |H_n|^2$ is an exponentially distributed random variable with probability density function expressed as:

$$f_X(x) = \frac{1}{\sigma^2} \exp(\frac{-x^2}{2\sigma^2}), \; x \geq 0. \tag{14.2}$$

This model is called Rayleigh fading model.

We refer to X_n as channel state in slot n. Note that the channel state X_n may change from slot to slot either in an independent and identically distributed (i.i.d.) fashion or in a correlated fashion (e.g. may follow a Markov model). Moreover, the channel state X_n is a continuous (exponential) random variable. However, for the scheduling problems considered later in this chapter, we assume that the channel state X_n takes values from a finite discrete set $\mathbb{X}$. This discretization can be achieved by partitioning the channel state into equal probability bins with preselected thresholds. For example, let $x(1) < \ldots < x(L)$ be these thresholds. Then the channel is said to be in state x_k if $x \in [x(k), x(k+1))$, $k = 1, \ldots, L$. The channel state space $\mathbb{X}$ can be represented as $\mathbb{X} = \{x_1, \ldots, x_L\}$.

For a wireless channel, capacity analysis can be performed both in the presence as well as absence of Channel State Information (CSI) at the transmitter. Throughout this chapter, we assume that the transmitter has the knowledge of perfect CSI. In a Time Division Duplex (TDD) system, due to channel reciprocity, it may be possible for the transmitter to obtain the CSI through channel estimation based on the signal received on the opposite link. In a Frequency Division Duplex (FDD) system, the receiver has to estimate the CSI and feed this information back to the transmitter. In practice, e.g., in IEEE 802.16 [26], the channel related information can be conveyed using ranging request (RNG-REQ) messages. In this chapter, we do not take into account the specific feedback mechanisms; rather, we assume that the transmitter has the knowledge of perfect CSI.

Different notions of capacity of fading channels have been defined in the literature. The classical notion of Shannon capacity defines the maximum information rate that can be achieved over the channel with zero probability of error. This notion involves a coding theorem and its converse, i.e., that there exists a code that achieves the capacity (information can be reliably transmitted using this code at a rate less than or equal to the capacity) and that reliable communication is not possible if information is transmitted at a rate higher than the capacity.

In this chapter, we consider *ergodic* capacity (also termed as *throughput* capacity or *expected* capacity in the literature). This notion of capacity measures the rates achievable in the long run averaged over the channel variations. Here, the channel is assumed to vary sufficiently fast, yet the variations are slow enough such that 'reasonably' long codes can be transmitted. We first derive an expression for the ergodic capacity of a single user point-to-point link under an average power constraint and then extend the notion to multiuser scenario.

14.2.1 Point-to-Point Capacity with Perfect Transmitter CSI

Consider a single user wireless channel depicted in Figure 14.1. We assume transmitter has perfect knowledge of CSI. Let P_n denote the transmission power in slot n. Let $X_1 = x_1, \ldots, X_M = x_M$ be a given realization of channel states. We assume that the transmitter has an average power constraint of $\bar{P}$.

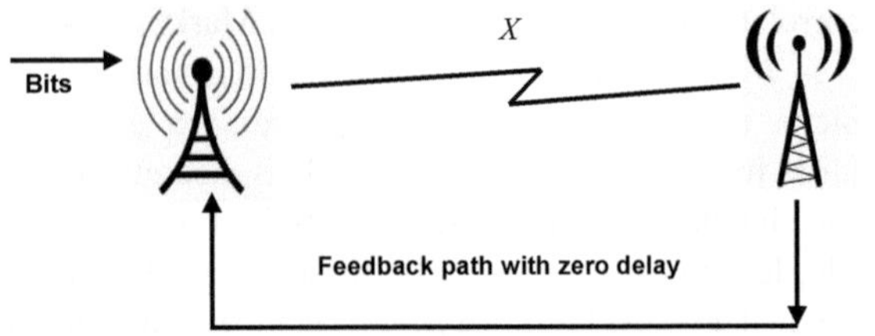

Fig. 14.1 Point-to-point transmission model

The restriction on the average power expenditure makes the problem of achieving capacity to be a power allocation problem. Specifically, the problem is to determine the transmission power in each slot that maximizes the information rate (and hence achieves capacity), while keeping the average power expenditure below the prescribed limit. The problem can be stated as:

$$\max_{P_1,\ldots,P_M} \frac{1}{M} \sum_{n=1}^{M} \log\left(1 + \frac{P_n X_n}{N_0}\right), \tag{14.3}$$

subject to,

$$\frac{1}{M} \sum_{n=1}^{M} P_n = \bar{P}. \tag{14.4}$$

This is a constrained optimization problem and can be solved using standard Lagrangian relaxation [8]. Let x^+ denote $\max(0, x)$. A solution to the optimization problem stated in (14.3) and (14.4) is a policy that determines the optimal power in n^{th} slot to be:

$$P_n^* = \left(\frac{1}{\lambda} - \frac{N_0}{X_n}\right)^+, \tag{14.5}$$

where λ is a Lagrange Multiplier which satisfies:

$$\frac{1}{M}\sum_{n=1}^{M}\left(\frac{1}{\lambda} - \frac{N_0}{X_n}\right)^+ = \bar{P}. \tag{14.6}$$

As $M \to \infty$, by ergodicity,

$$\lim_{M\to\infty}\frac{1}{M}\sum_{n=1}^{M}\left(\frac{1}{\lambda} - \frac{N_0}{X_n}\right)^+ = \mathbf{E}\left[\left(\frac{1}{\lambda} - \frac{N_0}{X}\right)^+\right] = \bar{P}, \tag{14.7}$$

where the expectation is taken with respect to the stationary distribution of the channel states. For a given realization of the channel state $X = x$, the optimal power allocation can be expressed as:

$$P^*(x) = \left(\frac{1}{\lambda} - \frac{N_0}{x}\right)^+. \tag{14.8}$$

Figure 14.2 provides a pictorial description of the power allocation (termed as *waterfilling*) policy. It can be observed that the transmitter allocates *more* power when the channel is *good* and *less* power when the channel is *poor*. This insight has been used later while designing scheduling schemes at network layer in Section 14.5. Note that the waterfilling power allocation is in contrast with the traditional power control policy which attempts to invert the channel.

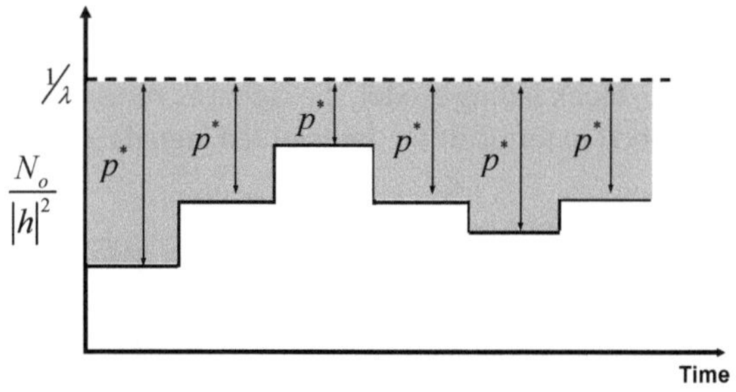

Fig. 14.2 Waterfilling power allocation

Once the optimal power is known, the channel capacity with perfect CSI at the transmitter can be expressed as:

$$C = \mathbf{E}\left[\log\left(1 + \frac{P^*(X)X}{N_0}\right)\right]. \tag{14.9}$$

Though a point-to-point or single user transmission system offers significant insight for transmission over a fading channel, it represents a somewhat restricted scenario. In the next section, we consider a more realistic multiuser scenario where we review generalization of the single user waterfilling power allocation.

14.2.2 Multiuser Capacity with Perfect Transmitter CSI on the Uplink

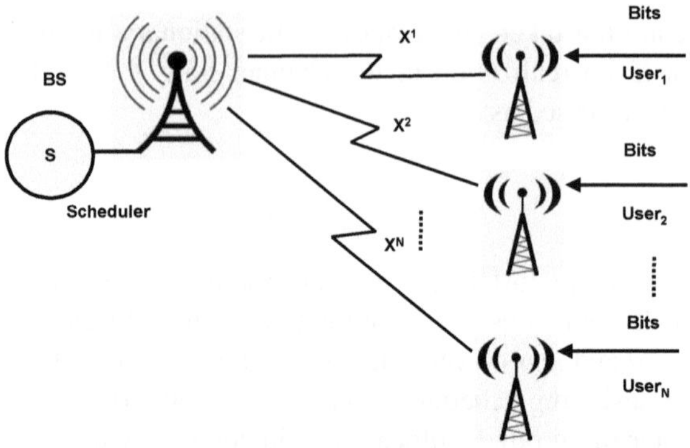

Fig. 14.3 Uplink transmission model, infinite backlog of bits at transmitters

In this section, our objective is to determine the ergodic capacity for a multiuser (uplink) fading channel as depicted in Figure 14.3 where N users communicate with a base station. With the block fading model, the signal Y_n received by the base station in slot n can be described in terms of the transmitted signals χ_n^i, $i = 1,\ldots,N$ as:

$$Y_n = \sum_{n=1}^{N} H_n^i \chi_n^i + Z_n, \tag{14.10}$$

where H_n^i is the channel gain for user i in slot n. Let $X_1^i = x_1^i, \ldots, X_M^i = x_M^i$, $i = 1,\ldots N$, be a given realization of the channel states. User i has an average power constraint of $\bar{P}^i$ ($\bar{\mathbf{P}} = [\bar{P}^1,\ldots,\bar{P}^N]^T$ being the average power constraint vector), the problem is to determine optimal power allocation for each user that maximizes the sum capacity subject to maintaining each user's average power expenditure below its prescribed limit. This problem can be stated as:

$$\max_{P_n^i, i=1,\ldots,N, n=1,\ldots,M} \frac{1}{M} \sum_{n=1}^{M} W \log\left(1 + \frac{\sum_{i=1}^{N} P_n^i X_n^i}{W N_0}\right), \tag{14.11}$$

subject to the per user power constraint:

$$\frac{1}{M}\sum_{n=1}^{M} P_n^i = \bar{P}^i,\ i = 1,\ldots,N. \tag{14.12}$$

We first consider *symmetric* scenario where all users have identical channel statistics and power constraints ($\bar{P}^i = \bar{P}, \forall i$). For simplicity, instead of individual power constraints as in (14.12), we consider the total power constraint:

$$\frac{1}{M}\sum_{n=1}^{M}\sum_{i=1}^{N} P_n^i = N\bar{P}. \tag{14.13}$$

It turns out that subject to the constraints expressed in (14.13), the sum capacity in (14.11) is maximized by allowing only one user with the best channel state to transmit in a slot. The power allocation is expressed as:

$$P^{i_n^*}_n = \begin{cases} \left(\frac{1}{\lambda} - \frac{WN_0}{x_n^i}\right)^+ & \text{if } x_n^i = \max_j x_n^j, \\ 0 & \text{otherwise,} \end{cases} \tag{14.14}$$

where λ again is a Lagrange Multiplier and is chosen to satisfy the sum power constraint (14.13) and i_n^* is the index of the best user in slot n. Taking $M \to \infty$ and by ergodicity of the fading process, we obtain the capacity-achieving power allocation policy that allocates power $P^{i^*}(\mathbf{x})$ to user i as a function of the joint channel state vector $\mathbf{x} = (x^1,\ldots,x^N)$ where:

$$P^{i^*}(\mathbf{x}) = \begin{cases} \left(\frac{1}{\lambda} - \frac{WN_0}{x^i}\right)^+ & \text{if } x^i = \max_j x^j, \\ 0 & \text{otherwise,} \end{cases} \tag{14.15}$$

where λ is chosen to satisfy the power constraint:

$$\sum_{i=1}^{N} \mathbf{E}\left[P^{i^*}(\mathbf{X})\right] = N\bar{P}. \tag{14.16}$$

The resulting sum capacity is:

$$C_{sum} = \mathbf{E}\left[W\log\left(1 + \frac{P^{i^*}X^{i^*}}{WN_0}\right)\right], \tag{14.17}$$

where X^{i^*} is the channel state of the best user indexed by i^*. Note that this result is derived by imposing a total power constraint (14.13). However, because of symmetry and independence between the user channel state processes, the power consumption of all users is same under the optimal power allocation policy. Hence, the per user power constraints in (14.12) are also satisfied.

The above scheduling policy where the user with the best channel state is scheduled in a slot is called *opportunistic* scheduling. It takes advantage of *multiuser diversity* in order to improve the sum rate (throughput), i.e., in a system with large number of users having independent and diverse channel states, there exists a user having good channel state with a high probability. Moreover, this probability increases with the number of users. The implications of opportunistic scheduling have been investigated in further detail in Section 14.4.

Let $\mathbb{C}_g(\mathbf{x}, \mathscr{P}(\mathbf{x}))$ denote the set of achievable rates under a policy allocation policy $\mathscr{P}(\mathbf{x})$. It can be expressed as:

$$\mathbb{C}_g(\mathbf{x},\mathbf{P}) = \left\{ \mathbf{R} : \sum_{i \in S} R^i \leq W \log\left(1 + \frac{\sum_{i \in S} x^i P^i}{W N_0}\right) \forall S \in \{1,\ldots,N\} \right\}. \tag{14.18}$$

A power allocation policy $\mathscr{P}$ is *feasible* if it satisfies the power constraints of all users, i.e., $\mathbf{E}[\mathscr{P}(\mathbf{X})] = \bar{\mathbf{P}}$. Let $\mathbb{F}$ be the set of all feasible power allocation policies. The *throughput capacity region* is defined as the union of the set of rates achievable under all power control policies $\mathscr{P} \in \mathbb{F}$, i.e.,

$$\mathbb{C}(\bar{\mathbf{P}}) = \bigcup_{\mathscr{P} \in \mathbb{F}} \mathbf{E}[\mathbb{C}_g(\mathbf{X}, \mathscr{P}(\mathbf{X}))]. \tag{14.19}$$

In a general case of asymmetric channels and power constraints, *weighted* rate maximization is a more appropriate metric. Let $\gamma = [\gamma^1, \ldots, \gamma^N]^T$ be a vector of weights assigned to the users. The weighted rate maximization problem can be expressed as:

$$\max \ \gamma \cdot \mathbf{R}, \tag{14.20}$$

subject to the constraint that the rate vector lies in the capacity region:

$$\mathbf{R} \in \mathbb{C}(\bar{\mathbf{P}}). \tag{14.21}$$

Using a Lagrangian formulation [8], it can be shown that the optimal power allocation policy can be computed by solving, for each channel state vector $\mathbf{X} = \mathbf{x}$, the following optimization problem:

$$\max_{\mathbf{R},\mathbf{P}} \gamma \cdot \mathbf{R} - \lambda \cdot \mathbf{P}, \tag{14.22}$$

subject to:

$$\mathbf{R} \in \mathbb{C}_g(\mathbf{X},\mathbf{P}). \tag{14.23}$$

The optimal solution to (14.22) thus provides a power allocation $\mathscr{P}(\mathbf{x})$ and a rate allocation $\mathscr{R}(\mathbf{x})$ at channel state vector $\mathbf{X} = \mathbf{x}$. If the choice of $\lambda = [\lambda^1, \ldots, \lambda^N]^T$ ensures that the power constraint is met then $\mathbf{R}^* = \mathbf{E}[\mathscr{R}(\mathbf{X})]$ is an optimal solution to (14.22). It can be shown that the optimal solution to (14.22) is a greedy *successive decoding* scheme where the users are decoded in an order that is dependent on the interference experienced by them.

14.3 A Framework for Cross Layer Scheduling

In the above sections, we have discussed information theoretic capacity notions for multiuser wireless system. These results indicate that significant performance gains can be obtained at the link and network layers by exploiting physical layer information. In this section, we discuss how information from physical layer can be exploited and opportunities created for making scheduling decisions in order to satisfy certain QoS measures.

14.3.1 Opportunistic Scheduling

As we have already studied, for the symmetric case, sum capacity is maximized by scheduling the user with the best channel state in a time slot. This suggests that users should transmit at opportunistic time. This leads to the foundation of *opportunistic scheduling*. Since the radio channel conditions vary independently for each user, in a given slot, there is a high probability of having a user whose channel state is near its peak. Scheduling such a user leads to high sum throughput. The gains are larger if the channel variations are larger which in turn are indeed larger if the number of users is large. Thus, the traditional view that rapid variations in the wireless channel pose a significant challenge for efficient communication has been converted into an opportunity for exploiting multiuser diversity.

Thus, we have a scheduling scheme where the scheduler picks up the user i_n in slot n such that

$$i_n = \arg\max_j x_n^j. \tag{14.24}$$

This 'pure' opportunistic scheduling, though, maximizes overall sum throughput, is not necessarily fair. It may starve the users who have poor average channel states. For example in Figure 14.4, user 2 is starved. This problem can be addressed by imposing fairness constraints in (14.24). We will review such algorithms later in the chapter.

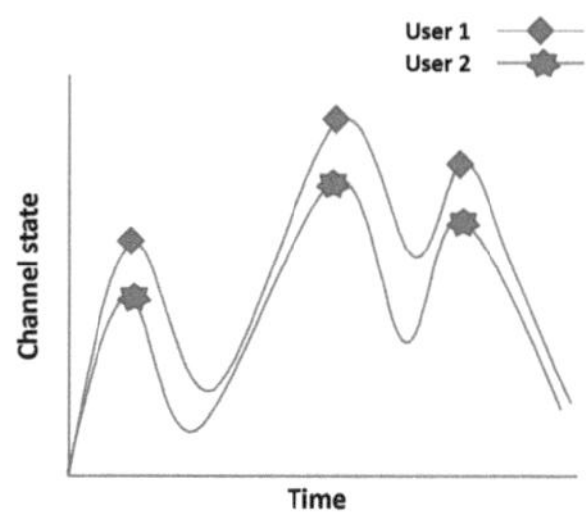

Fig. 14.4 User with perennially poor channel condition may be starved by a pure opportunistic scheduler

14.3.2 Energy Efficient Scheduling

Apart from maximizing throughput, energy efficiency is also an important concern in wireless communications. As is evident from the foregoing discussion on capacity, the power required for error free communication in Shannon's sense at a rate u when the channel state is x is given by

$$P(x,u) = \frac{N_0}{x}(e^u - 1), \tag{14.25}$$

where N_0 is the thermal noise power spectral density. Note that for a giver x, the transmission power is an increasing and strictly convex function of u.

Even for practical digital communication systems, the power required to transmit at a rate u for a specified bit error rate (BER) is given by

$$P(x,u) = \frac{N_0}{\Gamma(BER)x}(e^u - 1), \tag{14.26}$$

where $\Gamma(BER)$ denotes the signal to noise ratio (SNR) gap corresponding to a practical modulation and coding setting. Even in this case, the power is convex and strictly increasing function of u.

The convex power-rate relationship implies that if we want to, say, double the rate, we may have to transmit at more than double the power. Thus to transmit u packets, the scheduler can transmit $\frac{u}{2}$ packets in one slot and $\frac{u}{2}$ packets in the next slot instead of u packets in one slot (if the rate permits). This strategy saves the power, albeit, at a cost of delay of one slot. This suggest that the scheduler should transmit the data in opportunistic chunks for energy efficiency.

Note that packets (hence bits) arrive randomly and may be subjected to buffering. Hence to maximize throughput or to minimize energy, opportunistic or energy efficient scheduling has to contend with network layer issues like fairness, packet delay and queue stability. This leads to formulating the scheduling problem as a control problem which exploits fading state information to maximize (or minimize) a given utility function subject to some constraints such as fairness, delay, stability. In the rest of the chapter, we focus on such formulations. We begin by discussing the system model in more detail in the next section.

14.3.3 System Model for Centralized Scheduling

We consider a multiuser wireless system where N users communicate with a base station. On the uplink, as depicted in Figure 14.5, users communicate with the base station using TDMA, i.e., time is divided into slots of equal duration and only one user can transmit in a slot. We assume that the slot duration is normalized to unity. The base station is the centralized entity that makes the scheduling decision and the user scheduled by the base station transmits in a slot.

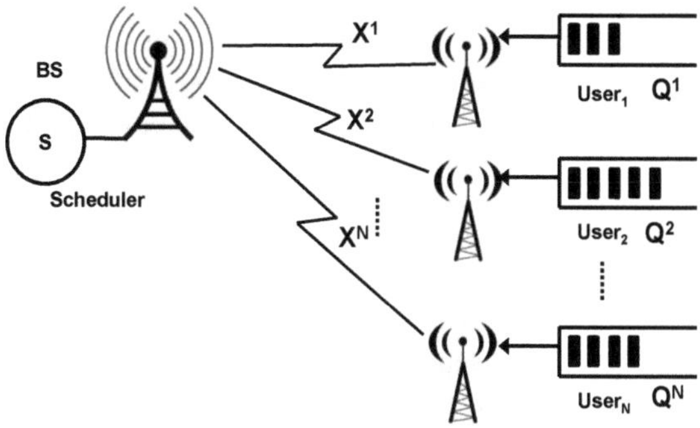

Fig. 14.5 Uplink transmission model, finite buffer at each user

The channel state as discussed in Section 14.2.2 is assumed to remain constant for the duration of a slot and to change in an i.i.d. manner across slots. We assume that packets arrive randomly into the user buffer and are queued in the buffer until they are transmitted.

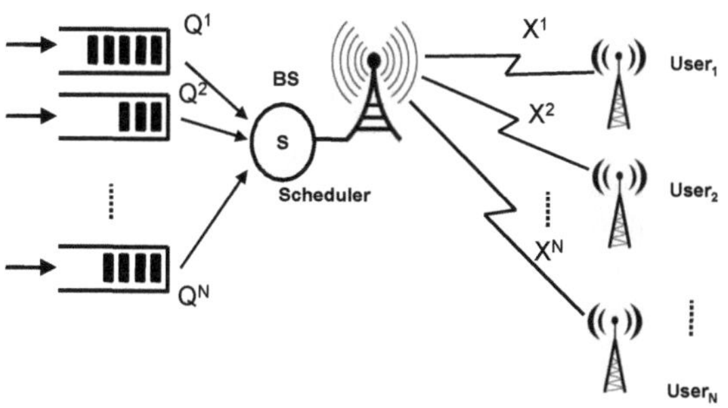

Fig. 14.6 Downlink transmission model, finite buffer for each user at base station

On the downlink, as depicted in Figure 14.6, we assume that the base station multiplexes the transmissions corresponding to N users using Time Division Multiplexing (TDM). The base station maintains a queue for each user.

Let $A_n^i \in \mathbb{A} = \{0,\ldots,A\}$ denote the number of arrivals into user i buffer at the *beginning* of slot n[1]. We also assume that the user's packets are of equal size, say, ℓ bits. The packet arrival process is assumed to be stationary and is independent of the channel fading and noise processes.

Let $Q_n^i \in \mathbb{Q}$ denote the queue length corresponding to user i at the beginning of slot n. Let U_n^i be the number of packets transmitted from user i buffer in slot n. Since the slot duration is normalized to 1, $U_n^i \in \mathbb{U}$ also denotes the transmission rate of user

[1] Random variables are denoted with capital letters while their values are denoted with small letters.

i in slot n. Let $R^i_n \in \mathbb{U}$ denote the number of packets that user i should transmit in slot n if it is scheduled. Then $U^i_n = I^i_n R^i_n$, where I^i_n is an indicator variable that is set to 1 if the user i is scheduled in slot n, otherwise it is set to 0.

The queue dynamics for user i can be expressed as (for U^i_n):

$$Q^i_{n+1} = \max(0, Q^i_n + A^i_{n+1} - I^i_n R^i_n). \tag{14.27}$$

Since the scheduler can at most schedule all the bits in a buffer in any slot, $U^i_n \leq Q^i_n$. Moreover, $Q^i_{n+1} \geq A^i_{n+1}$, $\forall n$. We assume that the scheduler can choose U^i_n based on the queue length Q^i_n, the channel state X^i_n and the number of arrivals (source arrival state) A^i_n. More generally, the scheduler can determine U^i_n based on entire history of queue lengths, channel states and source arrival states.

Different formulations make various assumptions on the arrival process $\{A^i_n\}$ and control action process $\{U^i_n\}$. We state these assumptions later while formulating different scheduling problems.

The average queue length of a user i can be expressed as:

$$\bar{Q}^i = \limsup_{M \to \infty} \frac{1}{M} \sum_{n=1}^{M} Q^i_n. \tag{14.28}$$

Average delay $\bar{D}^i$ can be treated to be equivalent to the average queue length $\bar{Q}^i$ because of the Little's law (Chapter 3, [9]) as follows:

$$\bar{Q}^i = \bar{a}^i \bar{D}^i, \tag{14.29}$$

where $\bar{a}^i$ is the average packet arrival rate of user i. Due to this relationship, queue length measure is usually considered to be synonymous with delay measure.

Similarly, the sum throughput over a long period of time can be expressed as:

$$\bar{T} = \liminf_{M \to \infty} \frac{1}{M} \sum_{n=1}^{M} \sum_{i=1}^{N} I^i_n R^i_n. \tag{14.30}$$

The average power consumed by a user i over a long period of time can be expressed as:

$$\bar{P}^i = \limsup_{M \to \infty} \frac{1}{M} \sum_{n=1}^{M} P(X^i_n, I^i_n R^i_n), \tag{14.31}$$

where $P(X^i_n, I^i_n R^i_n)$ is the power required by i when the channel state is X^i_n and user transmits $I^i_n R^i_n$ packets.

A number of scheduling algorithms have been proposed in the literature that focus on the above measures or variations of these measures. Broadly, these algorithms can be classified into three types:

1. *Maximize sum throughput subject to fairness constraint.*
2. *Maximize sum throughput subject to delay and queue stability constraint.*

3. *Minimize average power subject to delay constraint.*

We begin by first discussing fair scheduling algorithms.

14.4 Fair Scheduling

Exploiting multiuser diversity in an opportunistic manner by scheduling the user with the best channel state might introduce unfairness. Users who are closer to the base station might experience perennially better channel conditions and thereby obtain a higher share of the system resources at the expense of users who are farther away from the base station. On the other hand, scheduling users with poor channel states results in a reduction in the overall throughput. Thus, there exists a fairness-sum throughput tradeoff. One of the earliest systems to exploit this tradeoff in order to improve the sum throughput is the Code Division Multiple Access/High Data Rate (CDMA/HDR) system.

Different scheduling algorithms provide fairness over different time intervals. A scheduling algorithm is *long term* fair if it provides a *fair share* of a certain quantity such as fraction of time slots or throughput to all users over a long period of time. As outlined earlier, the average throughput achieved by a user i over a long period of time can be expressed as:

$$\bar{T}^i = \liminf_{M \to \infty} \frac{1}{M} \sum_{n=1}^{M} I_n^i R_n^i. \tag{14.32}$$

The fraction of slots allocated to a user in the long run can be expressed as:

$$\bar{I}^i = \liminf_{M \to \infty} \frac{1}{M} \sum_{n=1}^{M} I_n^i. \tag{14.33}$$

Long term fair algorithms allocate the quantities such as $\bar{T}^i$ and $\bar{I}^i$ $\forall i$ in a fair manner over a long period of time.

On the other hand, a scheduling algorithm is *short term* fair if it provides a fair allocation of a certain quantity such as fraction of time slots or throughput to all users in an interval of M slots. The average throughput by a user i over a window of M slots can be expressed as:

$$T^i(M) = \frac{1}{M} \sum_{n=1}^{M} I_n^i R_n^i. \tag{14.34}$$

The fraction of slots allocated to a user i in a window of M slots can be expressed as:

$$I^i(M) = \frac{1}{M} \sum_{n=1}^{M} \sum_{i=1}^{N} I_n^i. \tag{14.35}$$

Short term fair algorithms allocate the quantities such as $T^i(M)$ and $I^i(M)$ $\forall i$ in a fair manner over a window of M slots.

14.4.1 Notions of Fairness

There are various fairness measures that have been considered in the literature. Let $\phi = [\phi^1, \ldots, \phi^N]^T$ be a weight vector associated with the users indicating their relative priorities.

- Minimum Allocation: Under this notion of fairness, the scheduling scheme attempts to provide a certain minimum throughput or fraction of time slots to each user. Let $\bar{\Psi} = [\bar{\Psi}^1, \ldots, \bar{\Psi}^N]^T$ be a vector indicating certain minimum throughput that must be achieved by the users. Let $\bar{\varepsilon} = [\bar{\varepsilon}^1, \ldots, \bar{\varepsilon}^N]^T$ be a vector indicating minimum fraction of time slots that must be allocated to a user. Then the scheme is said to be fair if $\bar{\mathbf{T}} \geq \bar{\Psi}$ (minimum throughput allocation) or $\bar{\mathbf{I}} \geq \bar{\varepsilon}$ (minimum time slot allocation).
- Fair Relative Throughput/Time Slot Allocation: The system attempts to provide equal weighted throughput/fraction of time slots to all users under this notion of fairness. The scheme is said to be fair if $\frac{\bar{T}^i}{\phi^i} = \frac{\bar{T}^j}{\phi^j}$, $\forall i, j$ (fair relative throughput allocation) or $\frac{\bar{I}^i}{\phi^i} = \frac{\bar{I}^j}{\phi^j}$, $\forall i, j$ (fair relative time slot allocation).
- Proportional Fair Allocation: The fraction of slots allocated to a user is proportional to the average channel state of that user. Better the channel state perceived by a user on an average, higher is the fraction of slots allocated to such a user. The proportional fair scheduling algorithm is discussed in the next section.

Note that each notion of fairness defined above can have a probabilistic extension, where the system is allowed to be unfair with a certain probability.

14.4.2 Fair Scheduling Algorithms

Let T_n^i be the average throughput of a user i in an exponentially averaged window of length t_c. The proportional fair scheduling algorithm schedules the user i in a slot n where:

$$i = \arg\max_j \frac{U_n^j}{T_n^j}. \tag{14.36}$$

The average throughput T_n^j is updated using exponential averaging:

$$T_{n+1}^j = \begin{cases} (1 - \frac{1}{t_c})T_n^j + (\frac{1}{t_c})U_n^j, & j = i, \\ T_n^j & j \neq i. \end{cases} \tag{14.37}$$

Users having the same channel statistics tend to have the same average throughput and consequently the scheduling policy reduces to the opportunistic policy, i.e., in each slot, the user with the highest rate is scheduled. On the other hand, if the channel statistics of the users are not identical, then the users compete for resources based on their rates normalized by their respective throughputs. Note that the algorithm schedules a user when its channel state is high relative to its own average channel state over the time scale t_c. The proportional fair scheduler has the following property: For large t_c, i.e., for $t_c \to \infty$, the algorithm maximizes $\sum_{i=1}^{N} \log \bar{T}^i$.

Another fair scheduling policy is to maximize the sum throughput while providing minimum fraction of time slots to the users. This optimization problem can be expressed as:

$$\max \bar{T} \tag{14.38}$$

subject to:

$$\bar{I}^i \geq \bar{\varepsilon}^i, \ i = 1, \ldots, N. \tag{14.39}$$

One possible approach to designing such a scheduling policy can be based on stochastic approximation where one determines the throughput maximizing time slot allocation in an iterative fashion.

The opportunistic scheduling problem with short term fairness constraints (under the minimum time slot allocation criterion) can be expressed as the following optimization problem: in any window of M slots,

$$\max \sum_{i=1}^{N} \bar{T}^i(M), \tag{14.40}$$

subject to:

$$\bar{I}^i(M) \geq M\bar{\varepsilon}^i. \tag{14.41}$$

14.5 Power Optimal Scheduling

In this section, we discuss the problem of scheduling such that the average transmit power is minimized under constraints on the delay. As already discussed, the power required to transmit packets is a convex and increasing function of the rate (i.e., number of packets being transmitted in a slot). From an energy efficiency point of view, the scheduler should transmit packets opportunistically, i.e. when the channel condition is favorable. This leads to buffering the packets and consequent delay. Average delay may be a QoS metric for some applications and the packets need to be scheduled intelligently. We thus have the average cost scheduling problem where the objective is to minimize average power subject to constraint on average delay.

The system model has been already discussed in Section 14.3.3. In this section, the packet arrival process for each user $\{A_n^i\}$ is assumed to be i.i.d. across users. A_n^i takes values from a finite and discrete set $\mathbb{A} \stackrel{\Delta}{=} \{0, \ldots, A\}$. Since we consider TDMA, only one user can be scheduled in a slot. The scheduling problem is to determine the user to be scheduled in each time slot and also the number of packets that the user should transmit (i.e., rate) such that the average power $\bar{P}^i$ for each user i is minimized subject to a constraint on individual queue length $\bar{Q}^i$. Thus the scheduling problem can be stated as:

$$\text{Minimize } \bar{P}^i \text{ subject to } \bar{Q}^i \leq \delta^i \text{ for } i = 1,\ldots,N. \tag{14.42}$$

Since the scheduling decision in a slot affects the buffer occupancy of all users, these N problems are not independent.

Before we proceed with the multiuser case, we first consider the case of a point-to-point link, i.e., there is only one transmitter.

14.5.1 Single User Scheduling

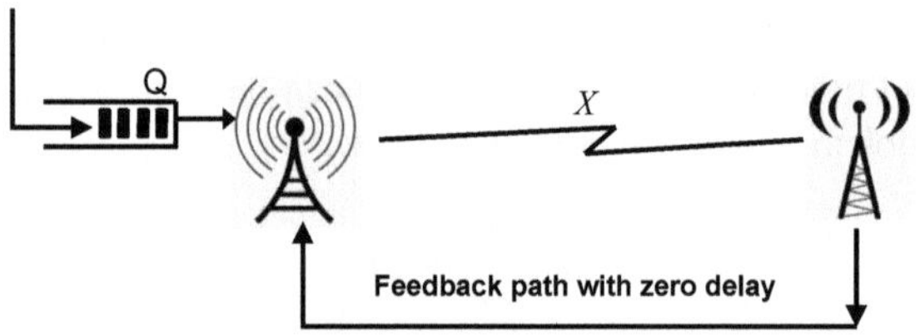

Fig. 14.7 Point-to-point transmission model with finite buffer

In this section, we assume that $\mathbb{Q} = \{0,\ldots,B\}$. For the point to point link, the queue evolution can be expressed as:

$$Q_{n+1} = Q_n + A_{n+1} - U_n. \tag{14.43}$$

Since we consider a single user, we drop the superscript i in the notation in this section. Let us define the state of the system in slot n by $S_n \stackrel{\Delta}{=} (Q_n, X_n, A_n)$, i.e., the state comprises of queue length, channel state and source arrival state. The system state space $\mathbb{S} = \mathbb{Q} \times \mathbb{X} \times \mathbb{A}$ is finite and discrete. In each slot, the control or scheduling action corresponds to the number of packets transmitted $U_n \leq Q_n$. U_n takes values from a finite action space $\mathbb{U} = \{0,\ldots,B\}$. Note that the channel state X_n is independent of the action, queue and arrival state. Similarly, the source arrival state is also independent of channel state and action. The control policy is a sequence of functions $\{\mu_1, \mu_2, \ldots\}$, where μ_n specifies U_n (or the probability of taking action U_n) given the past history of the system state and control actions.

We consider the arrivals to be i.i.d. across slots, in such a case, the state of the system simplifies to $S_n = (Q_n, X_n)$.

Let $c_p(S_n,U_n) = P(X_n,U_n)$ be the 'immediate' cost in terms of power required in transmitting U_n packets when the state is S_n. Let $c_q(S_n,U_n) \triangleq Q_n$ denote the 'immediate' cost due to buffering. Let $\mu = \{\mu_1,\mu_2,\ldots\}$ be the control policy. We would like to determine the policy μ that minimizes

$$\bar{P} = \limsup_{N\to\infty} \frac{1}{N}\mathbf{E}\sum_{n=1}^{N} c_p(S_n,U_n), \tag{14.44}$$

subject to

$$\bar{Q} = \limsup_{N\to\infty} \frac{1}{N}\mathbf{E}\sum_{n=1}^{N} c_q(S_n,U_n) \le \bar{\delta}. \tag{14.45}$$

It can be easily argued that this is a CMDP with average cost and finite state and action spaces. For average cost CMDP with finite state and action space, it is well known that a optimal stationary randomized policy exists. Let $\bar{P}^*$ denote the optimal cost i.e.,

$$\bar{P}^* = \min_{\mu} \bar{P}^{\mu}, \tag{14.46}$$

where $\bar{P}^{\mu}$ is the cost (14.44) under policy μ.

Let $\mu(\cdot|s) : s \in \mathbb{F}$ be the probability measure on $\mathbb{U}$. For each state s, $\mu(\cdot|s)$ specifies the distribution with which the control in that state is applied. We assume that $\{S_n\}$ is an ergodic Markov chain under such policies and thus has a unique stationary distribution ρ^{μ}.

Let $\mathbf{E}^{\mu}$ denote the expectation with respect to (w.r.t.) ρ^{μ}. Under a randomized policy μ, the costs in (14.44) can be expressed as:

$$\bar{P}^{\mu} \triangleq \mathbf{E}^{\mu}\Big[c_p(S_n,\mu(S_n))\Big] = \sum_{u,s} \rho^{\mu}(s)\mu(u|s)c_p(s,\mu(s)), \tag{14.47}$$

and,

$$\bar{Q}^{\mu} \triangleq \mathbf{E}^{\mu}\Big[c_q(S_n,\mu(S_n))\Big] = \sum_{u,s} \rho^{\mu}(s)\mu(u|s)c_q(s,\mu(s)), \tag{14.48}$$

respectively. Then the scheduler objective can be stated as:

$$\text{Minimize } \bar{P}^{\mu} \text{ subject to } \bar{Q}^{\mu} \le \delta. \tag{14.49}$$

We now demonstrate that the optimal average cost and policy can be determined using an unconstrained Markov Decision Process (MDP) problem and Lagrangian approach.

14.5.1.1 The Lagrangian Approach

Let $\lambda \ge 0$ be a real number. Define $c : \mathbb{R}^+ \times \mathbb{S} \times \mathbb{U} \to \mathbb{R}$ as follows,

$$c(\lambda,s,u) = c_p(s,u) + \lambda(c_q(s,u) - \delta). \tag{14.50}$$

Note that the function $c(\cdot,\cdot,u)$ is a strictly convex function of u (as the power required to transmit u packets is a strictly convex function of u). The unconstrained problem is to determine an optimal stationary policy $\mu^*(\cdot)$ that minimizes

$$L(\mu,\lambda) = \mathbf{E}^{\mu}\left[c(\lambda, S_n, \mu(S_n))\right], \tag{14.51}$$

for a particular value of λ called the Lagrange Multiplier (LM). $L(\cdot,\cdot)$ is called the Lagrangian.

Let $p(s,u,s')$ be the probability of reaching state s' upon taking action u in state s. Let $V(s)$ denote the optimal value function (i.e. expected cost) for a state s. The following dynamic programming equation provides the necessary condition for optimality of the policy.

$$V(s) = \min_u \left[c(\lambda,s,u) - \beta + \sum_{s'} p(s,u,s')V(s')\right], \; s' \in \mathbb{S}, \tag{14.52}$$

where $\beta \in \mathbb{R}$ is uniquely characterized as the corresponding optimal cost (power) per stage. If we impose $V(s^0) = 0$ for any pre-designated state $s^0 \in \mathbb{S}$, then V is unique. Furthermore, an optimal policy μ^* must satisfy,

$$\text{support}(\mu^*(\cdot|s)) \subseteq \arg\min \left[c(\lambda,s,u) - \beta + \sum_{s'} p(s,u,s')V(s')\right] \; \forall s \in \mathbb{S}. \tag{14.53}$$

It follows that the constrained problem has a stationary optimal policy which is also optimal for the unconstrained problem considered in (14.51) for a particular choice of $\lambda = \lambda^*$ (say). In general, this optimal policy may be a randomized policy. In fact, it can be shown that the optimal stationary policy is *deterministic* for all states but at most one s, i.e., there exists a unique $u^*(s)$ such that $\mu^*(u^*(s)|s) = 1$ and u^* is the solution to the following equation,

$$u^*(s) = \arg\min \left[c(\lambda^*,s,u) - \beta + \sum_{s'} p(s,u,s')V(s')\right] \; \forall s \in \mathbb{S}. \tag{14.54}$$

Furthermore, for the single (if any) state s for which this fails, $\mu(\cdot|s)$ is supported on exactly two points. The optimal average cost β gives the minimum power consumed $\bar{P}^*$ subject to the specified queue length constraint δ. Moreover, the following *saddle point condition* holds:

$$L(\mu^*,\lambda) \leq L(\mu^*,\lambda^*) \leq L(\mu,\lambda^*). \tag{14.55}$$

14.5.1.2 Structural Properties of the Optimal Policy

Before we discuss the computational issues in determining the optimal policy through dynamic programming equation (14.52), we discuss some structural properties of the optimal policy. The result for i.i.d. arrival and channel state processes can be stated as:

Theorem 14.1. *The optimal policy $\mu^*(s) = \mu^*(q,x)$ is non decreasing in channel state x and non-decreasing in queue length q.*

The proof of this theorem is based on supermodularity and increasing differences properties of the value function. These properties establish the monotonicity of the optimal policy in channel state and buffer occupancy.

The structural results imply that the optimal decision is to transmit a certain number of packets in a given slot where this number is an increasing function of the current queue length and channel state. Thus for a fixed channel state, the greater the queue length, the more will be the number of packets that will be transmitted. Similarly, for a fixed queue length, the better the channel, the more will be the number of packet transmissions. Thus, the optimal policy always transmits at the highest rate when channel condition is the most favorable and queue length is the largest.

Similar structural results also hold for Markovian arrivals and Markovian channel state and continuous state processes. However, for these general state spaces, the dynamic programming equation (14.52) for the long run average cost problem requires to be rigorously justified. See bibliographic notes and [1] for a discussion of this.

14.5.1.3 Learning Algorithm for Scheduling

In this section, we discuss the computation of optimal policy discussed in the preceding section. For a fixed λ, the Relative Value Iteration Algorithm (RVIA) can be used for solving the dynamic programming equation in an iterative fashion. The average cost RVIA for determining the value function such that (14.52) is satisfied can be written as:

$$V_{n+1}(s) = \min_{u \in U(s)} [c(\lambda, s, u) + \sum_{s'} p(s, u, s') V_n(s')] - V_n(s^0), \tag{14.56}$$

where $s, s', s^0 \in \mathbb{S}$ and s^0 is any fixed state. $V_n(\cdot)$ is an estimate of the value function after n iterations for a fixed LM λ.

Due to the large size of the state space $\mathbb{S}$, solution of this algorithm is computationally expensive. One approach for addressing this issue is to utilize the structural properties of the optimal policy outlined above to develop efficient heuristics that are computationally less expensive and hence can be implemented in a practical system. In [54], one such approach has been discussed. Techniques based on function approximation can also be employed for dimensionality reduction. The structural results of the policy may be used to choose the basis functions in function approximation.

Another technique is to use 'primal-dual' type approach with conventional iteration schemes for the value function (primal) and Lagrange multiplier (dual) [13]. However, even with this approach (and for that matter, with all other approaches discussed above for computing (14.56)), one major issue is that it requires the knowledge of transition probabilities $p(s, u, s')$. Note that $p(s, u, s')$ depends upon the channel state and packet arrival distributions. Both these distributions are unknown. In practice, it is usual to assume packet arrivals to be Poisson and channel state to be exponential (assuming Rayleigh fading as discussed in Section 14.2) but these are only modeling assumptions and it is difficult to obtain accurate information even about the parameters of these distributions. Reinforcement Learning (RL) algorithms are useful in such scenarios. We present here one such approach.

Let ζ be the law for the arrivals and $\kappa(\cdot|\cdot)$ the transition probability function for the channel state process[2]. To address the issue of these unknown probability laws, we can employ stochastic approximation to perform averaging in real time w.r.t unknown laws. Unfortunately, however, RVIA (14.56) is not amenable to real time implementation because of the occurrence of *min* operator outside the *averaging* operation. This problem is addressed by rewriting RVIA in a novel manner as discussed below.

We define *post-decision* state to be the virtual state of the system immediately *after* taking a decision but *before* the action of the noise (arrivals). If the transmitter transmits $U = u$ packets in a slot, then the post-decision state denoted by $\tilde{s}$, $\tilde{s} \in \mathbb{S}$ is $(q-u,x)$. We rewrite RVIA (14.56) in the form of post-decision state. Thus $\tilde{V}$ (i.e, the value function based on the post decision state) satisfies the following dynamic programming equation: for $\tilde{s} = (q,x)$,

$$\tilde{V}(\tilde{s}) = \sum_{a,x'} \zeta(a)\kappa(x'|x)(\min_{u \le q+a}[c(\lambda,(q+a,x'),u) + \tilde{V}(q+a-u,x')]) - \beta. \quad (14.57)$$

From (14.56) and (14.57), we get the following RVIA:

$$\tilde{V}_{n+1}(\tilde{s}) = \sum_{a,x'} \zeta(a)\kappa(x'|x)(\min_{u \le q+a}[c(\lambda,(q+a,x'),u) + \tilde{V}_n(q+a-u,x')]) - \tilde{V}_n(\tilde{s}^0);$$
$$\tilde{V}_{n+1}(\tilde{s}'') = \tilde{V}_n(\tilde{s}'') \ \forall \tilde{s}'' \neq \tilde{s}. \quad (14.58)$$

The important thing to note here is that we update only the $\tilde{s}$-th component, not the rest. Note (14.58) has a useful structure in the sense that the averaging operation has been moved outside of the min operator. We can now employ stochastic approximation to perform averaging in real time.

We set up 'primal-dual' iterations for the value function (of RVI (14.58)) and LM. To solve both value function and LM iteratively, let us first choose the sequences $\{f_n\}$ and $\{e_n\}$ that have the following properties,

$$\sum_n (f_n)^2, \ \sum_n (e_n)^2 < \infty, \quad (14.59)$$

$$\sum_n f_n = \infty, \ \sum_n e_n = \infty, \quad (14.60)$$

$$\lim_{n\to\infty} \frac{e_n}{f_n} \to 0. \quad (14.61)$$

The complete primal-dual RVI algorithm can be expressed as: for $S_n = \tilde{s} = (q,x)$,

[2] Here we allow the channel state process to be Markovian also.

$$\tilde{V}_{n+1}(\tilde{s}) = \tilde{V}_n(\tilde{s}) + f_n\Big\{\min_u[c(\lambda_n,(q+A_{n+1},X_{n+1}),u) + \tilde{V}_n((q+A_{n+1}-u,X_{n+1}))] - \tilde{V}_n(\tilde{s}) - \tilde{V}_n(\tilde{s}^0)\Big\}, \quad (14.62)$$

$$\tilde{V}_{n+1}(\tilde{s}'') = \tilde{V}_n(\tilde{s}) \ \forall\ \tilde{s}'' \neq \tilde{s}, \quad (14.63)$$

$$\lambda_{n+1} = \Lambda[\lambda_n + e_n(Q_n - \delta)], \quad (14.64)$$

where we use the projection operator Λ to project the LM onto interval $[0,L]$ for large enough $L > 0$, to ensure boundedness of the LM. These iterations can be performed in real time at every slot.

The sequences f_n and e_n are chosen appropriately to ensure that the sequences converge to 0 neither too fast nor too slow. Since we have assumed $\lim_{n\to\infty}\frac{e_n}{f_n} \to 0$, it induces two time scales, a fast one for (14.62) and a slow one for (14.64). Using the theory of two time scale stochastic approximation, it can be proved that these iterates indeed converge to optimal values.

The convergence of value function and LM is analyzed by first freezing $\lambda_n \approx$ a constant λ^i and then proving that the value function converges to its optimal value $\tilde{V}$, i.e., $\tilde{V}_n \to \tilde{V}$.

Let $h(\tilde{V}) = [h_{q,x}(\tilde{V})]$ be given by:

$$h_{q,x}(\tilde{V}) = \sum_{a,x'} \zeta(a)\kappa(x'|x) \times \min_u[c(\lambda,q+a,x,u) + \tilde{V}(q+a-u,x') - \tilde{V}(q^0,x^0)],$$

where (q^0,x^0) is any pre-designated state. The limiting o.d.e. for (14.62) is given by

$$\dot{\tilde{V}}(t) = \Lambda(t)(h(\tilde{V}(t)) - \tilde{V}(t)), \quad (14.65)$$

where $\Lambda(t)$ is a diagonal matrix with nonnegative elements summing to 1 on the diagonal. Then the following Lemma can be proved.

Lemma 14.1. *If the diagonal elements of $\Lambda(t)$ remain uniformly bounded away from zero, $\tilde{V}_n^i \to \tilde{V}$.*

Note that the above analysis treats $\lambda_n \approx$ a constant, so what this Lemma states is that $\{\tilde{V}_n\}$ closely tracks $\{\tilde{V}^{\lambda_n}\}$, where $\tilde{V}^{\lambda}$ is $\tilde{V}$ with its λ-dependence made explicit. Note that the $\tilde{V}_n$ and λ_n iterations are primal-dual iterations. The primal iterations perform relative value iteration and determine a minimum of the Lagrangian (14.51) with respect to the policy for an almost constant LM.

To prove the convergence of λ_n, we consider the limiting o.d.e of (14.64). It can be shown that the limiting o.d.e. for the λ_n's is a steepest ascent for the Lagrangian minimized over the primal variables. By standard results for stochastic gradient ascent for concave functions, it can be proved that this o.d.e converges to the optimal LM λ^*. We thus have the following lemma.

Lemma 14.2. *The LM iterates λ_n converge to optimal value λ^*.*

Lemmas 14.1 and 14.2 imply that $(\tilde{V}_n,\lambda_n) \to (\tilde{V},\lambda^*)$ as required.

In practice, the power optimal scheduling algorithm can be implemented using the above primal-dual online algorithm. In each time slot, the scheduler observes the channel state x and determines u, the number of packets to be transmitted, that minimizes the right hand side of (14.62). The value function and LM are updated as in (14.62), (14.64) and the algorithm proceeds. Though, the convergence results are asymptotic, for most practical systems, it has been observed that the algorithm converges in reasonable number of iterations (time slots). The readers are referred to references in bibliographic notes for details of the implementation.

In the next section, we consider multiuser scheduling where a centralized scheduler has the responsibility of determining the user to be scheduled in a slot in addition to determining the number of bits (or packets) to be transmitted and the corresponding transmission power.

14.5.2 Multiuser Uplink Scheduling

As outlined earlier in (14.42), the multiuser power optimal scheduling problem is to design an algorithm that minimizes the average power expenditure of each user subject to a constraint on the individual queuing delay. This problem is an optimization problem with N objectives and N constraints and can also be formulated within the framework of CMDP. Let $\mathbf{Q}_n$ and $\mathbf{X}_n$ denote the vectors $[Q_n^1, \ldots, Q_n^N]$ and $[X_n^1, \ldots, X_n^N]$ respectively. The system state is then defined by $\mathbf{S_n} = (\mathbf{Q}_n, \mathbf{X}_n)$. Since we consider TDMA where only one user is scheduled in a slot, control action comprises of the index of the user and its transmission rate.

However, since the system state consists of vector of queue lengths and channel states of all users, even for moderate number of users, the size of the state space is very large. Thus numerical approaches for determining the optimal policy are computationally infeasible. Moreover, to compute the optimal policy, the base station needs to know the queue length information of each user. This requires significant communication overhead on the uplink where each user needs to communicate its queue length. Finally, as discussed in point to point link case, even the packet arrival and channel state distributions are unknown. One, therefore, has to consider suboptimal, albeit, efficient approaches.

One such strategy to solve this problem consists of decomposing the problems into N dependent single user problems and a base station problem. In a user problem, each user i determines a rate at which it should transmit in a slot as if it were being controlled by a single user policy. Since the channel and arrival statistics are not known, each user employs learning algorithm described in Section 14.5.1 (rate determination algorithm). The users' rates are then conveyed to the base station. The base station schedules the user with the highest rate (user selection algorithm). The rationale behind choosing the highest rate is that this strategy will favor the user either with good channel condition or large queue length. The queue length of a user, who is not scheduled for a while, will keep increasing thereby increasing its rate requirement as well. This will ensure that the user will eventually be scheduled.

It can be shown that the algorithm indeed converges to equilibrium and the delay constraints are satisfied. The complete strategy is explained below.

14.5.2.1 Rate Determination Algorithm

Each user implements the single user algorithm ((14.62)–(14.64)) discussed in Section 14.5.1 Thus user i determines its rate R_n^i (or the number of packets it should transmit) that maximizes the r.h.s of primal value function iteration (14.62). Note that the value function iteration (14.62) considers the following power cost:

$$\bar{P}_e^i = \limsup_{M\to\infty} \frac{1}{M} \sum_{n=1}^{M} P(X_n^i, R_n^i). \tag{14.66}$$

This is the same cost as discussed for the point to point link (14.44) (in this case U_n^i is replaced by R_n^i since U_n^i in multiuser case now refers to the packets actually transmitted by user i). User i transmits at this rate if the channel is allocated to it by the base station otherwise it is unable to proceed with the transmission. Thus the queue evolution equation for user i is given by,

$$Q_{n+1}^i = Q_n^i + A_{n+1}^i - I_n^i R_n^i, \tag{14.67}$$

where $I_n^i = 1$ if user i is scheduled in a slot n by the base station, else, $I_n^i = 0$. The control variables are $\mathbf{I}_n$, $\mathbf{R}_n = [R_n^1, \ldots, R_n^N]$. Note that users independently choose the corresponding components of $\mathbf{R}_n$ and the base station chooses $\mathbf{I}_n$ subject to the constraints that $I_n^i \in \{0,1\}$ and $\sum_i I_n^i = 1$.

The actual average power consumed by user i is given by,

$$\bar{P}^i = \limsup_{M\to\infty} \frac{1}{M} \sum_{n=1}^{M} P(X_n^i, I_n^i R_n^i). \tag{14.68}$$

The relative value iteration is performed not for the actual power cost (14.68) but exactly as discussed in single user case. However, the difference is reflected in the queue length update (14.67) (and thereby the LM update (14.64)) that depends on whether a user is scheduled in a slot or not. It can be proved that value function iteration will converge to the single user optimum rate required for the current quasi-static value of the LM. The difference with the single user case comes from the fact that the relative value iteration for each user is coupled through LM iterations also and the LM updates are indeed affected by the actual transmission through queue length. The convergence of LM would imply that the delay constraints are satisfied.

14.5.2.2 User Selection Algorithm

The base station schedules the user with the highest rate R_n^i in a slot, i.e, $I_n^i = 1$ if $i = \arg\max_j R_n^j$ and all other $I_n^j = 0$. If more than one user has the highest rate, one of them is selected at random with uniform probability.

This strategy can be thought of as an auction where users bid their rates to the base station which then schedules the user bidding the highest rate. Note that it is not in the interest of the users to bid unnecessarily high rates as this might result in higher power consumption.

14.5.2.3 Algorithm Analysis

Convergence Analysis: The convergence of value functions and LM follow as in the single user case. Since the value function of each user is updated in each slot regardless of whether the user is scheduled in that slot or not; the value functions are decoupled across users. This is in the spirit of the decoupling of static formulations of network flow problems via the Lagrange multiplier as in [22], except that here it is *mandated* by our algorithm. The decoupling is facilitated by the fact that the users compute their value function as though the cost is (14.66) and not (14.68). Convergence of the LMs implies that the delay constraints are satisfied and vice versa. This implies that if there is sufficient capacity, the multiuser scheduling satisfies the delay constraints of all the users.

Queue Stability: Assuming that the users and base station have already learned their policies, one can establish the queue stability. A sketch of the proof is as follows.

$\bar{a}^i = \mathbf{E}[A_n^i]$. Let π denote the (unique) stationary distribution of the Markov chain $\{\mathbf{X}_n\}$. Let $R \stackrel{\Delta}{=} \min_i \sum_{\mathbf{x}} \pi(\mathbf{x}) \hat{R}^i(x^i)$. Recall that R_n^i depends on (Q_n^i, X_n^i). Suppose that $R_n^i = \ell^i(Q_n^i, X_n^i)$ for some $\ell^i(\cdot,\cdot)$. Write $\ell(\mathbf{Q}_n, \mathbf{X}_n)$ for $[\ell^1(Q_n^1, X_n^1), \ldots, \ell^N(Q_n^N, X_n^N)]$.

For proof of stability, we assume a more general scheme where the channel is allocated to user i with a probability,

$$F^i(\mathbf{R}_n) := g(R_n^i)^m / (\sum_j g(R_n^j)^m)$$

,

where g is a monotone increasing and smooth function and $m >> 1$. Note that with this user selection algorithm, the base station uses a smooth approximation $\mathbf{F} := [F^1, \ldots, F^N]$ of the maximum function for channel allocation, i.e., it allocates the channel to the user with the highest rate (say, ith) with a probability $F^i(\mathbf{R}_n)$ close to one, but does allocate it to others also with a small but nonzero probability $F^j(\mathbf{R}_n), j \neq i$. We also assume that the $\ell^i(\cdot,\cdot)$ above is continuously differentiable.

The queue evolution (14.67) for user i can be rewritten as:

$$Q_{n+1}^i = Q_n^i + (\bar{a}^i - F^i(\mathbf{R}_n) R_n^i) + \Big((A_{n+1}^i - \bar{a}^i) + (F^i(\mathbf{R}_n) R_n^i - U_n^i) \Big).$$

Let $\widetilde{\mathbf{F}} = [\widetilde{F}^1, \ldots, \widetilde{F}^N]$ be defined by: $\widetilde{F}^i(r) = F^i(r)r^i \ \forall i$. Note that this is continuously differentiable. Then the queue evolution is of the form:

$$Q^i_{n+1} = Q^i_n + (\bar{a}^i - \widetilde{F}^i(\mathbf{R}_n)) + M_{n+1}, \ n \geq 0, \tag{14.69}$$

where $\{M_n\}$ is a martingale difference sequence.

Let $\mathbf{q} = [q^1, \ldots, q^N]$. Define $\widehat{\mathbf{F}} = [\widehat{F}^1, \ldots, \widehat{F}^N]$ by:

$$\widehat{F}^i(\mathbf{q}) = \sum_{\mathbf{x}} F^i(\ell(\mathbf{q}, \mathbf{x}))\ell^i(q^i, x^i)\pi(\mathbf{x}).$$

Consider the scaled version of (14.69), given by

$$Q^i_{n+1} = Q^i_n + \eta[(\bar{a}^i - \widetilde{F}^i(\mathbf{R}_n)) + M_{n+1}], \ n \geq 0, \tag{14.70}$$

for a small $\eta > 0$. If we consider smaller and smaller time slots of width η with $\bar{a}^i$, and $\widetilde{F}^i$ being 'rates per unit time' rather than 'per slot' quantities, then we obtain the 'fluid' approximation of (14.69) , which is given by the o.d.e.

$$\dot{q}^i(t) = \bar{a}^i - \widehat{F}^i(\mathbf{q}(t)), \tag{14.71}$$

It can be proved that the trajectories of the o.d.e. in (14.71) converge to an equilibrium for all almost all initial conditions. This in turn proves that the actual queue lengths will concentrate near the equilibrium set of this o.d.e with very high probability. The readers are referred to [41] for more details.

The above strategy thus ensures a stable energy efficient scheduling algorithm that also satisfies the delay constraints of the users. In the uplink scenario, the base station does not have information about queue lengths of users. In the above approach, each user needs to communicate only the desired rate (instead of the queue length information). In a practical system, we may have few possible rates say 16. In such a case, we may need only 4 bits of communications overhead. This strategy does not require the knowledge of packet arrival and channel state distributions. It thus provides a powerful framework for implementing multiuser packet scheduling algorithms.

14.6 Scheduling Schemes that Maximize Throughput

In the preceding section, we have considered power optimal scheduling algorithms that minimize power cost under *delay constraints*. In this section, we consider scheduling algorithms under the *power constraints*. We first study scheduling algorithms that consider queue stability as a notion of QoS. While some of these algorithms are throughput optimal, they do not necessarily ensure small average queue lengths and hence small delays. Subsequently, we consider scheduling algorithms that address this issue while achieving high sum throughput.

14.6.1 Throughput Optimal Scheduling

We first review feasible rate and power allocation with an objective of stabilizing the queues of the users. We define the overflow function as follows:

$$f^i(\xi) = \limsup_{M\to\infty} \frac{1}{M}\sum_{n=1}^{M} I_{Q_n^i > \xi}, \tag{14.72}$$

where $I_{Q_n^i>\xi}$ is an indicator variable that is set to 1 if $Q_n^i > \xi$, else it is set to 0. We say that the system is *stable* if $f^i(\xi) \to 0$ as $\xi \to \infty$ for all $i = 1,\dots,N$. Let $\bar{\mathbf{a}} = [\bar{a}^1,\dots,\bar{a}^N]^T$ denote the arrival vector, $\bar{a}^i$ being the average arrival rate for user i. In this section, in addition to the average power constraints, we also consider the peak power constraints, i.e., a user i can transmit at a maximum power $\hat{P}^i$ in any slot. Let $\hat{\mathbf{P}} = [\hat{P}^1,\dots,\hat{P}^N]^T$ denote the peak power constraint vector.

Note that, since the objective is to keep the queues stable, the power and rate allocation policies have to be cognizant of the queue lengths of the users in each slot. A power allocation policy $\mathscr{P}$ is a mapping from the joint channel state and queue length vector $(\mathbf{x},\mathbf{q})$ to a power allocation vector $\mathbf{P}$. A rate allocation policy $\mathscr{R}$ is a mapping from the joint channel state and queue length vector $(\mathbf{x},\mathbf{q})$ to a rate allocation vector $\mathbf{R}$. As noted previously in Section 14.2.2, a feasible rate allocation policy allocates rates within the multi-access capacity region $\mathbb{C}_g(\mathbf{x},\mathbf{P})$. The *stability region* of the multi-access system is the set of all arrival vectors $\bar{\mathbf{a}}$ for which there exists some feasible power allocation policy and rate allocation policy under which the system is stable. The stability region of a multi-access system can be shown to be given by:

$$\mathbb{C}_s(\bar{\mathbf{P}},\hat{\mathbf{P}}) = \bigcup_{\mathscr{P}\in\mathbb{F}} \mathbf{E}\left[\mathbb{C}_g(\mathbf{X},\mathscr{P}(\mathbf{X}))\right]. \tag{14.73}$$

Note that the power control policy $\mathscr{P}(\mathbf{X})$ depends only on the channel state vector $\mathbf{X}$. More importantly, this stability region of the multi-access system is same as the throughput capacity region under power control defined in Section 14.2.2.

If the joint arrival process $\{\mathbf{A}_n\}$ and joint channel state process $\{\mathbf{X}_n\}$ are ergodic Markov chains, then the system can be stabilized by a power and rate allocation policy if $\bar{\mathbf{a}} \in \mathbb{C}_s(\bar{\mathbf{P}},\hat{\mathbf{P}})$. In practice, one does not have a knowledge of the arrival vector $\bar{\mathbf{a}}$ and this can only be estimated over time. Power and rate allocation policies that do not assume knowledge of the arrival vector $\bar{\mathbf{a}}$ and stabilize the system as long as $\bar{\mathbf{a}} \in \mathbb{C}_s(\bar{\mathbf{P}},\hat{\mathbf{P}})$ are referred to as *throughput optimal* policies. Throughput optimal scheduling policies have been explored in several papers in literature, see bibliographic notes for further details. Longest Connected Queue (LCQ), Exponential (EXP), Longest Weighted Queue Highest Possible Rate (LWQHPR) and Modified Longest Weighted Delay First (M-LWDF) are some well known examples of throughput optimal scheduling policies. We now review some of these scheduling rules that are throughput-optimal under a power allocation policy $\mathscr{P}$.

- LWQHPR: Let $\alpha = [\alpha^1,\dots,\alpha^N]^T$ be a vector of weights. The throughput optimal rate allocation policy is obtained by maximizing $\sum_{i=1}^N \alpha^i Q_n^i U_n^i$ over $\mathbb{C}_g(\mathbf{x},\mathscr{P}(\mathbf{x}))$.

The solution $\mathbf{r}^*$ is obtained by successively decoding the users in an increasing order of their weights $\alpha^i Q_n^i$, i.e., shorter queues are decoded before longer queues. This implies that longer queues are given preference over shorter queues.

- M-LWDF: Let $\bar{\Upsilon}^i$ and $\bar{D}^i$ be the delay requirement and achieved delay for user i respectively. The M-LWDF scheduler attempts to satisfy the delay constraints of the form,
$$Pr(\bar{D}^i > \Upsilon^i) \leq \bar{\rho}^i, \tag{14.74}$$
where $\bar{\rho}^i$ is an upper bound on the probability with which $\bar{D}^i$ is allowed to exceed Υ^i. The M-LWDF scheme achieves this by scheduling a user i in a slot n where:
$$\frac{-\log(\bar{\rho}^i) \times Q_n^i \times U_n^i}{\bar{\Upsilon}^i} = \max_j \frac{-\log(\bar{\rho}^j) \times Q_n^j \times U_n^j}{\bar{\Upsilon}^j}. \tag{14.75}$$
Note that higher the queue length and better the channel state (and hence higher the rate) of a user in a slot, higher is the probability of scheduling the user in the slot.
- EXP: Let $\gamma = [\gamma^1, \ldots, \gamma^N]^T$, $\mathbf{b} = [b^1, \ldots, b^N]^T$ be an arbitrary set of positive constants. Let α and $\eta \in (0,1)$ be fixed. The Exponential (EXP) rule schedules a user i in a slot n where:
$$i = \arg\max_j \gamma^j U_n^j \exp\left(\frac{b^j Q_n^j}{\alpha + [\hat{Q}_n]^\eta}\right), \tag{14.76}$$
where $\hat{Q}_n \triangleq \frac{1}{N}\sum_{i=1}^N b^i Q_n^i$. Thus, a user with better channel state and hence higher rate and higher queue length has a higher probability of being scheduled.

14.6.2 Delay Optimal Scheduling

While throughput optimal scheduling policies maintain the stability of the queuing system, they do not necessarily guarantee small queue lengths and consequently lower delays. Delay optimal scheduling deals with optimal rate and power control such that the average queue length and hence average delay are minimized for arrival rates within the stability region under average and peak power constraints. Due to the nature of the constraints, there is no loss of optimality in choosing the rate and power control policies separately. Hence, to simplify the problem, one can choose any stationary power control policy that satisfies the peak and average power constraints. The delay optimal policy, therefore, deals with optimal rate allocation for minimizing delays under a given power allocation policy. The objective is to maximize a weighted combination of the rates expressed in (14.20), while at the same time minimizing the achieved delay $\bar{Q}^i$, $i = 1, \ldots, N$. Note that this problem is a multi-objective optimization problem.

We now study a scheme that is throughput optimal and delay optimal under certain assumptions on the arrival and channel state processes for both multi-access (uplink) and broadcast (downlink) channels. Before outlining these assumptions, we define a symmetric channel state process. The channel state process is called *symmetric* or *exchangeable* if for all n and $\mathbf{x} = [x^1,\ldots,x^N]^T$ in the channel state space $\mathbb{X}^N$,

$$Pr\left(X_n^1 = x^1,\ldots,X_n^N = x^N\right) = Pr\left(X_n^1 = x^{\pi(1)},\ldots,X_n^N = x^{\pi(N)}\right), \tag{14.77}$$

for any permutation $\pi \in \Pi$, where Π is the set of all permutations on the set $\{1,\ldots,N\}$. A power control policy $\mathscr{P}$ that is a function of the channel state vector only is symmetric if for all $\mathbf{x} \in \mathbb{X}$,

$$\mathscr{P}^i\left(x^1,\ldots,x^N\right) = \mathscr{P}^{\pi^{-1}(i)}\left(x^{\pi(1)},\ldots,x^{\pi(N)}\right). \tag{14.78}$$

Intuitively, under a symmetric power control policy, the power allocated to a given user is determined by the channel state perceived by that user relative to the channel states perceived by the other users and not on the identity of that user. In [56], the authors consider symmetric channel state and power control. Moreover, they assume Poisson arrivals and exponentially distributed packet lengths. Under these assumptions, they prove that the Longest Queue Highest Possible Rate (LQHPR) policy, besides being throughput optimal, also minimizes delay.

The problem of maximizing the sum throughput subject to constraints on the individual user delays has the structure of a Constrained Markov Decision Process (CMDP). However, the primary difficulty in computing optimal policy lies in large state space size that increases exponentially with the number of users. A simple heuristic would be to compute indices for all users in each slot based on their channel state and queue length. The base station schedules the user with the largest index. The indices are carefully updated so that the delay constraints of the users are satisfied and the system achieves a high sum throughput.

In this chapter, we have considered scheduling algorithms that take advantage of the opportunities provided by the fading wireless channel. These channel aware algorithms are termed as cross layer scheduling algorithms. Most of these algorithms can be considered as control problem where the objective is to optimize a given utility such as throughput, energy, queue stability subject to some constraints. These policies can be computed using dynamic programming tools. However, the curse of dimensionality is a major impediment in determining optimal solutions within this framework. MDP framework also needs knowledge of transition probability mechanism of the underlying Markov chain. In practice, this depends on the fading process and arrival process. In most cases, this knowledge is difficult to possess thereby requiring learning approaches. While significant progress has been made in the scheduling literature, multiuser scheduling still continues to be a challenging problem. In this chapter, we have given few formulations that outlined the nature of problems being considered in the literature.

Bibliographic Notes

An overview of fading in wireless channels has been provided in [44, 45]. The block fading channel model has been suggested in [33]. The Markov model for modeling a Rayleigh channel has been suggested in [55]. An excellent review of information theoretic limits over fading channels can be found in [11]. The waterfilling power allocation policy has been suggested in the work by Goldsmith and Varaiya [19]. Further insights on transmission in multiuser wireless channel has been provided in the works by Knopp and Humblet [23] and Tse and Hanly [50, 51]. The treatment in Section 14.2.2 borrows from [50].

For the multiuser scheduling with symmetric channel fading and symmetric constraints, it has been shown in [23] that the optimal policy schedules the user with the best channel condition in each slot. A generalized version of the problem has been considered in [50] where the authors show that the optimal policy is a greedy *successive decoding* scheme where the users are decoded in an order that is dependent on the interference experienced by them. 'Opportunistic scheduling' has been considered in [30]. Multiuser diversity has been discussed in a lucid way in [52]. Expressions for multiuser capacity with perfect transmitter CSI on the downlink have been derived in [49, 28].

The single user power optimal scheduling with delay constraints discussed in Section 14.5.1 has been formulated in a seminal paper by Berry and Gallager [6]. The paper has also quantified power-delay tradeoff and demonstrates the convexity of the power delay curve. Structural properties of the optimal policy under various assumptions on the channel and arrival processes have been proved in [5, 20, 18, 1, 37]. These problems have the structure of CMDP. [2] is an excellent monograph on CMDP. While there have been a number of studies to explore structural properties of the power optimal policy, numerical computation using dynamic programming is hard. Heuristic algorithms have been developed in [54]. Interestingly, Delay constrained power optimal scheduling has also been considered under non-fading gaussian noise channel [38] since the power-rate convexity relation also holds for additive white gaussian channel.

The learning scheme of Section 14.5.1.3 is from [39]. General references on reinforcement learning algorithms are [10, 47] while [7, 36] can be referred for dynamic programming. The result bounding the number of states at which the optimal stationary policy for a constrained problem takes randomized decisions can be found in [12]. The post-decision state concept used in this chapter was introduced in [53]. Similar ideas have been around for quite some time, see, e.g., [32]. For further details on stochastic approximation, refer to [46, 14, 25]. Techniques used to prove convergence of stochastic approximation schemes as well as multi-time scale and asynchronous stochastic approximation (which can be employed to prove the convergence of a learning scheme of this chapter) have been discussed in [14].

Multiuser power optimal scheduling under delay constraint is a somewhat difficult problem. This problem has been considered for sum power minimization subject to delay constraints in [34]. However, this scenario corresponds to the downlink scheduling. For an uplink case, weighted power minimization with a special case of two users has been studied in [5]. Power minimization of each user on the uplink subject to individual delay constraints has been explored in [41]. The discussion in Section 14.5.2 borrows from [41]. For a general discussion on theory of multiobjective optimization see [17, 42].

Various notions of fairness have been explored (See Chapter 8 [27] and the references therein). The proportional fair scheduler has been proposed in [16]. For a discussion on 'proportional fairness' and associated properties see [21]. Long term sum throughput maximization subject to providing minimum throughput or fraction of slots to users has been variously considered in [29, 15, 4, 31]. Formulation (14.38) has been considered in [29]. Short term fairness has been investigated in [24]. Formulation (14.40) in this chapter is from [24].

Throughput optimal policies have been considered in [50, 35]. LCQ has been suggested in [48], EXP in [43], LWQHPR in [56] and M-LWDF is discussed in [3]. Delay optimality of LQHPR policy has been proved in [56]. The indexing heuristic of Section 14.6.2 has been proposed in [40].

References

1. Agarwal, M., Borkar, V.S., Karandikar, A.: Structural Properties of Optimal Transmission Policies over a Randomly Varying Channel. IEEE Transactions on Automatic Control **53**(6), 1476–1491 (2008)
2. Altman, E.: Constrained Markov Decision Processes. Chapman and Hall/CRC Press, Boca Raton, FL (1999)
3. Andrews, M., Kumaran, K., Ramanan, K., Stolyar, A., Whiting, P., Vijayakumar, R.: Providing Quality of Service over a Shared Wireless Link. IEEE Communications Magazine **39** (1996)
4. Berggren, F., Jantti, R.: Asymptotically Fair Transmission Scheduling over Fading Channels. IEEE Transactions on Wireless Communication **3**(1), 326–336 (2004)
5. Berry, R.: Power and Delay Trade-offs in Fading Channels (2000). PhD Thesis, Massachusetts Institute of Technology
6. Berry, R.A., Gallager, R.G.: Communication over Fading Channels with Delay Constraints. IEEE Transactions on Information Theory **48**(5), 1135–1149 (2002)
7. Bertsekas, D.P.: Dynamic Programming and Optimal Control, vol. 1. Athena Scientific, Belmont, MA (1995)
8. Bertsekas, D.P.: Nonlinear Programming. Athena Scientific, Belmont, MA (1999)
9. Bertsekas, D.P., Gallager, R.: Data Networks. Prentice Hall (1987)
10. Bertsekas, D.P., Tsitsiklis, J.N.: Neuro-Dynamic Programming. Athena Scientific, Belmont, MA (1996)
11. Biglieri, E., Proakis, J., Shamai, S.: Fading Channels: Information-Theoretic and Communications Aspects. IEEE Transactions on Information Theory **44**(6), 2619–2692 (1998)
12. Borkar, V.S.: Convex Analytic Methods in Markov Decision Processes. In: E. A. Feinberg, A. Schwartz (Eds.) Handbook of Markov Decision Processes, pp. 347–375. Kluwer Academic Publishers, Dordrechet (2001)
13. Borkar, V.S.: An Actor-Critic Algorithm for Constrained Markov Decision Processes. Systems and Control Letters **54**, 207–213 (2005)
14. Borkar, V.S.: Stochastic Approximation - A Dynamical Systems Viewpoint. Hindustan Publishing Agency and Cambridge University Press (2008)
15. Borst, S., Whiting, P.: Dynamic Channel-Sensitive Scheduling Algorithms for Wireless Data Throughput Optimization. IEEE Transactions on Vehicular Technology **52**(3), 569–586 (2002)

16. Chaponniere, E.F., Black, P., Holtzman, J.M., Tse, D.: Transmitter directed Multiple Receiver System using Path Diversity to Equitably Maximize Throughput (2002). U.S. Patent No. 6449490
17. Coello, C.A.C.: A Comprehensive Survey of Evolutionary-Based Multiobjective Optimization Techniques. Knowledge and Information Systems **3**(1), 269–308 (1999)
18. Djonin, D.V., Krishnamurthy, V.: Structural Results on Optimal Transmission Scheduling: A Constrained Markov Decision Process Approach. The IMA Volumes in Mathematics and its Applications **143** (2007)
19. Goldsmith, A., Varaiya, P.: Capacity of Fading Channels with Channel Side Information. IEEE Transactions on Information Theory **43**(6), 1986–1992 (1997)
20. Goyal, M., Kumar, A., Sharma, V.: Power Constrained and Delay Optimal Policies for Scheduling Transmissions over a Fading Channel. In: Proceedings of IEEE INFOCOM, vol. 1, pp. 311–320 (2003)
21. Kelly, F.: Charging and Rate Control for Elastic Traffic. European Transactions on Telecommunications **8**, 33–37 (1997)
22. Kelly, F.P., Maulloo, A.K., Tan, D.K.H.: Rate Control for Communication Networks: Shadow Prices, Proportional Fairness and Stability. The Journal of the Operational Research Society **49**(3), 237–252 (1998)
23. Knopp, R., Humblet, P.A.: Information Capacity and Power Control in Single-Cell Multiuser Communications. In: Proceedings of IEEE ICC, vol. 1, pp. 331–335. Seattle, USA (1995)
24. Kulkarni, S., Rosenberg, C.: Opportunistic Scheduling Policies for Wireless Systems with Short Term Fairness Constraints. In: Proceedings of IEEE GLOBECOM, vol. 1, pp. 533–537 (2003)
25. Kushner, H.J., Yin, G.G.: Stochastic Approximation Algorithms and Applications. Springer-Verlag, New York (1997)
26. LAN/MAN Committee: IEEE 802.16-2004: IEEE standard for Local and Metropolitan Area Networks - Part 16: Air Interface for Fixed Broadband Wireless Access Systems. IEEE Computer Society (2004)
27. Lau, V.K.N., Kwok, Y.K.R.: Channel-Adaptive Technologies and Cross-Layer Designs for Wireless Systems with Multiple Antennas. Wiley-Interscience (2006)
28. Li, L., Goldsmith, A.: Capacity and Optimal Resource Allocation for Fading Broadcast Channels: Part I: Ergodic Capacity . IEEE Transactions on Information Theory **47**(3), 1083–1102 (2001)
29. Liu, X., Chong, E., Shroff, N.: Opportunistic Transmission Scheduling with Resource-Sharing Constraints in Wireless Networks. IEEE Journal on Selected Areas in Communication **19**(10), 2053–2064 (2001)
30. Liu, X., Chong, E., Shroff, N.: Transmission Scheduling for Efficient Wireless Resource Utilization with Minimum-Performance Guarantees. In: Proceedings of IEEE VTC'2001, vol. 2, pp. 824–828 (2001)
31. Liu, Y., Knightly, E.: Opportunistic Fair Scheduling over Multiple Wireless Channels. In: Proceedings of IEEE INFOCOM, vol. 2, pp. 1106–1115. San Francisco, USA (2003)
32. Magirou, V.F.: Stockpiling under Price Uncertainty and Storage Capacity Constraints. European Journal of Operational Research pp. 233–246 (1982)
33. McEliece, R., Stark, W.: Channels with Block Interference. IEEE Transactions on Information Theory **30**, 44–53 (1984)
34. Neely, M.J.: Optimal Energy and Delay Tradeoffs for Multi-user Wireless Downlinks. In: Proceedings of IEEE INFOCOM, pp. 1–13 (2006)
35. Neely, M.J., Modiano, E., Rohrs, C.: Power and Server Allocation in Multi-beam Satellite with Time Varying Channels. In: Proceedings of IEEE INFOCOM, pp. 138–152. San Francisco, USA (2002)
36. Puterman, M.: Markov Decision Processes. Wiley, New York (1994)
37. Rajadhyaksha, G., Borkar, V.S.: Transmission Rate Control over Randomly Varying Channels. Probability in the Engineering and Informational Sciences **19**(1), 73–82 (2005)
38. Rajan, D., Sabharwal, A., Aazhang, B.: Delay-Bounded Packet Scheduling of Bursty Traffic over Wireless Channels. IEEE Transactions on Information Theory **50**(1), 125–144 (2004)

39. Salodkar, N., Bhorkar, A., Karandikar, A., Borkar, V.S.: An On-Line Learning Algorithm for Energy Efficient Delay Constrained Scheduling over a Fading Channel. IEEE JSAC, Special Issue on Control and Communications **26**(4), 732–742 (2008)
40. Salodkar, N., Karandikar, A.: An Indexing Scheduler for Delay Constrained Scheduling with Applications to IEEE 802.16. In: Proceedings of IEEE WCNC'2008, pp. 1471–1476 (2008)
41. Salodkar, N., Karandikar, A., Borkar, V.S.: A Stable On-line Algorithm for Energy Efficient Multi-user Scheduling. URL http://www.ee.iitb.ac.in/ karandi/pubs_dir/submitted/uplink_tmc_revised_july2009.pdf. To appear in IEEE Transactions on Mobile Computing, 2010
42. Sawargi, Y.: Theory of Multiobjective Optimization. Academic Press, Orlando (1985)
43. Shakkottai, S., Stolyar, A.: Scheduling for Multiple Fows Sharing a Time-varying Channel: the Exponential Rule. AMS Translations Series 2 **207** (2002)
44. Sklar, B.: Rayleigh Fading Channels in Mobile Digital Communication Systems. I. Characterization. IEEE Communications Magazine **35**(7), 90–100 (1997)
45. Sklar, B.: Rayleigh Fading Channels in Mobile Digital Communication Systems. II. Mitigation. IEEE Communications Magazine **35**(7), 102–109 (1997)
46. Spall, J.C.: Introduction to Stochastic Search and Optimization. Wiley (2003)
47. Sutton, R.S., Barto, A.G.: Reinforcement Learning: An Introduction. MIT Press, Cambridge MA, USA (1998)
48. Tassiulas, L., Ephremides, A.: Dynamic Server Allocation to Parallel Queues with Randomly Varying Connectivity. IEEE Transactions on Information Theory **39**(2), 466–478 (1993)
49. Tse, D.: Optimal Power Allocation over Parallel Gaussian Broadcast Channels. In: Proceedings of IEEE ISIT, p. 27 (1997)
50. Tse, D., Hanly, S.: Multi-access Fading Channels - Part I: Polymatroid Structure, Optimal Resource Allocation and Throughput Capacities. IEEE Transactions on Information Theory **44**(7), 2796–2815 (1998)
51. Tse, D., Hanly, S.: Multi-access Fading Channels - Part II: Delay-Limited Capacities. IEEE Transactions on Information Theory **44**(7), 2816–2831 (1998)
52. Tse, D., Viswanath, P.: Fundamentals of Wireless Communication. Cambridge University Press (2005)
53. W. B. Powell: Approximate Dynamic Programming: Solving the Curses of Dimensionality. Wiley-Interscience (2007)
54. Wang, H., Mandayam, N.: A Simple Packet Transmission Scheme for Wireless Data over Fading Channels. IEEE Transcations on Communications **52**(7), 1055–1059 (2004)
55. Wang, H.S., Moayeri, N.: Finite-State Markov Channel – A Useful Model for Radio Communication Channels. IEEE Transactions on Vehicular Technology **44**(1), 163–171 (1995)
56. Yeh, E., Cohen, A.: Delay Optimal Rate Allocation in Multiaccess Fading Communications. In: Proceedings of IEEE Workshop on Multimedia Signal Processing, pp. 404–407 (2002)

Chapter 15
Major Trends in Cellular Networks and Corresponding Optimization Issues

Khiem Le

Abstract Cellular networks are entering a new phase, driven by some major evolutionary trends. With the boom in mobile data access, driven by attractive mobile Internet applications and more capable mobile devices, global mobile data traffic is expected to grow by orders of magnitude in the next five years (projections range from a ten-fold increase to a hundreds-fold increase). The Mobile Network Operators (MNOs) will incur costs to upgrade their networks to cope with that growth, but the revenues will not scale in the same proportion. To remain profitable, the MNOs have to reduce their cost per byte while increasing their revenues. There are many possible tradeoff points, depending on such considerations as the MNO's business strategy, and operating environment. Even for a given tradeoff point, the problem does not lend itself to a tractable formulation, due to the sheer number of contributing factors, the complexity of their interactions, and unavailability of complete data. However, it is possible to identify meaningful sub-problems that are easier to solve. This chapter focuses on some of the approaches and techniques that could be part of a toolbox to tackle these sub-problems. Another major trend is the expansion of mobile access (to voice and data, not necessarily broadband) to low income population segments. While this trend significantly overlaps with the one described above, the emphasis here is to provide affordable service rather than providing mobile broadband data. Optimization in the low income space is also intractable in general, and this chapter describes approaches for some particular cases.

15.1 Introduction

Cellular networks are now entering a new phase, driven by some major evolutionary trends. Mobile technologies such as GSM (Global System for Mobile Communications) have truly altered people's lifestyles and expectations all over the world.

Khiem Le
Research In Motion, 5000 Riverside Drive, Irving, TX 75039, USA, e-mail: khiem.le@ieee.org

J. Kennington et al. (eds.), *Wireless Network Design: Optimization Models and Solution Procedures*, International Series in Operations Research and Management Science 158,
DOI 10.1007/978-1-4419-6111-2_15,

People now expect to be able to connect while on the move, everywhere. While the basic GSM type voice connectivity initially provided the convenience of mobility and ubiquity to people who already have voice access, it now has grown to be a lifeline to people who have never had access to voice connectivity, e.g., in some rural areas of developing countries. In addition, mobile data (not necessarily broadband) technologies enable first-time access to the Internet and help to bridge the digital divide. Socioeconomic benefits include better access to healthcare and education, greater ability to buy/sell services and goods and make financial transactions and greater entrepreneurial activity [23]. However, there is a significant fraction of the global population whose level of income is so low that affordability is still a challenge [23].

In parallel, in the developed markets, the convergence of Internet and mobility has become a reality, and mobile networks are transitioning very quickly to offer broadband data connections. This has been enabled by progress in wireless access technologies and mobile terminal implementations. Wireless access technologies have improved so much that their throughput and latency are now comparable to traditional fixed broadband access technologies such as cable or DSL. With advances in implementation technologies, mobile terminals are capable of providing a viable Internet experience for traditional Internet applications and services in a portable form factor. In addition, a new category of uniquely mobile applications, synergistically blending the phone's camera, GPS, display, various sensors and Internet connectivity has emerged. Examples include location-based services, context-aware applications (e.g., where is the nearest ATM? Is one of my friends currently in my geographical area?) and even social augmented reality, where the virtual world is superimposed on the physical world. For example, a user can tag a location or thing in the physical world (e.g., a restaurant) with a marker to share information or personal experience (e.g., a restaurant review) with friends. When the friend is in the same location, the marker can point to email, photos, URLs, etc. Many of these new applications provide value added addressing specifically the needs of users on the move, and contribute to driving the subscriber population growth.

Looking forward, the broadband wireless Internet is expected to spread to ever larger populations as the technologies become more capable and terminal prices drop. Additionally, ubiquitous and lower cost wireless connections will likely be a catalyst for new applications and usage scenarios, such as connection of appliances and machines leading to machine-to-machine communications and the " Internet of things". The wireless network operators will clearly have to upgrade their access and network infrastructures to cope with the anticipated explosive growth in data traffic volume, which translates into increased capital and operating expenses. At the same time, operators will experience enormous pressure on the revenue side; with data applications having a traffic volume orders of magnitude higher than for voice, they cannot scale linearly. That is, they cannot simply charge the same amount per byte for data and voice subscriptions. In some markets, such as the U.S., users even expect a flat rate billing based on the fixed Internet model rather than volume-based billing from the voice model.

Thus if we take the network operator's profit as a performance metric, there are two main optimization spaces which are somewhat overlapping. One space is first-time access to voice and Internet for low income populations. Another space is optimization in the context of broadband wireless Internet. Although the general problems are intractable, they can be decomposed into optimization sub-problems that are easier to solve. The profit metric in many cases is correlated with a higher service penetration rate, i.e., more widespread access to mobile communications with ensuing socioeconomic benefits. For example, in its Global Mobile Tax Review 2006-2007, the GSMA found that "penetration boosts economic activity. In developing countries a 10% increase in penetration leads to a 1.2% increase in the annual growth rate in GDP." [25]

This chapter focuses on some of the optimization issues, approaches and techniques that could be part of a toolbox to tackle the general problem. Section 2 reviews the current situation in wireless networks, and Section 3 provides an overview of some solution approaches. Sections 4 through 8 describe in more detail some selected approaches. Section 9 concludes the chapter.

15.2 Current Situation

The two main factors impacting an operator's profit are the costs and revenues. Cost is more precisely total cost of ownership (TCO) consisting of Capital Expenses (CAPEX) and Operating Expenses (OPEX). CAPEX are one-time expenses and include items such as cost to acquire the spectrum, and cost to buy and deploy the network infrastructure. OPEX are recurring expenses and correspond to items such as staff salary, rental of facilities, and electricity bills. Revenues mostly stem from providing wireless voice and data access services. They could also come from commercial scale/grade products such as bundled voice, web hosting, VPNs, mail, conference and messaging facilities offered to government and corporate clients.

Many, if not most operators, have to deal with multiple generations of equipment and technologies from different vendors that coexist and interoperate, and this makes the network costly to operate and manage. Capacity expansion and growth becomes more and more complex and costly as each new wave of expansion has to take into account the previous systems. On top of that, current network architectures and technologies still carry the legacy from the voice centric, circuit switched networks. That is, they are well suited to carry voice traffic, which is low volume, and incur the limitations of older technologies. A good example is the backhauling technology, based on T1/E1 trunks, used to transport the traffic between the base stations and the core network. Since the T1/E1 are circuit switched, they are not well optimized for the packet traffic of Internet access as there can be no efficiency gain from statistical multiplexing. Furthermore, the network operators commonly lease the T1/E1, and the leasing fees are proportional to the bandwidth. Thus, it does not scale well with the envisioned broadband data traffic which is orders of magnitude higher than voice. Clearly, cost of mobile backhauling is an issue and a new paradigm is needed to address it. Another long standing cost contributor is the spectrum. As licenses cost hundreds of millions or billions of dollars, operators make every effort to use the spectrum as efficiently as possible.

Convergence with the Internet is a major trend that affects the MNO's revenues [26]. The "Web2.0" with players such as YouTube, Facebook and MySpace is growing fast, and using business models based on indirect revenues rather than revenues from direct charging for usage. These business models challenge Communication Service Providers (CSPs) traditional user fee based revenue model. (In this chapter, CSPs is a generic term that also includes wireline and hybrid operators.) The Web 2.0 Internet players are diversifying their offerings and are starting to cannibalize not only CSPs core business, such as voice services, but also new growth opportunities in areas such as IPTV. According to estimates, in some markets about half of users time spent on communication services is already occupied by Internet based messaging and VoIP services. Web video and TV platforms are posing as competition to IPTV. As the boundary between fixed and mobile access gets more fuzzy with the technical advances of wireless technologies, the cannibalization trend also takes place for MNOs. The appeal and variety of usage scenarios for Instant Messaging (IM) in a Web 2.0 world is a threat to profitable Short Message Service (SMS) and voice business. Google's mobile operating system "Android" is an attempt to gain direct access to mobile consumers, bypassing MNOs. Google has introduced Google Voice, a service offering cheap voice calls that supports multiple features including allowing a user to access his voicemail online, read automatic transcriptions of his voicemail, and create personalized greetings based on the caller.

Some MNOs have been very successful at gaining new subscribers by offering mobile devices with an attractive menu of applications and user friendly interface for which they have exclusive rights. An example is AT&T with the iPhone. The downside is that these applications tend to generate a disproportionate amount of traffic on the MNO's network. AT&T has experienced a 50-fold mobile data traffic increase from 2006 to 2009 [19]. This example illustrates the challenge the MNOs are facing as the traffic explodes, but the revenues do not grow in the same proportion.

Network operators' core assets include the subscribers identity infrastructure [5] and the capability to issue billing to the users. The subscribers identity infrastructure corresponds to the subscribers' identities and associated security credentials in the operators' databases along with the existing network mechanisms to authenticate subscribers and control their access to the network services, and the Subscriber Identity Modules (SIMs). SIMs are by far the most widely deployed secure devices in the world, and MNOs master the process of managing the lifecycle of SIMs. Additionally, roaming agreements between operators provide a globally interconnected "network of trust" [5]. The subscribers identity infrastructure is truly an asset unique to the mobile operators, considering that the population of mobile subscribers now exceeds 4 Billion [4] and is growing. No other industry can boast such a large network of trust. Closely linked to the subscribers identity infrastructure is the network operators' capability to record a subscriber's usage down to a fine granularity (e.g., minutes or seconds of use), consolidate and generate a bill.

For low income populations, affordability is key. The relevant metric is user TCO, which is the total cost to buy and use a mobile device. This includes the price of the mobile device and the price of the service. An analysis of global income [25] shows that 2.7bn people earn 2 USD or less per day, and have a budget of 3 USD per month or less to spend on Information and Communications Technology (ICT). Slightly higher up the income scale, we find that 2.8bn people have an income of between 4 USD and 40 USD per month. Yet today the average TCO for the lower income consumer is about 13 USD [21]. Clearly, the operators need to lower their own TCO to be in a position to offer lower-priced services and still be profitable. Innovation in the business model and cost structure is called for.

15.3 Overview of Solution Approaches

This section provides an overview of the most significant solution approaches, but is not intended to be exhaustive. Subsequent sections provide more details on a selected subset of solutions. In general, solutions attempt to reduce costs and/or increase revenues.

The first challenge is the explosive growth in data traffic. As mobile data traffic is expected to grow from 1.3 Exabytes in 2008 to 19.2 Exabytes in 2014 [11], and users expect higher throughput and lower latency, an incremental upgrade will not be enough. Operators will have to acquire new spectrum and overhaul the network infrastructure from the base station all the way to the core network, including the transport infrastructure. Furthermore, new technologies such as LTE (Long Term Evolution) will be needed to achieve a quantum jump in the throughput and latency compared to 3G [3].

A near term solution to alleviate the spectrum capacity problem is to use femtocells. Femtocells are miniature cellular base stations, with relatively short radio range, deployed in the home to provide regular cellular service to the users in the household. Traffic to/from the femtocell is backhauled through the home's fixed broadband connection (xDSL, cable, etc.) to the cellular core network [16]. It should be noted that femtocells offer other benefits such as better radio quality for the user, and lower power drain on the mobile terminal. To take the U.S. as an example, operators such as AT&T [10] and Verizon [36] are starting to offer femtocells.

Unlicensed mobile access (UMA) is another near term approach to alleviate the spectrum capacity problem by offloading the cellular traffic to a WiFi access point in the home [2]. As with Femtocells, UMA traffic is backhauled on the home's fixed broadband connection to the cellular core network. UMA-based service is currently offered in the U.S. by T-Mobile.

A generalization of Femtocells and UMA is Smart Connectivity (SC). SC enables seamless access to services in heterogeneous networks across a variety of access systems (2G, 3G, LTE, WIMAX, WiFi, xDSL, PON, etc.). As the user roams, Smart Connectivity determines what accesses are available and provides the ability to select the optimal one for the best user experience and most efficient use of resources without requiring user intervention [28]. SC adds intelligence in the network and in the terminal and can be initially implemented as an overlay. The general architecture is shown in Fig. 15.1. SC can help to offload the traffic from cellular access to other access such as WiFi. This will help conserve the spectrum and mitigate the cellular network capacity upgrade requirements. SC also helps to improve the user experience by providing better ubiquity of access and continuity of communication in a manner transparent to the user.

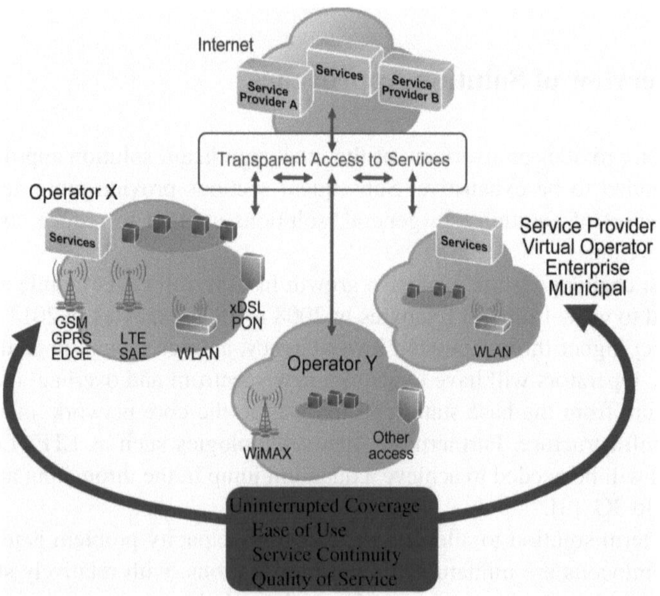

Fig. 15.1 Smart Connectivity (Courtesy of Nokia Siemens Networks)

Concurrently with expanding capacity for traffic growth, operators have to drastically decrease their cost per byte to be able to make a profit as the billing model for data is based more on an "all-you-can-eat" flat rate. New approaches and paradigms are needed. In the following sections we will look at the specific issues of mobile backhauling and reduction of CAPEX and OPEX through automation and built-in efficiency in the technology design.

Some operators have started to outsource the day-to-day management of their network to vendors with the objective of reducing their OPEX. Examples are Embarq with Nokia Siemens Networks, and Sprint with Ericsson. That trend will likely continue with the expansion to mobile broadband. MNOs can protect their traditional voice revenues and establish their role in the Internet value chain by exploiting their assets. In addition to the subscriber identity infrastructure, MNOs typically own the access and core infrastructure. They also have direct control of the network's quality parameters and know a customer's actual status and profile. Combining these assets can produce a massive improvement in convenience and personalization for the user [26]. Many providers offer multi-play packages consisting of high speed Internet access, voice, which can be either classic voice or VoIP, mobile telephony, and TV services. The packages usually give the user the advantage of one-stop-shopping, a perceived discount on his entire monthly telecommunication and TV spending, and cost control due to a single bill. Going beyond single billing, some operators are leveraging their control of fixed and mobile infrastructures to offer fixed-mobile integrated services, such as Converged TV. With a Converged TV subscription, a user can access TV services anywhere, any time, from any device (TV set, PC, Mobile device). Integration enables features such as session continuity, e.g., stopping a video session on the mobile device and continuing on the TV set, and common user interface. Following the success of the Apple App Store (two billion downloads less than 14 months after its commercial launch in July 2008 [34]), a growing number of MNOs are hoping to cash in on the application storefront model as a new revenue source.

Some operators also want to get in the mobile cloud computing game — where the computing and storage are moved away from the mobile devices into the network cloud — making the apps and user data available to a wide range of user devices. Cloud computing is the basis of popular services such as web mail and Blackberry's email services, hosted photos (Flickr, Photobucket), media streaming (YouTube), social networking sites (Facebook, MySpace), etc. The concept has been extended in new services like Apple's MobileMe, Microsoft's MyPhone and Google Sync, which allow users to synchronize the information on the mobile device and the PC by hosting it on the cloud which is accessible from any device. An example of an operator's mobile cloud service is Vodafone's 360 which lets users sync contacts, emails, photos, conversations, music and location-based services via mobile device and PC [37].

A more expansive view of the cloud is "platform as a service (PaaS)", where customers can build and run their own apps on a cloud platform. PaaS is about to go mobile with the advent of mobile broadband access as the key enabler. Smartphone versions of Google's App Engine and Amazon Web Services (AWS) are expected in 2010 [34].

According to ABI Research, "By 2014, mobile cloud computing will become the leading mobile application development and deployment strategy, displacing today's native and downloadable mobile applications." The number of subscribers will be close to 1 Bn [7].

With mobile cloud, applications and content are no longer dependent on the mobile device's operating system (OS). Application developers do not have to face the dilemma of writing an application for only one OS, or writing multiple versions of the same application. Furthermore, less-capable handsets can have access to sophisticated applications running in the network. Mobile cloud also offers new service opportunities: location-based services, business apps such as collaborative document sharing, scheduling, and sales force management apps, and remote control.

One of the most promising avenues for operators is to leverage their intimate knowledge of the customer (preference profile, location, etc.), their trust relationship with the customers, and control of network capabilities to provide optimized user experience, through personalized services, based on excellent navigation and location-based information. The operator can capitalize on the trust relationship to offer services related to payment (e.g., the operator can consolidate a third party provider's bill with the mobile subscription bill) or security (e.g., provide parental control of access to applications).

To be able to offer an attractive menu of services, operators have to open up their network and present APIs to third party application providers or application developers. Efforts are under way to define standard APIs, such as the GSM Association's OneAPI [30] reference implementation that attempts to create common web-friendly APIs for integrating apps with MNO network capabilities (charging, messaging, location and user context).

Many operators are adopting IT-style concepts and techniques such as Service Delivery Framework (SDF), Service Delivery Platform (SDP) and Service Oriented Architecture (SOA), to more efficiently in provide interfaces to applications, and open software development kits for third party application and service development, through support of distributed computing and easy reuse of components and functions.

In Section 15.7, we will look more in depth at how the MNOs can leverage their unique subscriber identity infrastructure asset to go beyond connectivity services and generate value added in the Internet space. For the low income population case, the key is to reduce the operator's TCO. With lower costs, the CSP is in a stronger position to offer services more affordable to its customers and still remain profitable in its business. Complementary to reducing the cost of connectivity services, the operator could develop value added services on top of connectivity specifically geared towards the needs of low income people. We will describe some approaches based on a mix of innovative technology and business models to minimize the CSP's OPEX and CAPEX, and/or offer new services.

15.4 New Wireless Access Technologies

New technologies are designed to better address many of the above challenges. In particular, they attempt to improve user experience (higher data rates and lower round trip time), spectral and network architecture efficiency. As an example, LTE, the evolution of the GSM technology family towards 4G, boasts the following characteristics [1, 9]:

- Much improved user experience: Peak data rates in the range of 100-300 Mbps and 50-80 Mbps for downlink and uplink respectively, and packet round trip time reduced to 10 msec

- Reduction of OPEX and CAPEX: Flattening of the architecture and reduction of the number of network elements. Maintenance and operation of fewer elements will translate to lower cost
- Improved spectral efficiency through advanced techniques like Orthogonal Frequency Division Multiple Access (OFDMA), higher order modulation, channel dependent scheduling and time/frequency diversity
- Scalability of the frequency band also contributes to efficient usage of the spectrum, by enabling to choose the frequency bandwidth that best fits the available spectrum
- Maximized reuse of the infrastructure. LTE provides a packet core common to the various wireless access, as well as built-in mechanisms for inter-working and handover between these access systems. In particular, included are "legacy" access systems such as 2G and 3G. A common core will also help to reduce the OPEX and CAPEX.

In addition, an integral part of LTE design is Self Organizing Networks (SON). SON encompasses Self Configuration, Self Optimization, and Self Healing. The target of SON is to minimize human intervention in the tasks of configuration (e.g., when a new element such a as base station is put in service), optimization (e.g., when network element configuration parameters are readjusted during network operation to account for imperfections in the network planning and traffic evolution) and healing (outage detection and compensation). Besides its direct OPEX lowering benefit, SON's indirect benefit is to minimize the chance of human errors and the associated indirect costs to the network operator's image. However, SON's target is very ambitious, and it remains to be seen how much automation can be achieved, for example in the area of optimization. The next major step in evolution is LTE-Advanced, which aims at peak data rates around 1Gb/sec in the downlink, and 500 Mbps in the uplink, by using advanced techniques such as carrier aggregation, and coordinated multi-point transmission reception [6].

Another example is WIMAX 802.16. WIMAX, initially developed by IEEE as a wireless substitution to fixed broadband access (802.16-2004), was enhanced to be able to handle mobile users (802.16-e). The various versions of WIMAX are following the same evolutionary trend as the GSM family towards higher data rates, lower round trip time, improved spectral efficiency, frequency scalability and reduced OPEX and CAPEX [38]. IEEE 802.16m is the next major milestone in the WiMAX standards evolution beyond IEEE 802.16e-2005 [39].

15.5 Backhauling and Transport

Mobile backhauling is traditionally based on TDM E1/T1 lines, each providing 1.5 or 2.0 Mbps capacity. With mobile broadband such as LTE, a speed higher than 100 Mbps is required on the backhaul transport. Such speed can in principle be achieved by bundling a large number of E1/T1 lines, but as the leasing costs are proportional to the number of lines, that does not scale well. Technologies that can achieve speeds sufficient for mobile broadband are point-to-point Carrier Ethernet, Passive Optical Networks (PON) or microwave radio (MWR) [14, 17, 18, 20].

Microwave radio links today already account for more than 50% of wireless access transport solutions. Their advantage is flexibility and fast rollout, and a low OPEX. New technologies enable microwave links to extend transport capacity to 180 Mbps. Further capacity improvements may be possible, e.g., by using additional available capacity in unlicensed frequency bands. Microwave remains the most viable option where geographical and economic constraints preclude the use of wireline transmission media. Wireless backhauling solutions are able to address both TCO and operating efficiency. This is particularly important in emerging markets where cost is the primary concern.

In-band relaying employs frequency bands that have been assigned already. It is an attractive option because it does not require additional frequencies, but the requisite amount of capacity has to be reserved for the backhauling. On top of that, packetized transport solutions can achieve higher efficiency, as they can benefit from statistical multiplexing (Fig. 15.2).

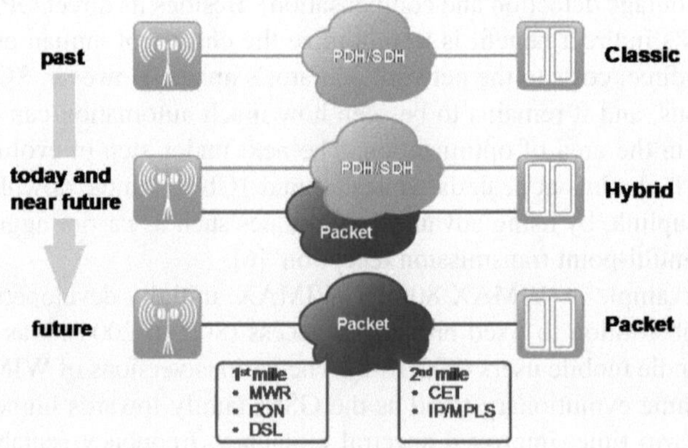

Fig. 15.2 Backhaul Evolution (Courtesy of Nokia Siemens Networks)

Backhauling is part of the more general area of transport. Many MNOs are also fixed network operators, and thus own a transport network carrying fixed traffic. Integration of all traffic, fixed and mobile, onto a single transport network, and streamlining of that network can significantly reduce the transport related costs for these operators. For historical reasons, there are four main infrastructures: the Internet for data transport, the telephone network for voice communication, the backbone network for mobile cellular communication, and the television content distribution network. As all media is digitized and packetized for transmission, these infrastructures are converging. Technologies such as adaptive optical core, Carrier Ethernet Transport, IP and Layer 2, multilayer networking can help bring down the transport costs [12].

With the massive increase in traffic, optical technology and fiber optics is clearly required in the backbone, where traffic is aggregated. In addition, a medium- to long term goal of many carriers is to deploy fiber in the last/first mile to enable the delivery of robust multimedia services, e.g., IPTV, Video on Demand, and Triple/Quad play. The main trends that can help improve the transport efficiency include [24]:

- The metro and access parts of today's networks will converge into a single network. An aggregation layer collects traffic from different sources and transfers it to the photonic core
- Hybrid architecture having transparent islands connected by opaque nodes that employ Optical/Electrical/Optical (OEO) conversion. Within the transparent islands packet headers may be processed electronically, but payloads are switched optically to minimize inefficient OEO conversions
- Further advances in the optical technology itself
- Flexible optical transport networks that can react quickly to external factors, such as (rapid) traffic changes or changes of the physical infrastructure. High traffic demand will result in the automatic addition of resources, either on a wavelength or on sub-wavelength level. In the event of failure in the network, the traffic will be automatically re-routed
- Packet-based architectures
- Long range transparent optical paths, enabled by more robust modulation techniques. The optical paths can potentially be end-to-end, which leads to new architectural approaches. For instance, network sites may be reduced, which can help to decrease the costs
- Automatic Switched Optical Network (ASON) will provide automatic and self-organized set-up and tear down of optical link requests.

While IP will be the common layer for applications, some of its functions and protocols are not needed for the efficient transport of data. Instead, simpler packet-based forwarding technologies, in combination with optical transport, will enable the cost-efficient transmission of large data volumes. Carrier Ethernet is the Layer 2 technology that will provide the common layer for packet-based transport, and the reduction of TCO will be achieved by an optimized split of functionality between different layers.

Network architectures are migrating to a three-layer model: Optical at Layer 1, Carrier Ethernet at Layer 2 and IP/MPLS at Layer 3. Optimizing the interaction of these three layers, i.e., by applying the concept of multi-layer optimization, can significantly reduce TCO. Multi-layer optimization means assigning functionality and traffic processing to the layer where it is technically and economically optimal [13].

In this target architecture, cost minimization is achieved since traffic is forwarded on the layer for which it is best suited, and unnecessary processing is avoided. For example, highly aggregated traffic can be carried by optical transport services using Reconfigurable Optical Add/Drop Multiplexer (ROADM) or Optical Cross- Connects (PXC) technology. The latter can switch without requiring optical/electrical/optical conversion. The layer that best meets the requirements of a service can be chosen. For example, 1+1 protection on the optical layer can be used for services needing fast recovery times below 50 msec [24].

In practical situations, one has to take into account the existing infrastructure, and the problem can be formulated as cost optimization to meet some traffic flow requirements under the constraints of existing infrastructure and operating constraints. The problem of determining what type of network element to place where, and in what quantity, lends itself to discrete optimization for which known algorithms apply.

15.6 Power Efficiency

Network energy efficiency means lower OPEX. In mature markets, power consumption can contribute as much as 10% of network OPEX, whereas in developing markets, it can contribute from 15% to 30%. According to [22], 86% of a MNO's energy consumption occurs in the network infrastructure.

In addition, networks are expanding into areas where the electricity grid is not available, or the grid quality is insufficient. In developing countries, almost a quarter of new cell sites will be off the power grid [22], and with the unpredictable cost of fuel as well as the location of these sites, there are significant financial benefits from renewable power solutions in addition to improved energy efficiency of equipment or technology. Solutions for "off grid" or "bad grid" sites can be based on solar panels and wind turbines [8, 15, 22].

The CSPs' energy OPEX can be optimized, generating energy savings and CO2 reductions, through approaches such as [22]

- Optimized field maintenance and use of batteries and generators
- Regular power source performance monitoring
- Energy control and management systems. They are connected to all power sources (grid, battery, generator, solar system, wind turbine) and smartly manage (through intelligent algorithms) the battery charging and discharging to optimize fuel consumption. They also enable remote monitoring, remote data collection and control
- Infrastructure lifecycle management
- Improved base station design, allowing fresh air cooling to be used instead of air conditioning. Software functions can also be used to balance energy consumption with traffic flows. Base stations can be set to save power during low usage times, such as at night, and by putting the downlink transmission into idle mode when possible.

In addition, energy efficiency will also help to reduce the emission of CO2 [15, 22].

15.7 Leveraging Assets to Protect and Increase Revenues

In saturated markets, MNOs cannot add new customers, and flat rate pricing for mobile data won't help much in increasing revenues. MNOs have to explore new revenue opportunities from services going beyond pure connectivity. Tomorrow's business models will continue to be based on interactions between two partners (people, machines or service providers) and identities are required for online interactions. Trust and security will be key, especially in market place transactions. Subscriber identity infrastructure, along with an intimate knowledge of the subscribers' status and profiles is a powerful combination of assets that can be leveraged by the MNOs to provide identity management services. There are three ways MNOs can position their services [5]:

- Provider of authentication services
- Full fledged identity provider
- Authentication and application enabler

As provider of authentication services, an MNO may offer economically attractive SIM-based user authentication solutions to other service or identity providers, leveraging the MNO's infrastructure and its circle of trust. A full fledged identity provider MNO will not only address the problem of user authentication, but, with the user consent, store and distribute the user's attributes to services providers. Attributes may be generated by the MNO, e.g., they can be location or presence attributes, or user defined. As an authentication enabler, an MNO rents SIM real estate and helps the service or identity provider provision its own security elements on the SIM card. This is more suited for cases where the provider needs to take more ownership of its users' authentication or authorization, due to high level of risk or significant liability. This type of business model is usually deployed with contact-less (Near Field Communication) applications. The MNO is responsible for providing a secure environment on the SIM card to host the NFC applications, and for partitioning the SIM card to isolate applications belonging to different providers.

However, a number of different identity management frameworks coexist today on the Internet: OpenID, CardSpace, SAML, WS-federation, etc., and to maximize the accessible market size, MNOs will have to support a large range of protocols. An idea to explore is to have a unified architecture combining multiple front-end Internet identity management technologies with a single backend identity management relying on the already, or soon-to-be, deployed cellular subscriber identity management infrastructure. Components of the Generic Authentication Architecture (GAA) can be serious candidates to be used as the common backend system. For example, with the Generic Bootstrapping Architecture (GBA), a MNO can leverage its security infrastructure including the AUC and Universal Integrated Circuit Card (UICC) to bootstrap the setup of a security association between a User Equipment (UE) and a relying party. Benefits of a unified architecture include:

- A secure (through strong 3G authentication mechanisms) , but simple sign-on end user experience
- Personalized and privacy preserved services to all subscribers on any device and any network (mobile and fixed)
- Common authentication/authorization functionality across network and application layers. For example, common user role definitions and privacy policies can be shared by the federation and roaming partners.

Identity management plays a crucial role in user convenience by eliminating the multiple login procedures, passwords and user names to remember for a variety of third party provider applications across multiple devices, access and networks. A common identity also facilitates consolidated billing.

MNOs can also enhance their identity infrastructure and use it for their own internal needs. As stated earlier, MNOs have to explore new revenue opportunities from services going beyond pure connectivity. To be able to quickly introduce services, MNOs need an open network architecture, that can easily integrate third party developers and applications. In this open, Internet-like model, there are new security risks. For example, hackers may try to attack and steal confidential information from users.

15.8 Low Income Populations

Around half the world's population lives in villages and many of these people need access to communications to reap the socioeconomic benefits of being connected. Mobile access technologies inherently have lower implementation costs than technologies for wireline access, and have already enabled billions of people to be connected for the first time. However, there is still a sizable segment of the global population that cannot afford mobile access.

For some segments in rural markets, the main challenge is potential users who only have a budget for two or three USD per month [29]. Mobile operators need to adopt innovative solutions to keep the user TCO below USD 3 per month to provide access for this segment. To achieve such low TCO, an innovative approach is required to minimize the CSP's OPEX and CAPEX.

Networks and their support functions typically account for more than 50% of net OPEX in emerging markets. Other significant cost factors are power consumption of base station sites, site infrastructure, and backhaul. Reducing these costs will enable the CSP to offer services to the end-user at a lower TCO.

End-user costs are also widely reduced by the sharing of devices. Recent studies have shown that 81% of lower income consumers in India do not own a handset, yet 55% can get to a phone within five minutes [25]. However, in this chapter, we focus on network-based solutions. Value, the other factor in the cost/benefit ratio, can also be increased through services uniquely tailored to the needs and constraints of low income populations.

15.8.1 Cost Reduction

In this section, we describe two examples of innovative approaches to provide low cost connectivity to rural areas. The Village Connection (VC) initiative [29], through a novel approach, lowers the CAPEX and OPEX of GSM to provide voice and SMS service to villages. WiLDNET [31, 35], which has been deployed in India, Ghana, Guinea Bissau and the Philippines, enhances the inherently low cost WiFi technology to extend its range and provide low cost data connectivity to villages.

In VC, the main innovation is in the business model, however technology and regulation are needed to support it. Operating costs are significantly reduced compared with conventional network roll-outs, as responsibility for local network and business functions is transferred to a local level, building cost-effective connectivity village by village.

VC has a network structure different from traditional GSM networks (Fig. 15.3). It comprises village-level GSM access points (GAPs) and regional access centers (ACs). GAPs interface with subscriber terminals according to the standard GSM air interface, while the link between the GAPs and ACs is based on IP. Call control has been shifted to the edge of the network, with a GAP handling call completion within each village. Multiple GAPs aggregate to an AC, which handles call completion within its area (known as a VC domain). As this call control is now taking place in the village, the AP host is able to provide subscriber management related tasks for the village area. ACs connect to an MSC via a standard A interface for those calls which require connection to other networks. The CSP aggregates traffic from various villages, channels the traffic between different villages, and provides external world connectivity to VC subscribers.

Each GAP comprises RF, power, and IT hardware and software, bundled into a compact unit. The GAP is plug-and-play and can be installed and powered up easily, without requiring commissioning or operations support from trained personnel. Each village typically hosts one GAP. The plug-and-play nature of the GAPs, coupled with support for local subscriber management (provisioning, billing, and customer support, for instance), enables a distributed management model where network operation and subscriber management functions are handled at the village level. This makes possible a new business model where a local entrepreneur can host the GAP. The entrepreneur would be responsible for the operation and maintenance of the equipment, village-level demand aggregation (subscriber acquisition), sale/lease of terminals to local subscribers, and handling basic customer management functions, including provisioning and billing. The village entrepreneur is an entirely new player in the value network of mobile networks. Alternatively, an operator can have personnel at the village level to handle the functions locally. This distributed management model is critical to the viability of rural roll-outs, since it eliminates the OPEX traditionally associated with wireless networks (network operation and maintenance, billing, customer care, and so on).

The IP connection between the GAP and the AC can be supported by a variety of transmission media, including point-to-multipoint WiMAX, microwave, fiber, or satellite. Backhaul and interconnection costs are optimized through the hierarchical handling of calls and the IP connection.

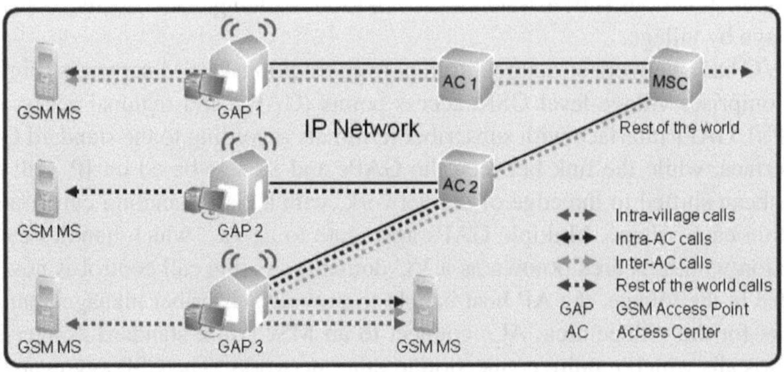

Fig. 15.3 Village Connection Architecture (Courtesy of Nokia Siemens Networks)

Another approach is based on extending the range of 802.11 WiFi technology to provide a low cost alternative to traditional connectivity solutions for rural areas [31, 35]. In [32], the authors argue that rural areas tend to be low density environments, where people are usually clustered around small localities (e.g., villages), with large distances among these clusters. Thus, the topology assumed is one where long range point-to-point links are suitable to meet the connectivity needs. In addition to the physics of power loss, other issues that need to be addressed are the poor performance of 802.11 Medium Access Control (MAC) over long distances, and severe throughput limitations caused by high and variable loss characteristics. WiLDNet attempts to overcome these issues by using high gain directional antennas with Line of Sight (LOS), enhanced MAC protocol, and adaptive loss-recovery mechanism using Forward Error Correcting Code (FEC) and bulk acknowledgements. According to [31], with these enhancements distances from 10 to 100 km can be achieved, and a 2 to 5 fold improvement in TCP/UDP throughput has been observed in a real-world deployment compared to the best achievable throughput by conventional 802.11 MAC.

The primary cost benefit stems from the use of low cost and low power single board computers, high-volume off-the-shelf 802.11 wireless cards originally intended for industrialized markets, and low cost towers. In addition, the typically lower incomes lead users to share computer terminals (e.g., Internet kiosks) to amortize the relatively high cost of the devices and network connection. Thus, the cost barrier of data terminals capable of leveraging the WiFi is lowered.

The authors also report a variety of non-technical problems encountered in the deployments. The installation, maintenance, and servicing costs turned out to be larger, due to lack of local technical expertise, equipment availability, and logistics. They identified the need for production-quality solutions, and the hardware and software must be robust, user friendly, and simple to install, maintain, and manage. Local partners must be trained as well. Another barrier is local telecommunication regulation, which is hindered by limited technical staff, "imperfect" government, and the presence of local incumbent monopolies. Problems related to the use of VoIP over unlicensed frequencies include restrictions on using VoIP (favoring local telecom monopolies), licensed or even restricted frequency bands that are unlicensed everywhere else in the world, and unregulated wireless usage resulting in significant same-band interference.

The WiLDNET approach is potentially more disruptive, as it relies on unlicensed frequency bands and therefore does not depend on a CSP in the traditional sense. It can be setup by grassroots organizations [32]. This is important for those rural areas which are less enticing to CSPs due to the low density and income of users.

15.8.2 Value

Another approach, not mutually exclusive of the previous one, is to provide higher value through services tailored to the needs and constraints of the village population. Examples include telemedicine, mobile commerce, mobile banking, and mobile access to public services.

Telemedicine can help to address the severe shortage of trained doctors and nurses in rural areas, by providing villagers with remote access to health care. This is enabled by affordable connectivity reaching the villages. The deployment of an intranet network in South India between hospitals and rural vision centers based WiLDNET is described in [33]. In the vision center model, doctors remain at the hospital but interact with patients over the communication network. A vision center is a room in the village equipped with basic ophthalmic equipment and a PC with a webcam, and staffed by two personnel who do not have a degree or a broad technical skill set. The local personnel perform some basic vision tests, present the results to the doctor at the base hospital via a videoconference, and then the patient interacts with the doctor. The personnel then follows up on the doctor's advice by handing out prescriptions, filling out referral forms, or creating glasses. They may also refer the patient to the base hospital for further examinations or treatments such as cataract surgery.

An example of eCommerce is the integrated solution developed by Nokia Siemens Networks (NSN), that enables businesses to reach rural consumers via their local village shops [27]. A key challenge experienced in many remote rural regions is the inability of businesses to reach consumers due to insufficient distribution and information systems. Many rural small and medium retail/distribution players cannot afford to be online, and so their business remains tied to the traditional form of local village shops with limited goods to sell. However, rural consumers have sufficiently high spending power to demand a wider choice of rich information and high quality products at good prices. The solution is proven in a joint project between NSN and Fujian Mobile in Fujian Province in China. The model enables a bigger business (product supplier) to use small local businesses (village retail shops) to promote and sell its products to rural consumers. It also offers CSPs the opportunity to tap into the potentially powerful new business model of Business to Business to Consumer (B2B2C) services. The solution provides a flow of up-to-date and accurate product information via a Catalog Sales Platform, a mobile means of cash payment on delivery, and cost-effective, reliable product delivery based on the existing rural logistics system. Mobile phones are an essential part of the Catalog Sales Platform service logic, even when only simple SMS notifications are used. Everyone (supplier, retail-point owners, and rural consumers) has a phone and can be contacted instantly. This is a new Internet service delivery approach with three key characteristics: (i) an end to end, integrated Information and Communications Technology (ICT) business solution based on shared Internet access that cuts the cost of using ICT by aggregating the needs and spending power of the local community, (ii) ICT empowerment of local entrepreneurs that links their existing local businesses into a new value creation ecosystem, and (iii) a sustainable business model for traditional operators that helps them to explore new revenue potential from B2b2C services.

In the mobile banking (or m-banking) case, the mobile phone, along with the associated trust relationship with the CSP, becomes a means of extending financial services to people without bank accounts [23]. M-banking allows these people to save, borrow, and remit safely, without having to rely on insecure cash transactions. Examples of services are person-to-person transfers, utility bill payments, and payments to purchase goods and services. M-banking can even remove the need for cash dispensing infrastructure (ATMs), or even retail banking offices, as every user with a mobile phone can act as an ATM, by exchanging cash in return for a mobile money transaction. The ecosystem involves a CSP and a bank, and requires the appropriate support of regulators and policy makers. The technology enablers are based on SMS and SIM Application Toolkit (ATK), which are available in most low end mobile devices. Since the marginal cost of m-banking on top of the existing infrastructure is low, banks can create new, viable business out of offering m-banking to low income groups. M-banking has been deployed in South Africa and the Philippines. This is a primary example of leveraging the trust relationship between the CSP and the subscriber, the subscriber identity infrastructure and the ubiquity of the mobile phone.

15.9 Conclusions

The advent of ubiquitous mobile broadband access will put tremendous pressure on the MNOs, who need to cope with the surge in traffic. They will need to reinvent themselves to tackle very challenging optimization problems related to decreasing the cost per byte and/or identifying new opportunities to increase revenues. There is an opportunity to extend services to the billions of the world's lowest income people, who mostly live in rural areas. Such an opportunity requires innovative approaches to dramatically reduce the user TCO and/or provide added value services geared towards the low income population.

Acknowledgements The author would like to thank the editors for their helpful comments and suggestions, and Nokia Siemens Networks for providing the figures.

References

1. Mobile broadband showcase. http://gsmworld.com/documents/LTE.pdf. Accessed on August 15, 2010
2. UMA technology. http://www.umatechnology.org/technology/index.htm. Accessed on August 15, 2010
3. 3G Americas: HSPA to LTE advanced. http://www.3gamericas.org/documents/3G_Americas_RysavyResearch_HSPA-LTE_Advanced_Sept2009.pdf. Accessed on August 15, 2010
4. 3G Americas: World cellular subscriber distribution. http://www.3gamericas.org/index.cfm?fuseaction=page&pageid=566. Accessed on August 15, 2010
5. 3G Americas: Identity management: Overview of standards and technologies for mobile and fixed internet. http://3gamericas.org/documents/3GAmericas_Unified_Identity_Management_Jan2009.pdf (2009). Accessed on August 15, 2010
6. 3GPP: Proposal for candidate radio interface technologies for IMT Advanced based on LTE release 10 and beyond. http://www.3gpp.org/IMG/pdf/2009_10_3gpp_IMT.pdf. Accessed on August 15, 2010
7. ABI Research: Mobile cloud computing subscribers to total nearly one billion by 2014. http://www.abiresearch.com/press/1484-Mobile+Cloud+Computing+Subscribers+to+Total+Nearly+One+Billion+by+201. Accessed on August 15, 2010
8. Alcatel-Luncent: Qatar's mobile service greatly expands thanks to nation's first green powered base station. http://webform.alcatel-lucent.com/res/alu/survey/alu2CustomForm.jsp?cw=alu2CorpDocDownload=&LMSG_CABINET=Docs_and_Resource_Ctr&LMSG_CONTENT_FILECase_Studies/vodafone_Qatar_Alternative_Energy_CS.pdf
9. 3G Americas: 3GPP mobile broadband innovation path to 4G: Release 9, release 10 and beyond: HSPA+, LTE/SAE and LTE-advanced.

http://www.3gamericas.org/documents/3GPP_Rel-9_Beyond%20Feb%202010.pdf. Accessed on August 24, 2010
10. at&t: AT&T 3G Microcell. http://www.wireless.att.com/learn/why/3gmicrocell/. Accessed on August 15, 2010
11. Cherry, S.: Forecast for Cloud Computing: Up, Up, and Away. IEEE Spectrum **46** (10), 69 (2009)
12. Cisco: Connecting customers with services: Cisco IP next-generation network carrier ethernet design. http://www.cisco.ag/en/US/solutions/collateral/ns341/ns524/ns562/ns577/net_brochure0900aecd806aec0d.pdf. Accessed on August 15, 2010
13. Engel, T., Autenrieth, A., J.-C.Bischoff: Packet layer topologies of cost optimized transport networks: Multilayer network optimization. In: International Conference on Optical Network Design and Modeling, pp. 1 – 7. IEEE, New York, NY, USA (2009)
14. Ericsson: High speed technologies for mobile backhaul. http://www.ericsson.com/res/docs/whitepapers/High-speed-technology-mobile-backhaul.pdf. Accessed on August 15, 2010
15. Ericsson: Sustainable energy use in mobile communications - key facts. http://www.ericsson.com/campaign/sustainable_mobile_communications/downloads/sustainable_energy_use_key_facts.pdf. Accessed on August 15, 2010
16. Femto Forum: http://www.femtoforum.org/femto/. Accessed on August 15, 2010
17. Lively, J.: Mobile backhaul landscape. http://www.huawei.com/perspective2/mobile_backhaul/mobile_backhaul_landscape.do. Accessed on August 15, 2010
18. Metro Ethernet Forum: Carrier ethernet access for mobile backhaul networks. http://metroethernetforum.org/MSWord_Documents/CarrierEthernetAccessforMobile%20BackhaulNetworksreleaseversion.doc. Accessed on August 15, 2010
19. Morgan Stanley: Economy + Internet Trends. http://www.morganstanley.com/institutional/techresearch/pdfs/MS_Economy_Internet_Trends_102009_FINAL.pdf. Accessed on August 15, 2010
20. Nokia Siemens Networks: Broadband access for all: A brief technology guide. http://www.nokiasiemensnetworks.com/portfolio/products/broadband-connectivity/broadband-connectivity. Accessed on August 15, 2010
21. Nokia Siemens Networks: Creating connectivity that everyone can afford. http://www.nokiasiemensnetworks.com/sites/default/files/document/TCO_NSN_flyer_A3_2811.pdf. Accessed on August 15, 2010
22. Nokia Siemens Networks: End to end energy solutions: Discover some surprising ways to save energy with nokia siemens networks end-to-end energy solutions (executive summary). http://www.nokiasiemensnetworks.com/sites/default/files/document/Energy_Solutions_Executive_Summary__301009.pdf. Accessed on August 15, 2010
23. Nokia Siemens Networks: Expanding horizons, 1/2009, special edition, connecting rural communities. http://www.nokia.com/NOKIA_COM_1/Corporate_Responsibility/Society_/Expanding_Horizons/Expanding_Horizons_NEW/pdf/Expanding_Horizons_Q1_2009.pdf Accessed on August 24, 2010
24. Nokia Siemens Networks: Future Transport Networks
25. Nokia Siemens Networks: Internet for the next billion. http://www.nokiasiemensnetworks.com/sites/default/files/document/Internet_for_the_next_billion_WP.pdf. Accessed on August 15, 2010
26. Nokia Siemens Networks: Monetizing Web 2.0: Embracing the internet. http://www.nokiasiemensnetworks.com/sites/default/files/document/Monetizing_WP.pdf. Accessed on August 15, 2010
27. Nokia Siemens Networks: On line shopping reaches remote rural regions in pioneering projects. http://www.nokiasiemensnetworks.com/sites/default/files/document/Online_shopping_reaches_regions.pdf. Accessed on August 15, 2010
28. Nokia Siemens Networks: Smart connectivity. http://www.nokiasiemensnetworks.com/about-us/company/innovation-and-technology. Accessed on August 15, 2010

29. Nokia Siemens Networks: Village connection, bringing the benefits of affordable mobile access to rural communities.
http://www.nokiasiemensnetworks.com/sites/default/files/document/Village_Connection_WP.pdf. Accessed on August 15, 2010
30. One API: http://www.gsmworld.com/our-work/mobile_lifestyle/oneapi.htm. Accessed on August 15, 2010
31. Patra, R., Nedevschi, S., Surana, S., Sheth, A., Subramanian, L., Brewer, E.: Wildnet: Design and implementation of high-performance wifi-based long distance networks. In: ACM/USENIX NSDI (2007)
32. Subramanian, L., Surana, S., Patra, R., Nedevschi, S., Ho, M., Brewer, E., Sheth, A.: Rethinking wireless for the developing world.
http://conferences.sigcomm.org/hotnets/2006/subramanian06rethinking.pdf. Accessed on August 15, 2010
33. Surana, S., Patra, R., Nedevschi, S., Brewer, E.: Deploying a rural wireless telemedicine system: Experiences in sustainability.
http://tier.cs.berkeley.edu/docs/TelemedicineSustainability.pdf. Accessed on August 15, 2010
34. Tanner, J.C.: Cloudy with a chance of $$$. http://www.telecomasia.net/content/cloudy-chance. Accessed on August 15, 2010
35. Technology and Infrastruture for Emerging Regions, University of California, Berkeley: TIER. http://tier.cs.berkeley.edu/wiki/Home. Accessed on August 15, 2010
36. Verizon: Verizon network extender. http://wirelesssupport.verizon.com/faqs/Equipment/network_extender.html. Accessed on August 15, 2010
37. Vodafone 360: http://login.vodafone360.com/en/web/home/index. Accessed on August 15, 2010
38. WIMAX: Mobile WIMAX - part 1: A technical overview and performance evaluation.
http://www.wimaxforum.org/technology/downloads/Mobile_WiMAX_Part1_Overview_and_Performance.pdf. Accessed on August 15, 2010
39. Yaghoobi, H.: Mobile WIMAX update and 802.16m. Accessed on August 15, 2010

MIX
Papier aus verantwortungsvollen Quellen
Paper from responsible sources
FSC® C105338

If you have any concerns about our products,
you can contact us on
ProductSafety@springernature.com

In case Publisher is established outside the EU,
the EU authorized representative is:
Springer Nature Customer Service Center GmbH
Europaplatz 3, 69115 Heidelberg, Germany

Printed by Libri Plureos GmbH
in Hamburg, Germany